International Conference on Electrical, Electronic, Informatic and Vocational Education (ICE-ELINVO 2019)

Advancing Green Technology Application in Industry and Vocational Education

Journal of Physics: Conference Series Volume 1413

Yogyakarta, Indonesia
14 September 2019

ISBN: 978-1-7138-1319-4
ISSN: 1742-6588

Printed from e-media with permission by:

Curran Associates, Inc.
57 Morehouse Lane
Red Hook, NY 12571

Some format issues inherent in the e-media version may also appear in this print version.

Institute of Physics
Dirac House, Temple Back
Bristol BS1 6BE UK

Phone: 44 1 17 929 7481
Fax: 44 1 17 920 0979

techtracking@iop.org

Additional copies of this publication are available from:

Curran Associates, Inc.
57 Morehouse Lane
Red Hook, NY 12571 USA
Phone: 845-758-0400
Fax: 845-758-2633
Email: curran@proceedings.com
Web: www.proceedings.com

TABLE OF CONTENTS

ELECTRICAL AND ELECTRONICS

INFORMATICS

Journal of Physics: Conference Series **1413** (2019) 011001 doi:10.1088/1742-6596/1413/1/011001

Preface

On behalf of the ICE-ELINVO organizing committee, I am honoured to welcome you to the 2nd International Conference on Electrical, Electronics, Informatics, and Vocational Education held today, 14th September 2019. This conference is organized by Department of Electrical Engineering Education in cooperation with Department of Electronic and Informatic Education, Faculty of Engineering, Universitas Negeri Yogyakarta.

This conference takes theme "ADVANCING GREEN TECHNOLOGY APPLICATION IN INDUSTRY AND VOCATIONAL EDUCATION". As we know, For the last two decades, the environmental deterioration caused by both human and industry activities has been increasing at alarming level. Numerous efforts have been carried out to alleviate the damage to the environment including the campaign of Green Technology (Greentech) or Clean Technology (Clentech) or Environmental Technology (Envitotech) utilization worldwide.

Greentech is an integration between modern technology and environmental science which is applied to preserve the fulfilment of future community needs without harming the environment. Advanced innovation and education are needed to develop the Greentech. Innovation is set as foundation on invention of environmentally-friendly technological alternatives to meet human needs, while education is channeled as an attempt to improve human awareness in implementing the Greentech which in the long run, leads to a sustainable environment. Therefore, the industries as the inventors of technological innovation and vocational education as the end users of technology must be able to work hand in hand in implementing and evolving the Greentech.

This conference provides a forum for knowledge and experience exchange and we encourage researchers, lectures, teachers, scholars, practitioners, decision maker including the government, and stakeholder to contribute cutting-edge ideas to the Greentech growth.

I realize that you are fully dedicated to the session that will follow but I do hope you will also take time to enjoy the fascinating Yogyakarta city with a vibrant history, friendly people, delicious cuisines, and traditions of Indonesia, which is also renowned as a center of education.

Finally, I would like to express our heartfelt thanks to the committee for enabling this to take place and to everybody who comes to this conference. I wish we will have productive and successful conference

Thank you

Yogyakarta, 14th September 2019
Conference Chairperson
Dr. phil. Nurhening Yuniarti, M.T

Peer review statement

All papers published in this volume of *Journal of Physics: Conference Series* have been peer reviewed through processes administered by the proceedings Editors. Reviews were conducted by expert referees to the professional and scientific standards expected of a proceedings journal published by IOP Publishing.

Development of Practicum Tools DC Motor Speed Control System as a Learning Media of Robotics Learning Practice

O Widiantama and H S Pramono
Electrical Engineering Education Department, Universitas Negeri Yogyakarta, Yogyakarta, Yogyakarta, Indonesia

Email: okky.widi.0@gmail.com, herlambangpramono@yahoo.com

Abstract. The purpose of this research is to: (1) Understand how the DC motor control systems trainer works, and (2) measure the feasibility of DC motor control systems trainer as a learning media. This research applies research and development approach and ADDIE (Analyze, Design, Develop, Implement, Evaluate) model. The research was conducted in mechatronic engineering education study program, Yogyakarta State University with the students of mechatronic engineering education as the subjects of the research. Based on this research, it is known that: (1) Vexta series of DC motor control systems trainer can control the speeds appropriate with the input speed, and (2) The result of feasibility assessment by material experts obtained an average score of 50.5, that included in the "feasible" category, while the media experts feasibility obtained an average score of 72, that included in the "very reasonable" category. For the students' assessment results based on the experiment on small group, 50 % of the students agreed that the material is "feasible" category, while another 50% agreed that it is "very reasonable" category. For the assessment on operational products, 65 % of the students agreed that it is "feasible" category, while another 30 % agreed that it is "very reasonable" category, and the other 5 % agreed that it is "good enough" category, to be applied as a learning media.

1. Introduction

Higher education is one of the government's solutions in the development of students who are able to advance domestic technology. Higher education is a continuation of secondary education that is held to prepare students to become members of the community who have academic and professional abilities that can apply, develop science, technology and arts (Law No. 2, 1989). In higher education there are many fields of expertise that are offered to students to learn. Each existing field of expertise makes students focus in their learning and has a clear direction for their future.

Department of Electrical Engineering Education (JPTE) Faculty of Engineering (FT) Yogyakarta State University (UNY) as one of the higher education institutions, including those affected by changes in DU/DI. Changes that occur in the area of controlling machines or tools in the DU / DI are not all and cannot always be followed by JPTE, so it becomes less conducive to learning [5].

Mechatronics Engineering Education Program, Department of Electrical Engineering Education, YSU, which studied industrial control and robotics. One of the main subjects of the Mechatronics Engineering Education Study Program is the Robotics Practice course. In the course of robotics practice, learning science related to robots, starting from the design of robots, sensors, actuators, mechanical engineering, and programming. From the science of designing robots, sensors, actuators, mechanical engineering, and programming can later be expanded to other fields in the field of robotics, namely robot kinematics including navigation systems, manipulation systems, image processing, intelligent robots and so on. Vexta series DC motors are actuators that are often used in the field of robotics. This Vexta series DC motor is a product of Oriental Motor, which is a type of brushless motor equipped with several sensors including a speed sensor and a current sensor. With this sensor it can be used as a feedback signal device to the controller to form a closed loop system.

Based on observations made in the Mechatronics Engineering Education study program, the Department of Electrical Engineering Education, Yogyakarta State University, in robotics practice courses, students are introduced to the basics of robotics that are still theoretical due to the small number of existing facilities and infrastructure. So students are less able to explore the capabilities and science of robotics more deeply. With the lack of available facilities and infrastructure, students have not been able to test the practice of robotics to develop their knowledge. This causes students' interest and motivation to learn is low.

Efforts to increase students' interest and motivation to learn need to develop adequate learning media. Through innovative learning media students can be more motivated to learn and improve their understanding in the field of robotics. In particular, the notion of media in the learning process tends to be interpreted as graphic tools, diagrams, slides (ppt) that serve to convey messages or visual or verbal information [3][5]. Gerlach and Ely explain the media when it is understood that it is broadly human, material and events that establish conditions that make students able to obtain knowledge, skills or attitudes [1]. In this sense, educators, textbooks, and the school environment are media. While Criticos explained that quoted by Daryanto the media is one component of communication, namely as a messenger from the communicator to the communicant [4]. Based on some of the opinions above it can be concluded that the media are all things or components that can be used to channel messages from the sender to the recipient so that they can stimulate the thoughts, feelings, attention and interests of students in the learning process.

The use of instructional media in the teaching and learning process can arouse new desires and interests, arouse motivation and stimulation of learning activities, and even bring psychological influences on students [5][6]. From the above problems the researcher intends to use learning media that is more applicable to the industrial world and is able to motivate students to study harder.

2. Method and Types of Research

The type of research applied is development research. This research produces a product in the form of a DC motor speed control system trainer as a learning medium for robotics practice. Product development applies the ADDIE model (Analyze, Design, Development, Implement, Evaluate) from Robert Maribe Branch [2].

Product development applying the ADDIE model includes analysis, product design, product development, implementation and evaluation. Analysis is carried out on teaching and learning activities then prepared appropriate learning media. Learning media in the form of a trainer applies a closed loop control system with a rotary encoder transducer. The product design begins with making the transfer function diagram and system block diagram. Making a gradual trainer from making hardware is continued by making software. Trainer that has been formed needs to pass validation from expert lecturers before applying it to teaching and learning activities.

This research began in July 2016 until completion. The research location is in the Department of Electrical Engineering Education, Faculty of Engineering, Yogyakarta State University. The research subjects were students of the Mechatronics Engineering Education Study Program, Faculty of Engineering, Yogyakarta State University.

Data obtained from respondents were processed and analyzed by quantitative descriptive statistics. The quantitative data of the feasibility instrument was changed into four categories of feasibility namely very feasible, feasible, sufficiently feasible and not feasible.

3. Results and Discussion

The product produced in this research is learning media in the form of Vexta series DC Motor Speed Control System Trainer and lab worksheet. The learning media of the Vexta series DC Motor Speed Control System was developed through the ADDIE development research method proposed by Robert Maribe Branch. This learning media uses several types of components including: ATMega16 Microcontroller, Rotary Encoder as speed sensor, one for Vexta series DC motors as actuators, keypads to provide input values for motor and LCD speeds to display input data and output speed of the Vexta DC series motors.

After being developed, this learning media took several tests. First is testing the blackbox to find out the learning media's functionality. The results obtained are the learning media trainer speed control system of this DC motor Vexta series can function properly according to the program that has been made.

When testing the instructor media of the Vexta series DC motor control system, there were some disadvantages. The learning media is still less informative for the speed unit value on the LCD, so an additional speed is needed to be displayed on the LCD. In addition, the absence of a stop button on the learning media hardware is also an input that, this learning media is less practical. Because there are only one tool, this learning media must be used interchangeably.

3.1. Result

This study aims to: (1) Determine the test results of the learning media functionality, and (2) Determine the feasibility of learning media for the Vexta series DC Motor Speed Control System. Discussion of the results of this study is explained as follows.

3.1.1. Learning Media Performance

The learning media performance is assessed from the blackbox test. The results of the analysis of the performance data is known by testing each function of the instructor learning media section of the Vexta series DC motor speed control system. The test data shows that all trainers of the Vexta series DC motor speed control system are functioning normally. After taking the blackbox test stage, the learning media are run according to the procedure and get the results that the program can control the motor speed according to the entered speed value.

3.1.2. The Feasibility Learning Media of Vexta series DC Motor Speed Control System Trainer

The feasibility of instructional media is assessed based on assessments by media experts, material experts and students. The feasibility assessment is explained as follows.

3.1.2.1. Expert Judgment of the Material

The assessment of the feasibility of learning materials by material experts is assessed based on two aspects, namely the relevance aspect of the material, and the technical aspects of learning media. The assessment data from media experts can be shown in Table 1.

Table 1. Data from Expert Material Evaluation Results

No	Aspect	Average Score	Category
1.	Relevance of Material	33,50	Worthy
2.	Technical Learning Media	17	Worthy
Average Total Score		**50,5**	**Worthy**

The conclusion of the feasibility of instructional media material based on data from the assessment of material experts obtained an average total score of 50.5 assessment with the category "feasible". Material experts assess learning media as "feasible" because the aspect of relevance of the material is in accordance with the subject of robotics practice being taught. Furthermore, from the technical aspects of instructional media, instructional media are considered to have good design quality, can be operated easily and easily in their maintenance.

3.1.2.2. Assessment of Media Experts

Assessment of the feasibility of instructional media by media experts is assessed based on three aspects, namely aspects of expediency, aspects of device engineering and aspects of visual communication. The assessment data from media experts can be shown in Table 2.

Table 2. Data from the Media Expert Assessment Results

No	Aspect	Average Score	Category
1.	Usefulness	27	Very decent
2.	Device Engineering	31	Worthy
3.	Visual Communication	14	Very decent
Average Total Score		**72**	Very decent

The conclusion of the feasibility of instructional media based on data from the results of the assessment of media experts obtained an average total score of 72 with the category "very feasible". Media experts consider the Android-based learning media to be "very feasible" because of the usefulness of the media that supports the explanation of learning material. Furthermore, from the engineering aspect of the device, learning media can be used for other learning materials. From the aspect of visual communication considered to be able to attract students' interest in learning and can help students to understand the learning material.

3.1.2.3. Student Assessment

Response assessment by students contains product assessments in terms of aspects of material quality, aspects of performance, and aspects of expediency. Student assessment of learning media is obtained from small group trial data and operational product tests.

3.1.2.3.1. Evaluation of Small Group Trials

Assessment in small group trials is carried out by 8 students to determine students' assessment responses on a small scale. Data on student assessment results in small group trials is explained in the frequency distribution table in Table 3.

Table 3. Frequency Distribution of Small Group Trial Results

Category	Frequency	Percentage (%)
Very decent	4	50,00
Worthy	4	50,00
Decent enough	0	0,00
Not feasible	0	0,00
Total	8	100,00

Based on the data in Table 3, the total frequency distribution of learners' scores in small group trials is presented in the form of a diagram in Figure 2.

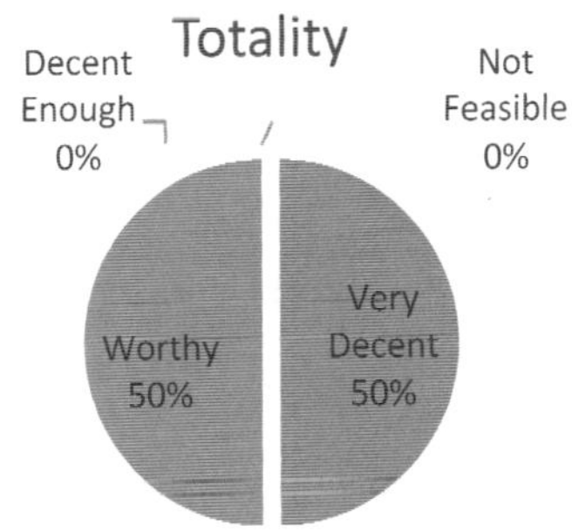

Figure 1. Diagram of Frequency Distribution of Small Group Trial Results

Based on the frequency distribution diagram in Figure 2 it can be seen that 50% of participants in a small group trial stated that learning media in the feasible category as learning media and 50% of other students rated the media in the category very feasible to be used as a learning madia. The conclusion was obtained that the Vexta series DC motor trainer speed control system learning media is "very feasible" to be used as a learning medium.

Learners assess learning media as "very feasible" to use because: (a) Aspects of the quality of the material, the portion of the material delivered is in accordance with the capacity of students, and is equipped with a labsheet; (b) Performance aspects, learning media are considered easy to operate and run well, the program is able to control the motor speed according to the input value; and (c) Aspects of expediency, learning media are considered to be able to attract students' interest in learning and make it easier for students to understand the learning material.

3.1.2.3.2. Assessment of operational product tests

The assessment on the operational product test is carried out by 20 students to determine the response of the end user assessment. Student assessment results data on the operational product test is explained in the frequency distribution table in Table 4.

Table 4. Frequency Distribution of Operational Product Test Results

Category	Frequency	Percentage (%)
Very decent	6	30,00
Worthy	13	65,00
Decent enough	1	5,00
Not feasible	0	0,00
Total	20	100,00

Based on the data in Table 4, the total frequency distribution of learners' scores on the operational product test can be presented in the form of a diagram in Figure 3.

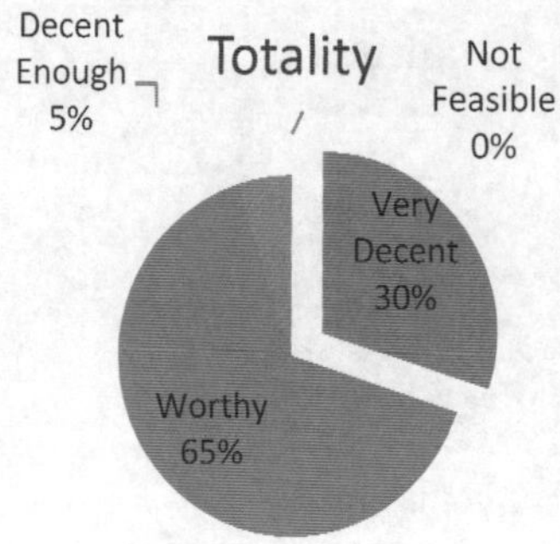

Figure 2. Frequency Distribution Diagram of Operational Product Test Results

Based on the frequency distribution diagram in Figure 3 it can be seen that 30% of students in the operational product test stated that learning media in the category are very feasible as learning media, while 65% of students rate learning media in the appropriate category as learning madia and 5% of students others considered learning media in the category quite feasible to be used as learning media. The conclusion was obtained that the learning media trainer speed control system DC motor Vexta series "feasible" is used as a learning medium.

Learners assess learning media as "very feasible" to use because: (a) Aspects of the quality of the material, the portion of the material delivered is in accordance with the capacity of students, and is equipped with a labsheet; (b) Performance aspects, learning media are considered easy to operate and run well, the program is able to control the motor speed according to the input value; and (c) Aspects of expediency, learning media are considered to be able to attract students' interest in learning and make it easier for students to understand the learning material.

On the other hand, students assess that there are deficiencies in learning media, namely the material delivered is less extensive, only one learning media, and the need for a GUI display on a computer for response graphs.

4. Conclusion and Recommendations

4.1. Conclusion

Based on the results of research and discussion of the development of instructor learning media trainer speed control system of DC motor Vexta series, the following conclusions can be drawn:

a. Development of learning media trainer speed control system of Vexta DC motor using several electronic components including: ATMega16 Microcontroller, LCD, Keypad, Vexta series DC Motor and Rotary Encoder. As for making programs for speed control using the Code Vision AVR software. The results of the trainer functionality test of the Vexta series DC motor control system can control the motor speed using a rotary encoder transducer by providing feedback to the microcontroller. With speeds ranging from 50 - 600 rpm, constants 0 to 20, and constants 0-10, constants 0 to 20.

b. The feasibility of learning media trainer speed control system of DC motor Vexta series based on the assessment by material experts obtained a score of 50.5 included in the category of "feasible" to be used as a learning medium. The feasibility assessment by the media experts obtained a score of 72 included in the category of "very feasible" used as a learning medium. Evaluation of students in small group trials obtained 50% of students stated that learning media included in the category of "feasible" as learning media, and 50% of other students stated that learning media included in the category of "very feasible" as learning media. The results of the assessment of students in the operational product test obtained 65% of students stated that learning media included in the category of "feasible" as learning media, while 30% of students stated that learning media included in the category of "very feasible" as learning media and 5% of participants students state that learning media is included in the category of "reasonable enough" as a learning medium.

4.2. Recommendation
Based on the research that has been done, the advice given for the next research related to the development of instructional media trainer speed control system of DC motor Vexta series, namely:
a. Making learning media needs to be duplicated so that learning robotics practices is more effective.
b. Developing other trainer's abilities so that they can be used for other learning.
c. Developing more applicable media as a follow up to the motor speed control system practicum.

5. References

[1] Azhar Arsyad 2011 Media Pembelajaran (Jakarta: PT. Raja Grafindo Persada)
[2] Bekti Wulandari dkk. 2015 Pengembangan Trainer Equalizer Grafis dan Parametris Sebagai Media Pembelajaran Mata Kuliah Praktik Sistem Audio (Vol.22. No.4. Hlm.375)
[3] Branch, M. Robert 2009 Instructional Design The ADDIE Approached. (New York: Springer)
[4] Daryanto 2010 Media pembelajaran: Peranannya Sangat Penting Dalam Mencapai Tujuan Pembelajaran (Yogyakarta: Gava Media)
[5] Haryanto & Moh Khairudin 2012 Pengembangan Model Pembelajaran Jaringan Syaraf Tiruan Tipe Supervised Learning Sebagai Media Pembelajaran (Vol.21. No.1. Hlm.84)
[6] Muhammad Munir 2014 Pengembangan Media Pembelajaran Interaktif Kompetensi Dasar Register Berbasis Inkuiri Terbimbing. Jurnal Pendidikan Teknik dan Kejuruan (Vol.22. No.2. Hlm.185)

Kinematics Simulation of Welding Manipulator Based on V-REPPRO EDU

Muslikhin[1], D Irmawati[1], F Arifin[1], A Nasuha[1] and N Hasanah[2]
[1]Electronics Education Department, Universitas Negeri Yogyakarta, Yogyakarta, Indonesia
[2]Informatics Education Department, Universitas Negeri Yogyakarta, Yogyakarta, Indonesia

Email: muslikhin@uny.ac.id

Abstract. As the primary tool in modern industrial welding, 6 DOF manipulator robots can weld in all positions. In this context, Danevit-Hartenberg parameters were developed to simplify kinematics equations for manipulator equations. Inverse kinematics is obtained based on the algebraic method. Simulation of kinematics manipulators based on V-REP PRO EDU is present. Kinematics simulations produce processing motion and relevant kinematics curves. The simulation results and kinematic equations of robotic manipulators are compared, showing that valid kinematic equations and weld trajectories can be passed by the end-effector successfully.

1. Introduction

Welding in manufacturing tasks has become an essential part, especially welding overhead positions, shape, material, and other variables. Welding using a robot manipulator is nothing new. However, so many variations V-REP (Virtual Robot Experimentation Platform) in the shape of the welded workpiece make it essential to be able to handle it quickly. Therefore, simulation is the necessary and primary step before using these robots in real duty [1].

The idea of welding in the welding industry is not only done on one workpiece but also more. Human welding operators devote themselves to more straightforward jobs and more natural positions, such as underhand welding [2]. While the welding robot will do some duties such as; uniform work, intricate welding patterns, high accuracy, and repetitive [3], in this way, the system can accommodate increased quality and welding production capacity.

In short, 6 DOF can be divided into two parts. The first three parts are used to specific positions, and the other three are the essential dexterity characteristics of robotic space manipulators [3]. Therefore, the focus of this paper is six degrees of freedom from specific permits and connections.

The manipulator is a critical part of robot welding. Vitally, kinematic analysis, and manipulator simulation is needed to analysis the welding results before being applied. This paper is organized as follows: the structure of the manipulator described in section 2 continues with kinematic analysis discussed in section 3. The procedure for developing the manipulator modelling in simulation is outlined in section 4, and the results are displayed in section 5. Definitely, conclusions are drawn in section 6.

2. Structure of Manipulator

The manipulator is an essential element of the whole system. The basis of manipulators is to imitate human hands designed with certain DOFs. The number of DOF is very influential on the system to be designed. The same number of DOF with different dimensions will have strange results even applied to the same system or vice versa [4] [5]. The application of manipulators to welding requires us to be able to work based on 3D fields. Whereas the 6 DOF constructively accomplish to work on 3D dimensions, as shown in Figure 1.

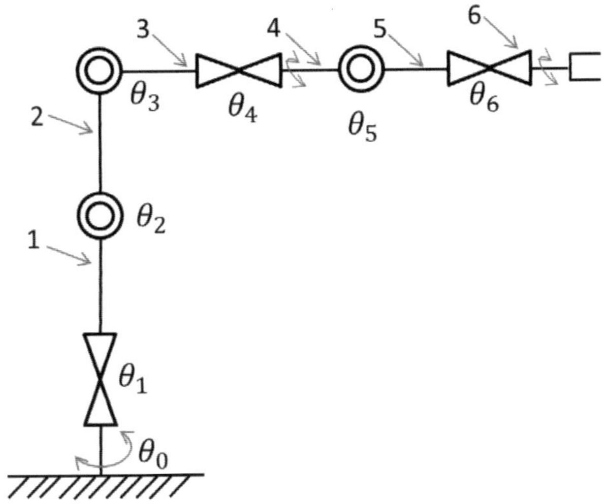

Figure 1. The structure of the manipulator

2.1. D-H Coordinate System

The idea of Denavit and Hartenberg (DH) proposed a matrix method for constructing a coordinate system attached to each link in a chain with a robot to describe the translation or rotation relationship between links [6] [7]. Fig. 2 depicts the rotation and translation relationships for each link.

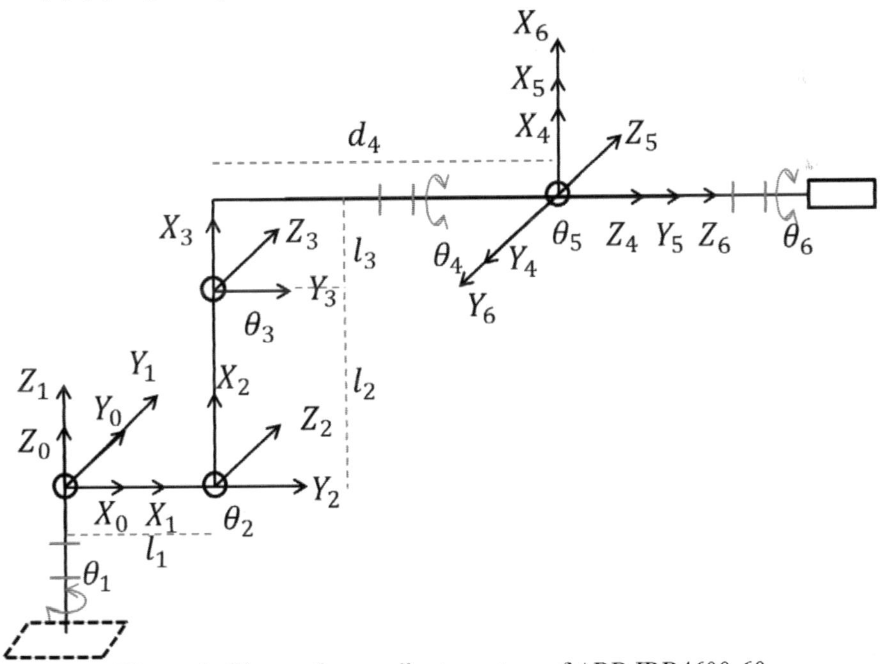

Figure 2. Kinematic coordinate system of ABB IRB4600-60

In this paper, the ABB IRB4600-60 robot industry is taken as a simulation object [8] [9]. DH coordinates are composed of four parameters. This notation describes the parameters of the relationship between one joint i and another joint i+1. The four parameters consisted of1) link l_i length is the distance between the common vertical line of joint axes l_i and i+1, 2) the angle of torsion connected with rod α_i

is the angle between of the joint axes i and i+1, 3) the distance between the two connection rodsd$_i$ is the distance between two vertical lines l_i and l_{i-1}, and 4) angle θ_i is the angles between l_i and l_{i+1}.Based on the DH parameters above, the structure can be written for the ABB IRB4600-60 manipulator robot, as shown in Table 1.

Table 1. A DH parameter for ABB IRB4600-series

joint	θ_i	l_{i-1}(mm)	d_i(mm)	α_{i-1}(deg.)	range
1	θ_1	0	0	0	-180~180
2	θ_2	l_1=175	0	-90	-90~150
3	θ_3	l_2=900	0	0	-180~75
4	θ_4	l_3=175	d_4=960	-90	-400~400
5	θ_5	0	0	90	-125~120
6	θ_6	0	0	-90	-400~400

Based on Table 1 of the DH parameter system, the homogeneous transformation of the link ${}^{i-1}_{i}T$ can be obtained.

$$ {}^{i-1}_{i}T = \begin{bmatrix} cos\theta_i & -sin\theta_i & 0 & \alpha_{i-1} \\ sin\theta_i cos\theta_{i-1} & cos\theta_i cos\alpha_i & -sin\alpha_{i-1} & -sin\alpha_{i-1}d_i \\ sin\theta_i cos\theta_{i-1} & cos\theta_i cos\alpha_i & cos\theta_{i-1} & cos\theta_{i-1}d_i \\ 0 & 0 & 0 & 1 \end{bmatrix} \quad (1) $$

Transformation of a robot manipulator matrix is given based on the DH parameters method, so for each transformation at each joint as follows.

$$ {}^{0}_{1}T = \begin{bmatrix} cos\theta_1 & -sin\theta_1 & 0 & 0 \\ sin\theta_i cos\theta_{i-1} & cos\theta_i cos\alpha_i & 0 & 0 \\ sin\theta_i cos\theta_{i-1} & cos\theta_i cos\alpha_i & 1 & 0 \\ 0 & 0 & 0 & 1 \end{bmatrix} \quad (2) $$

$$ {}^{1}_{2}T = \begin{bmatrix} cos\theta_2 & -sin\theta_2 & 0 & l_1 \\ 0 & 0 & 1 & 0 \\ -sin\theta_2 & -cos\theta_2 & 1 & 0 \\ 0 & 0 & 0 & 1 \end{bmatrix} \quad (3) $$

$$ {}^{2}_{3}T = \begin{bmatrix} cos\theta_3 & -sin\theta_3 & 0 & l_2 \\ sin\theta_3 & cos\theta_3 & 0 & 0 \\ 0 & 0 & 1 & 0 \\ 0 & 0 & 0 & 1 \end{bmatrix} \quad (4) $$

$$ {}^{3}_{4}T = \begin{bmatrix} cos\theta_4 & -sin\theta_4 & 0 & l_3 \\ 0 & 0 & 1 & d_4 \\ -sin\theta_4 & -cos\theta_4 & 0 & 0 \\ 0 & 0 & 0 & 1 \end{bmatrix} \quad (5) $$

$$ {}^{4}_{5}T = \begin{bmatrix} cos\theta_5 & -sin\theta_5 & 0 & 0 \\ 0 & 0 & -1 & 0 \\ sin\theta_5 & cos\theta_5 & 0 & 0 \\ 0 & 0 & 0 & 1 \end{bmatrix} \quad (6) $$

$$ {}_{6}^{5}T = \begin{bmatrix} cos\theta_6 & -sin\theta_1 & 0 & 0 \\ 0 & 0 & 1 & 0 \\ -sin\theta_6 & -cos\theta_6 & 0 & 0 \\ 0 & 0 & 0 & 1 \end{bmatrix} \quad (7) $$

2.2. Forward Kinematics Analysis

The forward kinematic analysis means that the position of the end-effector is unknown. In this case, it will set angle $\theta_{i...n}$to reach the intended end position. In case forward kinematic is a translation from joint space to Cartesian space [4]. It could be expressed by the following equation.

$$ {}_{6}^{0}T = {}_{1}^{0}T \cdot {}_{2}^{1}T \cdot {}_{3}^{2}T \cdot {}_{4}^{3}T \cdot {}_{5}^{4}T \cdot {}_{6}^{5}T = \begin{bmatrix} n_x & o_x & a_x & p_x \\ n_y & o_y & a_y & p_y \\ n_z & o_z & a_z & p_z \\ 0 & 0 & 0 & 1 \end{bmatrix} \quad (8) $$

where (n_x, n_y, n_z), (o_x, o_y, o_z), (a_x, a_y, a_z), and (p_x, p_y, p_z) are the normal vector, orientation vector, approach vector, and position of the end-effector, respectively.

2.3. Inverse Kinematics Analysis

Inverse kinematics is the opposite of the forward kinematic process. In this method, the position of the end-effector is known, and the joint angle will calculate how much it must rotate or translate. The solution for kinematic inverses can be done by algebraic, geometric, and iterative methods. In this paper, we use the algebraic method for inverse kinematics [3].

Based on formula (8) will be obtained ${}_{6}^{0}T^{-1}(\theta){}_{6}^{0}T = {}_{2}^{1}T \cdot {}_{3}^{2}T \cdot {}_{4}^{3}T \cdot {}_{5}^{4}T \cdot {}_{6}^{5}T$through the correspondence of the matrices element, the equation of the two segments can determine values from$\theta_1$to θ_6.In this simulation, we use the Algebraic method; thus, we can separate all joint angles as follows.

$$ {}_{1}^{0}T^{-1} \cdot {}_{6}^{0}T = {}_{6}^{1}T \quad (9) $$

$$ {}_{2}^{1}T^{-1} \cdot {}_{1}^{0}T^{-1} \cdot {}_{6}^{0}T = {}_{6}^{2}T \quad (10) $$

$$ {}_{3}^{2}T^{-1} \cdot {}_{2}^{1}T^{-1} \cdot {}_{1}^{0}T^{-1} \cdot {}_{6}^{0}T = {}_{6}^{3}T \quad (11) $$

$$ {}_{4}^{3}T^{-1} \cdot {}_{3}^{2}T^{-1} \cdot {}_{2}^{1}T^{-1} \cdot {}_{1}^{0}T^{-1} \cdot {}_{6}^{0}T = {}_{6}^{4}T \quad (12) $$

$$ {}_{5}^{4}T^{-1} \cdot {}_{4}^{3}T^{-1} \cdot {}_{3}^{2}T^{-1} \cdot {}_{2}^{1}T^{-1} \cdot {}_{1}^{0}T^{-1} \cdot {}_{6}^{0}T = {}_{6}^{5}T \quad (13) $$

Although DH parameters have become standard in robotic manipulators, there is no standard for optimal solutions in inverse kinematic applications. Practically, we can choose the solution closest to the position of the manipulator according to the range of motion, collision, and robotic joint axis requirements.

3. Kinematics Simulation Based on V-REPPRO EDU

V-REP is software that is used for robotic simulations. This software is quite powerful because it can facilitate ROS nodes, Remote API clients, Plugins, Embedded Scripts, add-ons, and BlueZero nodes. V-REP PRO EDU is a free professional education version. Because it has the capability of Remote API, this software can run scripts from C/C++, Lua, Python, Java, MATLAB, Octave, and Urbi.

3.1. Simulation Setting

In this paper, the simulation is done using V-REP PRO EDU version 3.5.0 software (rev.4) 64 bits. Manipulator robots using ABB IRB4600-40 will work with inverse and forward kinematics for the same welding task. The DH parameter values for robotic manipulators are as shown in Table 1. The inverse and forward kinematic calculations were done in MATLAB 2017. Computers used Intel Core i5 CPU @ 3.0 GHz (6 CPUs), 16 GB RAM, with Windows 10 Pro 64 bit operating system.

3.2. 3D Modeling for the 6 DOF Manipulator

The 3D robot manipulator modeling consists of four parts: base, shoulder (l_1) 175 mm, elbow (l_2) 175 mm, and wrist (l_3) 175 mm. The 3D simulation model was built using the V-REP PRO EDU software as shown in Figure 3.The welding torch is placed at the end of the end-effector with a length of 300 mm, while the yellow circle shows the welding path 1010 mm in diameters.

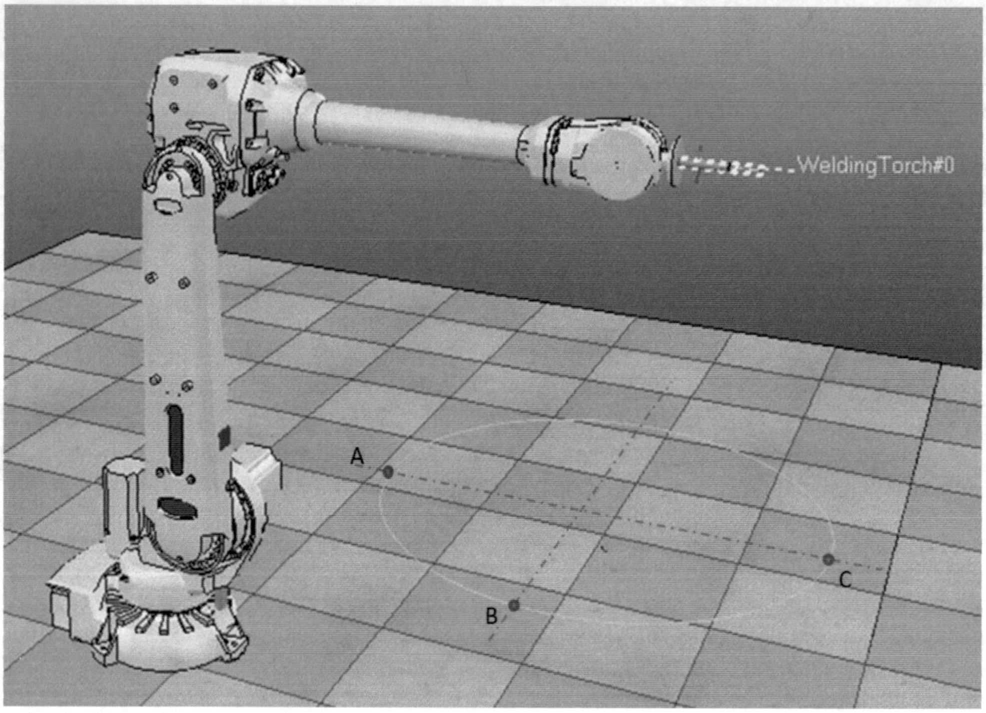

Figure 3. Virtual ABB IRB4600 with welding torch and yellow welding pathin V-REP PRO EDU

3.3. Trajectory Planning Expected

The weld trajectory planning in this simulation is set for flat-circular objects (Figure 3). The basis for planning a point to point (PTP) manipulator has three points. Point A ($\theta_{1...6}$= 0 90 0 0 0 0), as the prefix then moves down to point B ($\theta_{1...6}$= 0 120 60 0 0 0) as well as the starting point of the circle. Finally, at point C ($\theta_{1...6}$= 0 52 -13 0 0 0), which is 180 degrees from point A, in other words, it will be at the point after forming a half-circle. A series of data variations of all joints together with time has been obtained. Data is stored in the script and given to V-REP PRO EDU; data can be used for torch welding to move along the specified circle line.

4. Simulation Result

Simulation results are compared using the RoboAnalyzer formula. The aim is to verify the correctness of kinematic equations and simulation time is 50 seconds with 100 steps. From V-REP PRO EDU we can see the displacement of the end-effector in XYZ as described in Table 2 and all units in millimeters. Now based on the Eq. 8 above we can find out the values of n for each endpoint A, B, and C. Including for orientation (o) and approach vector (a) are obtained from the forward kinematics of formula too.

Table 2. The end-effector configurations

position	n	o	a	$p_{[simulation]}$	$p_{[RoboAnalyzer]}$	Δp
A	0	0	-1	1.135	1.135	0
	0	1	0	0	0	0
	1	0	0	1.075	1.075	0
	0	0	0	1	1	0
B	0.5	0.866	0	0.643	0.643	0.016
	0	0	1	0	0	0
	0.866	0.5	0	0.45	0.451	*0.094*
	0	0	0	1	1	0
C	0.454	0.89	0	1.654	1.654	0
	0	0	1	0	0	0
	0.89	-0.454	0	0.435	0.447	0.012
	0	0	0	1	1	0

The end-effector position in this paper compares the simulation results and the DH parameter from RoboAnalyzer software. There is a difference, especially in the column of position (p), where this column represents the end-effector position of the first three lines (X, Y, and Z). Initially, this simulation was given in flat 1G (groove) welding with a circular shape. Because it is flat, the position of Z should not change, but we can see the results of the highest DH analysis with a deviation of 0.094 mm. This condition is still acceptable in welding for objects with a diameter of 1010 mm [10].

The Δp value, which reaches up to 0.094 mm, is suspected to be influenced by the number of steps in RoboAnalyzer. The number of steps around 100 is considered still too rough to compare with this simulation. As an illustration, see Figure 4,the motion position of each step is always shorter than the 1/100 arc or 10.1 mm apart; the aforementioned is where the deviation occurs. Whereas to overcome this deviation could be done by increasing the number of steps in crossing the circle or welding path.

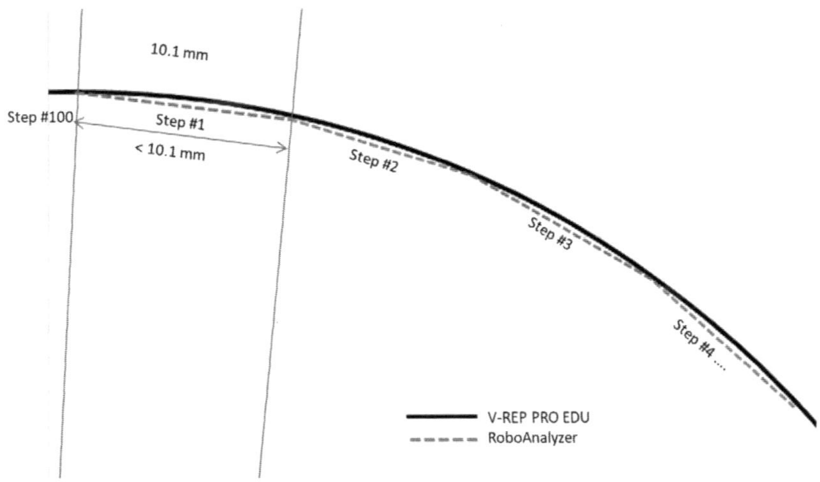

Figure 4. The deviation in the simulation and RoboAnalyzer

5. Conclusion

This paper takes ABB IRB4600-60 robot manipulator as a welding simulation object. Kinematic simulation is made in V-REP PRO EDU. DH parameters and trajectory welding solve modeling is obtained for a workpiece. The robot model was made, and RoboAnalyzer verified the truth of the robot kinematics model. Simulation results show that kinematic analysis is correct with acceptable errors.

6. References

[1] Erden M S and Billard A 2016 Robotic assistance by impedance compensation for hand movements while manual welding IEEE Transactions on Cybernetics 46 2478656

[2] Yueqiang Y and Zhiwei T 2014 The study of path planning of welding manipulator based on improved QPSO Proc. 6th Int.Conf. on Measuring Technology and Mechatronics Automation (Zhangjiajie) vol 1 (Zhangjiajie: China) p 774

[3] Linna X et al2017 Kinematics analysis and simulation of six - dof industrial robotInt. Journal of Computer Trends and Technology 49229

[4] Yujie Cet al2010 Kinematics simulation of an aided fruit-harvesting manipulator based on ADAMS Proc. Int. Conf. on Computer Design and Applications (Qinhuangdao) vol 5 (Qinhuangdao: China) p 222

[5] Velarde S J Aet al2010 5-dof manipulator simulation based on matlabsimulink methodology Proc. 20thInt. Conf. on Electronics Communications and Computers (Cholula) vol 1 pp 295

[6] Liu F et al 2017 Kinematic analysis and simulation of a 3-dof robotic manipulator3rd Int. Conf. on Computational Intelligence & Communication Technology (Ghaziabad) vol 1 (Ghaziabad: India) p 1

[7] Hong Set al2018 Realization of the fast real-time hierarchical inverse kinematics for dexterous full body manipulation Proc. 18th Int. Conf. on Control, Automation and Systems (Daegwallyeong),vol 1 (Daegwallyeong: South Korea) p 166

[8] Chapelle F and Bidaud P A 2001 Closed form for inverse kinematics approximation of general 6R manipulators using genetic programming Proc. IEEE Int. Conf. on Robotics and Automation (Seoul) vol4 (Seoul: South Korea) p3364

[9] Minh C N and Graefe V 2001 Object manipulation by learning stereo vision-based robots Proc. IEEE/RSJ Int. Conf. on Intelligent Robots and Systems (Maui) vol 1 (Hawaii: US) p 146

[10] Denault M et al2009 Low distortion welded joints for NCSX*23rd IEEE/NPSS Symposium on Fusion Engineering (San Diego) (California: US) p 1

Teaching Microcontrollers using Arduino Nano Based Quadcopter

E S B Santosa[1] and S Waluyanti[2]
[1]Electronics and Informatics Engineering Education Department,
Universitas Negeri Yogyakarta, Yogyakarta, Indonesia
[2]Technology and Vocational Education Department,
Universitas Negeri Yogyakarta, Yogyakarta, Indonesia

Email: ekasetia@2017.student.uny.ac.id

Abstract. Learning to use the Arduino Nano based quadcopter is an innovation that can improve students' motivation in learning. Microcontroller learning is one of the subjects in vocational high school, especially in electronics engineering. This study aims to motivate students in learning and knowing students' responses about the use of Arduino Nano based quadcopter learning media. The first stage of this learning is students learn about components that exist in the quadcopter and remote control. The second stage of students integrates between components to form a quadcopter and remote-control system. The last stage is students learn to program using Arduino. Data is obtained by using a questionnaire to determine students' responses to learning by using these media. Data are obtained from students' motivational questioner in 1 Magelang public vocational high school in Indonesia with a total of thirty-two students. The result showed that motivation students increased after participating in learning using Arduino Nano based quadcopter.

1. Introduction

The microcontroller has been used as a digital system that has been widely applied in household and industrial electronics. Microcontrollers are a choice because of the affordable prices. Microcontrollers are usually used as control systems, signal processing, instrumentation, and others [1].In vocational high school, the microcontroller is a productive subject that must be mastered by students. Microcontroller learning is very important because it is related to practical skills and programming [2]. Microcontroller learning is a basic knowledge of today's digital technology and knowledge [3].Therefore, the expertise of the microcontroller is very important in response to technological challenges [4].

Students at vocational high school must learn through practice in real life from only limited theoretical learning [5]. Simple microcontroller learning usually uses computer simulations but this has become less effective because it does not show the real application of the microcontroller [2][6]. Students will be bored and less interested. Several learning methods that can support more innovative learning. One of them is problem-based learning (PBL). The PBL is a learning model that emphasizes in real conditions to solve real problems [7]. PBL is one learning model that can increase student motivation [8][9].Implementation of the PBL model in microcontroller learning will motivate students more when carried out with interesting media, one of which is using Arduino Nano based quadcopter. Arduino is a prototype and a platform that is a combination of hardware and software [3]. Arduino uses an open-source library available on the Internet.

The use of Arduino is very diverse from very simple to complex. One of its uses in learning is quadcopter. Arduino becomes popular because it is simple and easy to use. Learning by using Arduino includes of electronic design, block diagram, programming, etc. Arduino Nano based quadcopter has a variety of devices and sensors: Arduino Nano, accelerometer sensor, gyroscope sensor, electronic control speed (ESC), brushless motor, liquid crystal display (LCD), transceiver, and joystick. Each devices and sensors are arranged together to form a quadcopter system. Each circuit has its own characteristics based on its settings. Data was obtained through research in 1 Magelang public vocational high school. Syllabus in microcontroller subjects are as follows: (1) concepts of the basics of designing hardware and software microcontrollers, (2) supporting components of microcontrollers, (3) input/output programming of microcontrollers, (4) liquid crystal display (LCD) programming, (5) DC Motor programming, (6) internal pulse width modulation (PWM), (7) microcontroller interruptions, (8) timer/counter, and (9) internal analog-digital converter (ADC).

1.1. Problem-Based Learning

Problem-based learning models are based on the learning process that helps students to solve real-world issues, communication, and teamwork skills [10][11]. PBL is one of the learning strategies that focuses on students learning. PBL is widely applied in learning methods because it includes a problem-solving process and high-level thinking [12]. PBL positions the teachers as the facilitator in charge of supervising and the students ' instruction in learning. Students use existing knowledge to learn to solve problems and to participate actively in the teaching process [13][14]. Through the problem-based learning model, students learn each part as a whole. Here are some things to be achieved in problem-based learning.

- Students will be motivated in learning, understanding the Arduino Nano architecture, describing the central processing unit, addressing input-output, and programming.
- Students will get used to Arduino Nano through practical practice on a quadcopter.
- Students can combine all parts of the quadcopter and programming between parts so that the parts can be connected properly.

1.2. Arduino Nano Based Quadcopter and Programming

Arduino is an open-source platform arranged in a simple microcontroller board [15]. The Arduino is designed to be easy to use anit has its programming language called as the Arduino programming language [16]. Arduino has the advantages of being able to connect with a computer easily through the USB port. Arduino can also combine with other components such as the transceiver NRF24L01, liquid crystal display, and others very easily. Along with the growing popularity of Arduino at this time so Arduino began to be used to facilitate the learning of embedded microcontroller systems. One of themicrocontroller systems that can be developed in learning using the Arduino is learning by utilizing the Arduino for Arduino based quadcopter. The quadcopter is a type of vehicle that enters the category of flying and landing in a vertical [17]. The quadcopter has the advantage of a helicopter which can fly and move in a small space. Tools and materials used in the learning are Arduino based quadcopters and personal computers for Arduino based programming. The quadcopter component consisted of: Arduino Nano, electronic speed control (ESC), accelerometer gyroscope sensor, NRF24L01 transceiver, 2x16 LCD, analog joystick, brushless motor, and propeller. Programming using Arduino language.

2. Research Method

This study applied a quantitative research method. Data was obtained using questioner which includes the student's motivation and interest in the use of Arduino Nano based quadcopter in learning. The questioner consisted of four alternative answers: strongly disagree, disagree, agree, and strongly agree. Instrument validation is done by the lecturer. The study conducted in the Electronics Engineering Department of 1 Magelang public vocational high school. The number of the subjects in the study was thirty-two grade XI students.

3. Topic Learning Materials

The topic plan was used by the teacher as a reference in microcontroller material topic. The stages of learning activities consist of ten topics to facilitate students in learning microcontroller. Each topic was delivered in sixty minutes.

Table 1. Learning Materials

Lesson	Topic
1	Introduction to electronic digital and embedded systems
2	Arduino embedded system
3	Arduino language programming
4	Internal timer
5	Internal analog digital converter (ADC)
6	Electronic speed control (ESC) and internal pulse width modulation (PWM)
7	Serial communication
8	Sensor accelerometer and gyroscope
9	Liquid crystal display (LCD)
10	Transceiver NRF24L01

The following is a description of each topic.

3.1. *Introduction to electronic digital and embedded systems*

In this topic, the teacher provided an overview of electronic digital and its application in the real world. The teacher also gave motivation to the students regarding to the goal to be achieved after learning activities. After that, the student was introduced to the embedded systems. The students were given with an overview of the kinds of embedded systems and its application in the real world. Then the teacher explained the advantages and disadvantages of the embedded systems.

3.2. *Arduino embedded system*

The students listened to the teacher's explanation of Arduino Nano architecture, block diagrams, memory addressing, and input/output ports.

3.3. *Arduino programming language*

The teacher explained to students about Arduino programming with the C language includes its structure, syntax, variables, mathematical operators, comparison operators, and structuring structures. The teacher presented the problems related to Arduino programming with C language then students tried to solve the problem with the guidance of the teacher.

3.4. *Internal Timers*

Students tried to learn the material by reading from microcontrollers book and articles from the internet. The material learned includes how to the timers' work, the typeof the timers, and the timer applications on the system. In the final stage, students tried to write a program in Arduino using a timer and then they need to observe the output using the oscilloscope.

3.5. *ADC*

In this material students were challenged to read data from the output of a joystick and potentiometer. The students tried to write a program to read the ADC input which is in an analog joystick form and then they tried to display it to the LCD.

3.6. *ESC and PWM*

The teacher explained the ESC function in a system includes: how ESC works, how program the ESC and how to control ESC with PWM. The students learned about ESC by reading books and article from internet. The teacher delivered a little description of the PWM and its application to a system. The students were also challenged with a case: how to regulate the speed of a brushless and ESC motorbike using PWM that supplied with DC voltage through the converter in the form of ESC [18].

3.7. Serial Communication

The teacher explained about serial communication in Arduino, how to program, and its application to the system. Students were also challenged by a case: how to program the accelerometer sensor and gyroscope to adjust the balance in a system.

3.8. Accelerometer and gyroscope sensor

The teacher explained about the accelerometer and gyroscope function of a system, then the students learned it more by reading books and articles from internet.

3.9. LCD

Students listened to the teacher's explanation of the 16x2 LCD port, how to install the system, and how to program. After that student challenged to write a program that can create the menu of a system and displayed on the LCD.

3.10. Transceiver NRF24L01

Students tried to write a program to the transmitter at the remote so they can send signals according to the pattern of changes in the joystick.

4. Task Discussion

The focus of task discussion was the assembly and the programming of the quadcopter. This section included remote and quadcopter systems. For the remote system, it was consisting of analog joystick reader, LCD, and a transceiver with NRF24L01. Meanwhile, for the quadcopter it was consisting of accelerometer/gyroscope sensor, reception with NRF24L01, and PWM on a brushless ESC motor.

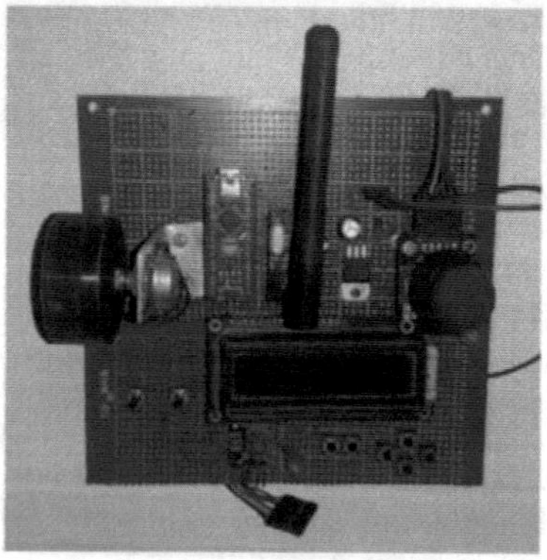

Figure 1. Remote control based Arduino Nano

Figure 2. Quadcopter based Arduino Nano

4.1. Remote Control System

In this project, students were challenged to assemble and write a program for an analog joystick so that it can be read by Arduino Nano through the ADC port and displayed on the LCD. The programming activities included of four ADC inputs which are used for up and down, back and forth, left and right controls, and turn left and right rotate. For make sure the ADC output is good, the calibration conducted to the ADC input. The programming activities started by making a simple LCD menu first. After that, the students proceed with the ADC program and displayed the result on the LCD. After all ADC inputs can be read properly, then the data transmitted to the quadcopter with NRF24L01.

Journal of Physics: Conference Series **1413** (2019) 012003 doi:10.1088/1742-6596/1413/1/012003

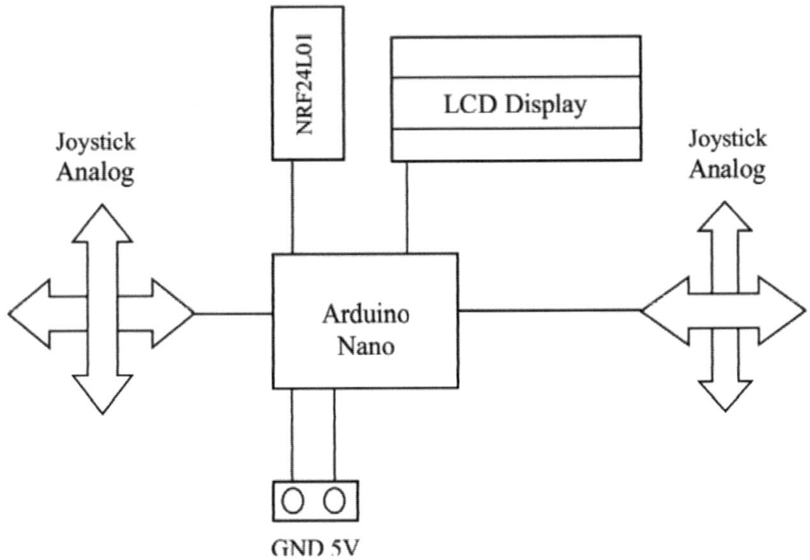

Figure 3. Remote Control Diagram

4.2. Quadcopter system

In this project, students tried to write a program for the Arduino so it able to read the data received by NRF24L01. After the data can be received by the NRF24L01 receiver, the students continued it by trying to use the data to control the brushless motor movement through ESC. The students also wrote a program to control one BLDC motor. After it worked properly, then they challenged to write a program to control two BLDC motors, and finally the students need to wrote the program to control all four BLDC motors. Motor speed adjusted by ESC program that written using PWM. After the BLDC motors can be programmed and the speed can be adjusted, then the students need to wrote a program to control the quadcopter according to the remote control's commands such as the up and down movements, back and forth, left and right, turn left and right rotate. Finally, a program also needs to write to control the balance and stability the quadcopter by utilizing the accelerometer and gyroscope.

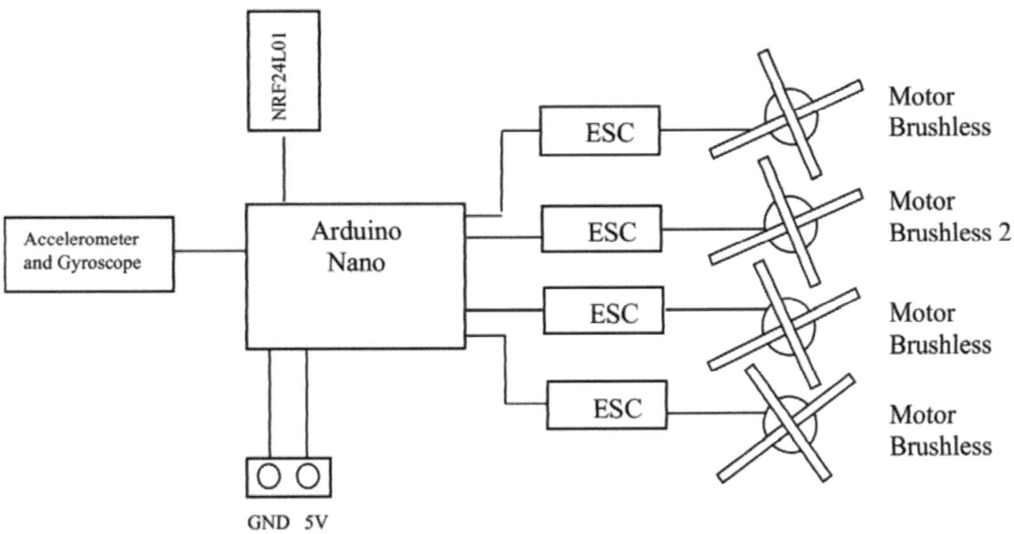

Figure 4. Quadcopter Diagram Based on Arduino Nano

5. Result of Methodology

In this section, an evaluation is carried out to determine students' responses to the learning process that has been carried out. The evaluation carried out was about students' motivation after following the learning. The following are the results of student scores.

Table 2. Survey Result

Question	Strongly Disagree	Disagree	Agree	Strongly Agree
My knowledge increased after studying with Arduino Nano based quadcopter	6%	14%	55%	25%
The ability to program with my Arduino increased after learning with Arduino based quadcopter	2%	16%	48%	34%
The use of Arduino Nano based quadcopter makes it easier for me to learn microcontroller material	5%	18%	38%	39%
The use of Arduino Nano based quadcopter supports in microcontroller learning	9%	19%	40%	32%
My motivation increased after learning to develop programs with Arduino based quadcopter	0%	21%	53%	26%
The Arduino Nano based quadcopter gives me an attraction	4%	22%	55%	19%
Learning with Arduino Nano based quadcopter is interesting because it fits the application in real life	3%	15%	24%	58%
My activity increases when learning with Arduino Nano based quadcopter	6%	11%	56%	27%
My creativity increases when learning with Arduino Nano based quadcopter	9%	16%	46%	29%

Based on the results of the study in table 2 the calculated data is the choice of agreeing and strongly agree. The results showed that most students felt the knowledge of microcontroller increased by 80%. A total of 82% of students felt the perfection of programming with the Arduino increased. As many as 77% of students find it easier to learn microcontrollers. 72% of students feel Arduino Nano based quadcopter supports in learning microcontroller. 79% of students answered that motivation increased in microcontroller learning. Most students view Arduino Nano based quadcopter as having a good attraction of 74%. A total of 82% of students are interested in learning microcontroller by using Arduino Nano based quadcopter. 83% of students felt that activeness in learning increased when participating in microcontroller learning. As many as 78% of students answered their creativity increased when microcontroller learning. Based on these data, the use of quadcopter based Arduino Nano is very supportive of student learning activities.

6. Conclusion

The use of Arduino Nano based quadcopter in microcontroller learning helps students in learning, increased their knowledge, and the skills in system development. Arduino Nano based quadcopter application also increased the motivation and activeness based on the results of survey responses from students in learning. The application of the problem-based learning model in the learning of microcontrollers using Arduino based quadcopter increased students' motivation and creativity because the students tried apply the concept they received in the class to real life.

7. References

[1] Bo S and Li W 2010 *International Conference on E-Health Networking Digital Ecosystems and Technologies (EDT)* Shenzhen pp 375-378

[2] Chao M, Qingli L, Zhongyuan L and Yu J 2010 *2nd International Conference on Education Technology and Computer* Shanghai pp V1-35-V1-38

[3] Winzker M and Schwandt A *2009 IEEE International Conference on Microelectronic Systems Education* San Francisco CA 2009 pp 89-92

[4] Purahong B 2013 *Proceedings of 2013 IEEE International Conference on Teaching, Assessment and Learning for Engineering (TALE)* Bali pp 29-32

[5] Felder R M, Woods D R, Stice J E, and Rugarcia A 2000 The future of engineering education II. Teaching methods that work *J. Chem. Eng. Educ.* 34: 1

[6] Lo D C, Qian K and Hong L 2012 *IEEE 2nd Integrated STEM Education Conference* Ewing NJ pp 1-4

[7] Li S C and Chun K K. 2011 *IEEE 11th International Conference on Advanced Learning Technologies* Athens GA pp 129-130

[8] Hernandez B A 2014 Teaching introduction to robotics: Using a blend of problem- and project-based learning approaches *IEEE SOUTHEASTCON 2014* Lexington KY pp 1-5

[9] Reck R M 2017 *IEEE Frontiers in Education Conference (FIE)* Indianapolis IN pp 1-8

[10] Pardo A 2014 Problem-based learning combined with project-based learning: A pilot application in digital signal processing *XI Tecnologias Aplicadas a La Ensenanza de La Electronica (Technologies Applied to Electronics Teaching) (TAET)* Bilbao pp 1-5

[11] Winarno S, Muthu K S and Ling L S 2017 *International Seminar on Application for Technology of Information and Communication (iSemantic)* Semarang pp 48-51

[12] Krishnan M and Ruhizan M Y R 2009 *International Conference on Engineering Education (ICEED)* Kuala Lumpur pp 122-124

[13] Qian X 2009 *First International Workshop on Education Technology and Computer Science* Wuhan Hubeipp 16-20

[14] Thamarai S S, Kaleel D and Chinnaiah V 2012 International *Conference on Recent Trends in Information Technology* Chennai Tamil Nadu pp 244-249

[15] Galadima A A 2014 *11th International Conference on Electronics Computer and Computation (ICECCO)* Abuja pp 1-4

[16] Novak M, Kalova J and Pech J 2018 *International Conference on Information Technologies in Engineering Education (Inforino)* Moscow pp 1-4

[17] Khavnekar A, Gondalia J and Shah D 2017 *International Conference on Inventive Systems and Control (ICISC)* Coimbatore pp 1-6

[18] Iosifidis D, Alvanos N, Yfoulis C, Papadopoulou P and Galatsopoulos C *2018 UKACC 12th International Conference on Control (CONTROL)* Sheffield pp 428-433

Journal of Physics: Conference Series **1413** (2019) 012004 doi:10.1088/1742-6596/1413/1/012004

A Heart Rate Measuring Trainer Kit as a Medical Electronics Practice Learning Media

D Anugrah and S Waluyanti
Electronics and Informatics Engineering Education Department,
Universitas Negeri Yogyakarta, Yogyakarta, Indonesia

Email: dena.anugrah@gmail.com, waluyanti@uny.ac.id

Abstract. This research was aimed to produce the learning media for a heart rate measuring trainer kit that can be used easily at relatively cheap price to support the learning process of Medical Electronics Practice. This research adapted the Alessi and Trollip's development model which consisted of three stages, namely planning, design, and development. The designs used to test the product was alpha testing and beta testing. The subjects of the trial were limited to four students from the Electronics Engineering Study Program (D3) at Yogyakarta State University. The techniques used to to collect the data were interviews, observation, and questionnaires. The technique used to analyze the data was qualitative descriptive analysis. The result of the research showed that in general the heart rate measuring trainer obtained the average score of 3.49 which was included in the very feasible category. The assessmentof trainer kit was seen from four assessment indicators, namely design, operation, usefulness, and suitability. A trainer kit with a very feasible category could be used in learning Medical Electronics Practice. A heart rate measuring trainer kit made the practical learning process more real.

1. Introduction

Health is a very important factor in human life. By a healthy body condition, human can work productively and it can be assured that all of their organs can function properly. Therefore, maintaining health is an effort that must be done by human at all times.

Globally, cardiovascular disease is the number one cause of death every year [1]. According to data from the Indonesia Health Ministry in 2008 an estimated 17.3 million deaths were caused by cardiovascular disease. Cardiovascular disease is a disease caused by impaired heart and blood vessel function. The heart organs, therefore, need to be monitored regularly.

Heart as one of the organs of the human body has a very important role in the cardiovascular system. Life and death of human can be known from the pulse of heart or not. Heart rate contains electrical signals originating from within the human body and is one of the health parameters that must be monitored at all times. When the heart rate starts to become irregular, it can be considered as a critical sign.

Measurement of heart rate by medical experts is used to diagnose the patient's condition. There are several methods that can be done to measure heart rate such as Palpation, Photoplethysmography (PPG), and Electrocardiography (ECG). All these three methods are clinical and often used by medical experts to measure heart rate. However, commercially available manufacturing instrumentation for this method is relatively expensive, so it can only be used by people or agencies who can afford it.

Palpation method is done by touching the body parts associated with the blood circulatory system such as the arteries to feel the presence or absence of pulses. In general, there are many arteries in the human body, but there are only a few arteries that can be sensed. The radial artery and brachial artery are the best arteries to examine because they can provide reliable heart rate information [2]. Method provides an accurate measure of heart rate [2]. Therefore, arterial pulses can represent heart rate because blood flows in one cycle in the body.

Photoplethysmography (PPG) method is a light radiation technique on the skin and measuring the amount of light scattered through the bloodstream that changes dynamically due to changes in heart rate or changes in blood volume [3]. Heart rate measurement by Photoplethysmography (PPG) method uses a light source from LED (Light Emitting Diode) and photodetector to detect the changes in blood volume in blood vessels [4]. The light source from the LED emits light to the tissue and the photodetector measures the light reflected from the tissue [5]. The reflected light is proportional to the variation in blood volume [5]. Broadly speaking, the sensor used in the Photoplethysmography (PPG) method can measure the amount of light absorbed or reflected by blood [6].

Electrocardiography (ECG) is an instrument used to record the electrical activity of the heart [7]. The electrical activity of the heart can be recorded using electrodes mounted on the surface of the body. The recording of the heart's electrical activity is depicted in a graph called an electrocardiogram. The term "electrocardiogram" is a combination of three words, namely "electro" relating to electrical signals, "cardio" meaning heart, and "gram" which means recording [8]. Based on the meaning of the three words, the electrocardiogram can be interpreted as a recording of heart electrical activity. This Electrocardiography (ECG) serves to monitor the condition of the heart and diagnose cardiovascular disease [9].

Learning about medical instrumentation is not only studied by the students of medical department, but also studied by the students of engineering department. The difference is in the learning focus, the students of medical department more focus on practical use of medical instrumentation while engineering students more focus on designing medical instrumentation. In Yogyakarta State University, Electronics Engineering students at the D3 program also study medical instrumentation in the Medical Electronics Practice course. The status of this new course was required in the year study of 2016/2017 which was previously an optional course. Given the importance of this course as one of the branches of electronics and many industries require the skilled technicians in the field of medical instrumentation as well, and because of no students who take this course every year, so the status of this course becomes a compulsory course. Change in the status of this course has not been balanced with the availability of learning tools that will be used to teach students. One of the learning devices that have not been ready is a trainer kit for practical activities.

There are two important elements in the learning process, namely learning media and teaching methods [10]. Learning media is needed in the Medical Electronics Practice course as a tool that can support the learning process. While, the trainer kit learning media has not yet been available for Medical Electronics Practice courses. Until now, the learning process still uses laptop as a learning media to find and analyze journals relating to the project to be made. The use of laptop in the Medical Electronics Practice course causes the students' learning activities become less active because they only sit in front of the laptop screen during the learning process. Even, most of them are more likely to explore other things beyond learning content. Whereas, the student learning activities can be a determinant factor in the success of learning if all or at least some students are actively involved either physically, mentally or socially in the learning process [11].

Each course certainly has a competency standard that must be achieved, and so a Medical Electronics Practice course does. Based on the fact, it shows that theres is only one standard competency in the course of Medical Electronics Practice achieved in the learning process, while the other one is not achieved yet. Competency standard that has not been achieved is measuring electrical signal originating in the human body. The learning process in this competency is abstract. Therefore, the students are required to think so logically that the learning process needs to be applied in the real life.

To be able to achieve all the competencies of the Medical Electronics Practice course, it needs a learning media trainer kit for measuring the heart rate with various measurement methods as performed by medical experts, in which the price is relatively cheaper than the manufacturer's medical instrumentation, and contains educational elements to support the learning process of Medical Electronics Practice. The availability of trainer kit learning media in the Medical Electronics Practice course is expected that the practical learning process will feel more real. This assumption is of course necessary to be proven through a series stages of research activities. Therefore, the author takes the research titled "A Heart Rate Measuring Trainer Kit as a Medical Electronics Practice Learning Media".

This research was aimed to produce a trainer kit which was feasibility tested so that it could be used as a learning media in the Medical Electronics Practice course. Theoretically, the benefit of this research was that it could add insight into the use of Medical Electronics Practice learning media and could also be used as reference material for relevant future research. Practically, the benefit of this research was that it could produce a trainer kit that had been tested for its feasibility so that it could be used as a supporting media in the learning process of Medical Electronics Practice.

2. Research Method

This research belonged to the type of Research and Development. The development model used in this research adapted the Alessi and Trollip's development model. The development procedure in this research consisted of three stages, namely planning, design, and development as shown in Figure 1.

Planning	Design	Development
• Determining the research scope • Identifying the problems • Collecting the initial data • Conducting a needs analysis through brainstorming	• Making flowchart • Arranging storyboard	• Preparing the trainer kit component • Conducting alpha testing • Conducting an initial revision of the product • Conducting beta testing • Conducting a final revision of the product

Figure 1. Research Procedures Adapted from The Alessi and Trollip's Development Model [12]

The design of product testing used alpha testing involving two media expert lecturers, two material expert lecturers, and beta testing involving some students. The subject of the trial involved four students from the Electronics Engineering Study Program (D3) of Yogyakarta State University. The techniques used to collect the data were interview, observation, and questionnaire. The technique used to analyze the data was qualitative descriptive analysis.

The trainer kit feasibility data obtained from the questionnaire used a combination of quantitative and qualitative data analysis techniques. Data obtained from media experts, material experts, and students in the form of qualitative values will be converted into quantitative values. The rules for giving scores to convert qualitative values into quantitative values were carried out by Likert scale with assessment ranges 1 to 4 as shown in Table 1.

Tabel 1. Scoring Rules [13]

Assessment	Score
Absolutely Agree (AA)	4
Agree (A)	3
Disagree (D)	2
Absolutely Disagree (AD)	1

The quantitative data that had been collected was calculated by the average score using the following formula:

$$X = \frac{\Sigma X}{n} \quad (1)$$

Description:
X = average score
ΣX = score amount
n = assessment amount

This score calculation data was then changed into qualitative values based on 4 categories as shown in Table 2. Rules for converting scores into several categories were used to determine the feasibility level of learning media products being developed.

Table 2. Rules for Conversion of Feasibility Category Scores [13]

Score Interval	Category
$X \geq X + 1.SBx$	Very Feasible
$X + 1.SBx > X \geq X$	Feasible
$X > X \geq X - 1.SBx$	Unfeasible
$X < X - 1.SBx$	Very Unfeasible

Description:
X = obtained score
X = overall score average calculated by the use of formula $\frac{1}{2}$ (maximal score + minimal score)
SBx = overall deviation score calculated by the use of formula $\frac{1}{6}$ (maximal score – minimal score)

3. Results and Analysis

3.1. Planning

Planning was the first stage carried out in this research. At this stage, the researcher determined the scope of the research namely about the learning of Medical Electronics Practice. This course was chosen to be researched because it had several problems in the teaching process. Having been identified, it was found that the problem of its process was an unavailability of trainer kit learning media for Medical Electronics Practice courses.

To support the process of problem identification, so the initial data relating to the learning of Medical Electronics Practice was then collected. Data collection was done by interviewing lecturers in charge of Medical Electronics Practice courses. The results of the interview showed that the competency standard in the Medical Electronics Practice course had not been achieved yet, namely the students' ability to measure electrical signals originating from the human body. Having conducted observation, it was found that there was no one trainer kit that had been used to teach students to practice the theories about Medical Electronics. During this time, the practicum learning process frequently only used laptops connected to the internet network to find knowledge and train students' skills in making a project. The practical learning process liked this seemed to be passive because the students merely sit in front of the laptop screen and tended to explore other things outside the learning content.

Based on the initial collected data, a need analysis was then conducted through brainstorming together with the lecturer in charge of Medical Electronics Practice course. The results of brainstorming showed that it was necessary to develop a trainer kit learning media for measuring the heart rate to support the learning process of Medical Electronics Practice. By the availability of trainer kit learning media, it was expeted that the practical learning process would be more real.

3.2. Design

Designing was the second stage carried out in this research. At this stage, a design concept for the product of Medical Electronics Practice learning media was made by creating a flowchart about the design of the trainer kit as shown in Figure 2.

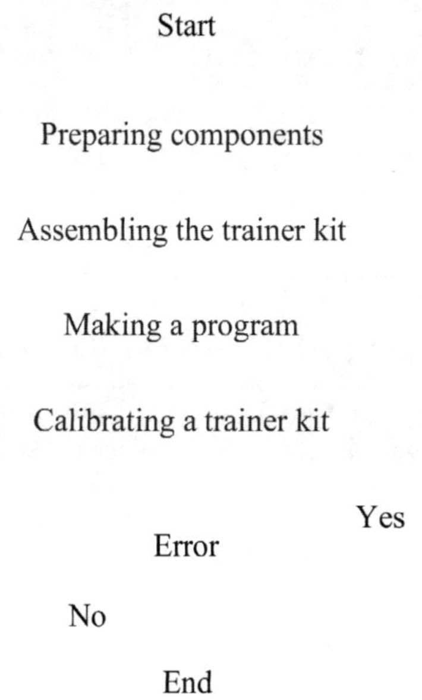

Figure 2. Flowchart of Trainer Kit Design

At this stage, a storyboard was made in the form of a trainer kit block diagram for measuring the heart rate using the Palpation method as shown in Figure 3. The trainer kit hardware design for measuring the heart rate using the Palpation method as shown in Figure 4.

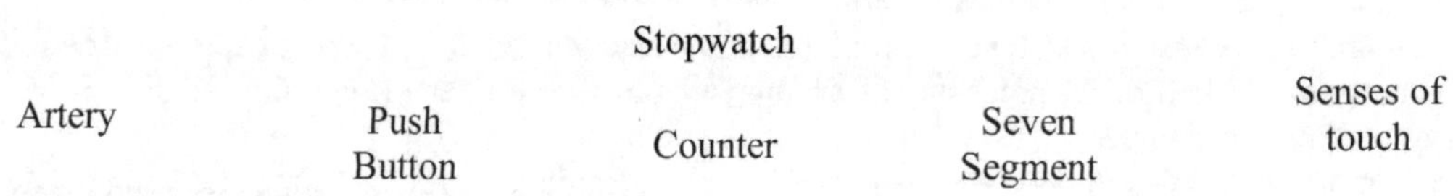

Figure 3. Trainer kit block diagram for measuring the heart rate using the Palpation method

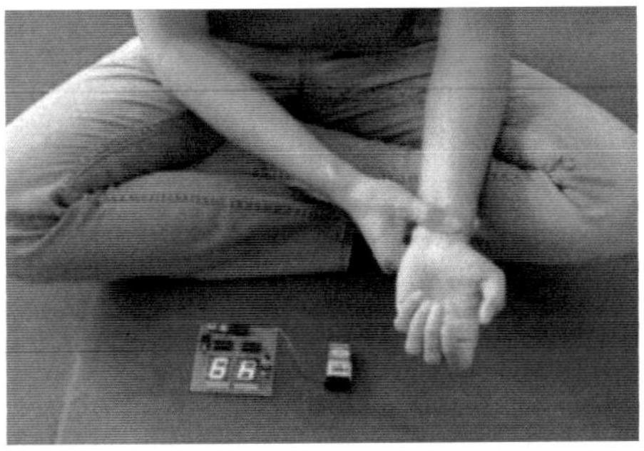

Figure 4. Trainer Kit Hardware Design for Measuring The Heart Rate using The Palpation Method

Besides, a storyboard was made in the form of a trainer kit block diagram for measuring the heart rate using the Photoplethysmography (PPG) method as shown in Figure 5. The trainer kit hardware design for measuring the heart rate using the Photoplethysmography (PPG) method as shown in Figure 6.

Light sensor

Amplifier

Arduino Uno

Figure 5. Trainer Kit Block Diagram for Measuring the Heart Rate using The Photoplethysmography (PPG) Method

Figure 6. Trainer Kit Hardware Design for Measuring The Heart Rate using The Photoplethysmography (PPG) Method

Besides, a storyboard was also made in the form of a trainer kit block diagram for measuring heart rate using the Electrocardiography (ECG) method as shown in Figure 7. The trainer kit hardware design for measuring heart rate using the Electrocardiography (ECG) method as shown in Figure 8.

Electrode

Arduino Uno

Figure 7. Trainer Kit Block Diagram for Measuring the Heart Rate using The Electrocardiography (ECG) Method

Figure 8. Trainer Kit Hardware Design for Measuring The Heart Rate using The Electrocardiography (ECG) Method

3.3. Development

Development was the last stage carried out in this research. At this stage, the product of a trainer kit of Medical Electronics Practice learning media was made based on the design concept that had been made previously. The process of making this product was begun by making a list of the required components and preparing them for making a heart rate measuring trainer kit using the Palpation method, Photoplethysmography (PPG), and Electrocardiography (ECG). Components needed for a heart rate measuring trainer kit using the Palpation method were push buttons, batteries, resistors, capacitors, ICs, diodes, connectors, seven segments, and other supporting components such as cables. Components needed for a heart rate measuring trainer kit using the Photoplethysmography (PPG) method were light sensors, pulse heart rate sensors, Arduino Uno, laptop, and other supporting components such as cables. Components needed for a heart rate measuring trainer kit using the Electrocardiography (ECG) method were electrodes, the AD8232 heart rate monitor, Arduino Uno, laptop, and other supporting components such as cables.

The available components were then assembled to become a trainer kit. This assembly consists of three parts, namely input devices, signal conditioning devices, and output devices. Measuring heart rate using the Palpation method was conducted manually, its input was obtained from arteries, then controlled using a stopwatch circuit, and its output was obtained through the sense of touch using the fingers. Measuring heart rate using the Photoplethysmography (PPG) method was carried out using a trainer kit, its input was obtained from the LED and photodetector as a light sensor, its signal conditioning was obtined by the use of a filter circuit and amplifier circuit available in the pulse heart rate sensor, and its output was displayed using a laptop through a serial monitor. Measuring heart rate using the Electrocardiography (ECG) method was carried out by using a trainer kit, its input was obtained from three electrodes, its signal conditioning was obtained by using a filter circuit and the amplifier circuit avaailable in the AD8232 heart rate monitor, and its output was displayed using a laptop through serial plotter.

Assembled trainer kit was then programmed in order that they could be operated well. Measuring heart rate using the Palpation method was done manually, therefore there was no special program made for measuring heart rate. The program for a heart rate measuring trainer kit using the Photoplethysmography (PPG) method and the Electrocardiography (ECG) method was created using the Arduino sketch.

The trainer kit which has been programmed was then calibrated. A heart rate measuring trainer kit using Palpation method was calibrated by using a stopwatch. The calibration results of a heart rate measuring trainer kit using Palpation method showed that the error rate was 0% for the measurement time of 15, 30, and 60 seconds. By an error rate of 0%, a heart rate measuring trainer kit using Palpation method showed a very good level of accuracy because the error rate value could still be tolerated in the range of measurement 0.6745% [14]. A heart rate measuring trainer kit using Photoplethysmography (PPG) method was calibrated by using Oximeter medical instrumentation. The calibration result of a heart rate measuring trainer kit using Photoplethysmography (PPG) method showcd that thc crror rate was 0.40%. By an error rate of 0.40%, so a heart rate measuring trainer kit using Photoplethysmography (PPG) method showed a good level of accuracy because the error rate value could still be tolerated in the range of measurement 0.6745% [14]. A heart rate measuring trainer kit using the Electrocardiography (ECG) method had not been calibrated because it was difficult to get permission to borrow its medical instrumentation for research purpose. Although a heart rate measuring trainer kit using the Electrocardiography (ECG) method had not been calibrated and the level of measurement accuracy had not been known yet, in general the measurement result could be seen quite clearly from the ECG waves displayed on the Arduino software, despite noise on the waves. The ECG waves pattern was shown in Figure 9.

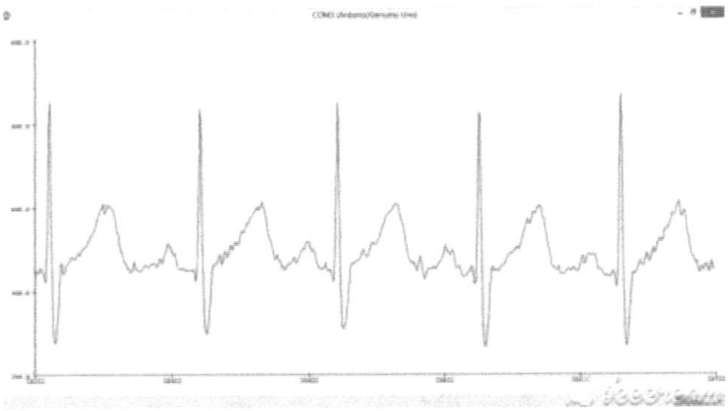

Figure 9. The ECG Waves Pattern in Arduino Software

The calibrated trainer kit was then used to measure the heart rate. Measuring heart rate using the Palpation method as shown in Figure 10. Measuring heart rate using the Photoplethysmography (PPG) method as shown in Figure 11. Measuring heart rate using the Electrocardiography (ECG) method as shown in Figure 12.

Figure 10. Measurement of Heart Rate using The Palpation Method

Figure 11. Measurement of Heart Rate using The Photoplethysmography (PPG) Method

Figure 12. Measurement of Heart Rate using The Electrocardiography (ECG) Method

The trainer kit was then tested using alpha testing by two lecturers of media expert and two lecturers of material expert. In general, the test results of alpha testing for a heart rate measuring trainer kit using the Palpation method, Photoplethysmography (PPG), and Electrocardiography (ECG) obtained the average score of 3.75 and were included into the very feasible category. These test results still required revisions on a small scale so that an initial revision needed to be made regarding to the supply of voltage inputs to the trainer kit from the PLN power source.

The trainer kit which had been tested by alpha testing and got initial revision was then tested again by beta testing in a limited way to the four students of Electronics Engineering (D3) Yogyakarta State University. In general, the results of beta testing for a heart rate measuring trainer kit using the Palpation, Photoplethysmography (PPG) and Electrocardiography (ECG) methods obtained the average score of 3.16 and were included into the very feasible category. Those testing results still required revisions on a small scale so that the final revision of the soldering on a heart rate measuring trainer kit was needed by the Photoplethysmography (PPG) method.

In general, the results of alpha and beta testings showed that a heart rate measuring trainer kit obtained the average score of 3.49 and were included into the very feasible category. A trainer kit with a very feasible category could be used in the learning of Medical Electronics Practice. The analysis of the trial results in the use of the learning media for a heart rate measuring trainer kit made the practical learning process more real.

A heart rate measuring trainer kit using Palpation method served as a timer for 60 seconds. The trainer kit could be controlled manually so that it was easy to adjust to your needs. The workings of this trainer kit was begun by pressing the on button and ended with pressing the off button. It could also be reset by pressing the reset button. The results of heart rate measurements were displayed on two seven segment. The trainer kit part used to measure the heart rate by the Palpation method was marked with a red line as shown in Figure 13.

A heart rate measuring trainer kit using the Photoplethysmography (PPG) method could be used to count the heart rate through the blood circulatory system in the body. The trainer kit used an LED light sensor and photodetector to detect blood flow in the digita arteries inside the finger skin. The trainer kit used a reflection mode where the LED and photodetector were put on the same area, while the finger was put on top of it. The LED functions as a light emitter, the photodetector functions as a light receiver, while the finger functions as a reflector that would reflect the light from the LED. The light from the LED would be spread through the skin tissue of the finger, then the light was absorbed by the blood so that it created a change in light intensity caused by the change in the volume of blood flowing in the digita arteries contained in the finger skin. The emitted light was received by the photodetector. The heart rate detected by the sensor circuit was filtered using High Pass Filter (HPF) and Low Pass Filter (LPF). The filtered signal was amplified by non-inverting operational amplifiers in which its data was then processed by Arduino Uno. The heart rate measurement results would be displayed on the laptop using the Arduino software on the monitor serial menu. The trainer kit part used to measure the heart rate using the Photoplethysmography (PPG) method was marked with a yellow line as shown in Figure 13.

A heart rate measuring trainer kit using the Electrocardiography (ECG) method could be used to count the heart rate by taking samples from the ECG waves per unit of time. The trainer kit used an electrode sensor to detect the heart rate and record the heart electrical activity. Three electrodes were attached on the right arm (RA), left arm (LA), and right leg (RL). The heart rate detected by the sensor circuit would be filtered using High Pass Filter (HPF) and Low Pass Filter (LPF). The filtered signal would then be amplified by non-inverting operational amplifiers in which the data would then be processed by Arduino Uno. The heart rate measurement results would be displayed on the laptop using the Arduino software on the serial plotter menu. The trainer kit part that was used to measure the heart rate using the Electrocardiography (ECG) method was marked with a green line as shown in Figure 13.

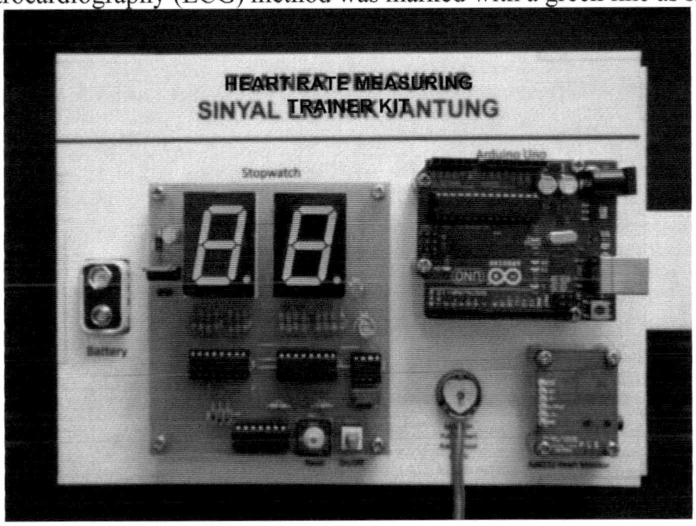

Figure 13. The Final Product of a Heart Rate Measuring Trainer Kit

4. Conclusion and Recommendation

In general, a heart rate measuring trainer kit obtained an average score of 3.49 which was included into the very feasible category. A trainer kit with a very feasible category could be used in learning Medical Electronics Practice. A heart rate measuring trainer kit could make the practical learning process more real.

A heart rate measuring trainer kit that had been considered very feasible needed to be implemented on a large scale practical learning activity. Furthermore, it was necessary to do an analysis of the process and student learning outcomes related to the use of a heart rate measuring trainer kit in practical learning.

5. References

[1] Kementrian Kesehatan Republik Indonesia. 2014. *Situasi Kesehatan Jantung*. Jakarta: Info Datin.
[2] Turner, R. W. D. & Gold, R. G. (1994). *Auskultasi Jantung*. Jakarta: Buku Kedokteran EGC.
[3] Utami, P. & Fajar, M. (2015). Pengembangan Trainer Indikator Denyut Jantung. *Elinvo (Electronics, Informatics, and Vocational Education)*, 1, 65-77.
[4] Yulian, R. & Suprianto, B. (2017). Rancang Bangun Photoplethysmography (PPG) Tipe Gelang Tangan untuk Menghitung Detak Jantung Berbasis Arduino. *Jurnal Teknik Elektro*, 6, 223-231.
[5] Castaneda, D., Esparza, A., Ghamari, M., et al. (2018). A Review on Wearable Photoplethysmography Sensors and Their Potential Future Applications in Health Care. *International Journal of Biosensors & Bioelectronics*, 4, 195-202.
[6] Moraes, J. L., Rocha, M. X., Vasconcelos, G. G., et al. (2018). Advances in Photopletysmography Signal Analysis for Biomedical Applications. *Sensors*, 18, 1-26.
[7] Khandpur, R. S. (2003). *Handbook of Biomedical Instrumentation*. New Delhi: Mc Graw Hill.
[8] Najarian, K. & Splinter, R. (2012). *Biomedical Signal and Image Processing*. Boca Raton: CRC Press.
[9] Rangayyan, R. M. (2002). *Biomedical Signal Analysis A Case Study Approach*. New York: IEEE Press and John Wiley & Sons.
[10] Arsyad, A. 2007. *Media Pembelajaran*. Jakarta: PT Raja Grafindo Persada.
[11] Qurniatush, D., & Sujono. 2015. Penerapan Model Problem Based Learning (PBL) untuk Meningkatkan Aktivitas dan Hasil Belajar pada Mata Pelajaran Teknik Listrik Siswa Kelas X Jurusan Teknik Elektronika Industri di SMK Negeri 3 Boyolangu Kabupaten Tulungagung. *Jurnal Tekno*, 23(1), pp. 1-6.
[12] Alessi, S. M., & Trollip, S. R. 2001. *Multimedia for Learning Methods and Development 3rd Edition*. United States of America: Allyn and Bacon.
[13] Mardapi, D. 2012. *Pengukuran Penilaian & Evaluasi Pendidikan*. Yogyakarta: Nuha Litera.
[14] Cooper, W. D. 1978. *Electronic Instrumentation and Measurement Techniques*. United States of America: Prentice Hall, Englewood Cliffs, New Jersey.

Acknowledgments
The author would like to thank Ms. Dr. Sri Waluyanti, M.Pd., as Chief of the D3 Electronics Engineering Study Program who had guided the author in compiling this paper. The author also thanked to Ms. Dessy Irmawati, M.T., as a lecturer in Medical Electronics Practice Course who had been a key sourcein this research.

Real-Time Audio Spectrum Analyzer to Analyze Loudspeaker

B J R Raharjo and M Zakaria
Electronics and Informatics Engineering Education Department,
Universitas Negeri Yogyakarta, Yogyakarta, Indonesia

Email: benijuniartorahmadraharjo@gmail.com

Abstract. This paper, the writer implements an Audio Spectrum Analyzer application that is used to see the frequency response of the loudspeaker in real-time. This application starts with the creation of the Fast Fourier Transform (FFT) function, which calculates the amplitude value of the captured audio sound. Furthermore, making the equalizer and normalization process to set the highest amplitude value of each spectrum so that it is more easily examined. This Spectrum Analyzer application was developed using the Matlab Guide User Interface (GUI) feature. Inputs are processed in the form of sound signals captured by an RTA microphone and then processed by an external soundcard and displayed in the software. From the test results and comparison of the Audio Spectrum Analyzer application, it can be concluded that this device can also show the exact location of the spectrum.

1. Introduction

The development of technology in the 21st century, especially in the field of electronics is required us to adapt quickly in every change which occurs. One of the development changes is in the field of digital signal processing. The field of digital signal processing addresses many sub-fields including frequency spectrum analysis devices. Spectrum analysis is one way of signal analysis techniques which translates the time domain into the frequency domain, which is usually implemented in industry and scientific research [1].

Spectrum analyzer needs are still in need of further development and depth. The audio field, this analyzer is used to carry out loudspeaker technical data or analyze loudspeakers. Besides, it is also used as a device to determine the frequency response in a loudspeaker [2]. Whereas, in the field of education, for example, audio spectrum analyzer devices are very necessary for the physics and electronics laboratory [3].

In general, Audio spectrum analyzer is currently paid. Thus, it is not easy to obtain. Besides the high prices offered, the audio spectrum analyzer is needed by audio engineering to analyze loudspeakers. Generally, audio technicians currently only use ear hearing. However, every human hearing is different, there will be different views from the analysis of a loudspeaker that is heard [2].

For one of the alternatives to overcome the problem, it is designed a spectrum analyzer audio software in Real-Time combined with hardware of external soundcard and RTA microphone. The software will be implemented using Matlab to generate functions from the audio spectrum analyzer. In the software built, it will be applied the fast Fourier transform algorithm method used to quickly and efficiently convert signals to the frequency domain [4].

2. Research Method

2.1. System Analysis

A system built in this research was a device which could analyze spectrum analyzer. The input from the audio spectrum analyzer was an audio signal, and the output was a real-time frequency spectrum. The design of this device consisted of a hardware and software design. The design of the hardware used to integrate with the software made included a Real-Time Analyzer (RTA) microphone, an external soundcard, and a loudspeaker as the audio source to be analyzed. Whereas software design was developed using Matlab by utilizing the Guide User Interface (GUI) feature by applying the Fast Fourier Transform (FFT) algorithm to convert the signal into a spectrum.

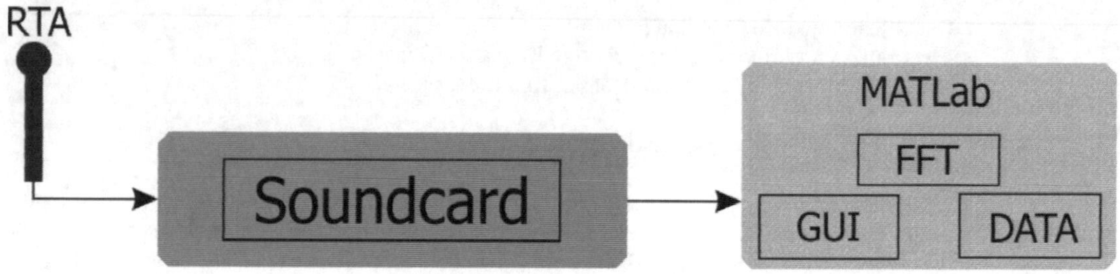

Figure 1. System Block Diagram

Figure 1 shows all system design diagram of the developed spectrum analyzer audio device. The explanation of parts from a diagram in figure 1 are as follow:

a. RTA Microphone
   A device used to receive audio wave signal emitted by the loudspeaker would be analyzed. RTA is a device that measures the amount of sound energy produced by a loudspeaker. RTA only captures sound pressure at certain frequencies and is only limited to its magnitude at the microphone location point [5].
b. External Soundcard
   A device used to process data was audio or sound.
c. Matlab Software
   1) Data
      Data which would be processed into spectrum was audio signal data. This data was obtained from the RTA microphone connected to the soundcard.
   2) FFT (Fast Fourier Transform)
      It was a method used to change signal into the spectrum. FFT Method consisted of some processes such as repairing bits of data taken.
   3) Matlab Guide User Interface (GUI)
      It was a software display consisted of a spectrum display and program menu. In an electronic device to implement FFT, one of which is Matlab. Matlab supports programming that can define sound in the form of FFT [6]. By using Matlab FFT, calculations will be simpler [7], [8].

The system design in figure 1 explains that between devices are integrated with the available ports and USB. After all, devices were connected properly and it could perform their work functions. Furthermore, it is the audio spectrum analyzer software. The software was created by applying the FFT algorithm, the data processed as a digital signal, and the appearance was in the form of a GUI. Fast Fourier Transform was the source of a mathematical algorithm to calculate Discrete Fourier Transform (DFT) quickly and efficiently. Complex calculations from FFT use equation (N/2) log2N for each multiplication and N log2N for complex addition[9]. Audio Spectrum Analyzer software was developed with three main components, such as the FFT calculation section, equalizer, and normalization. The overall design of the audio spectrum analyzer can be shown in Figure 2.

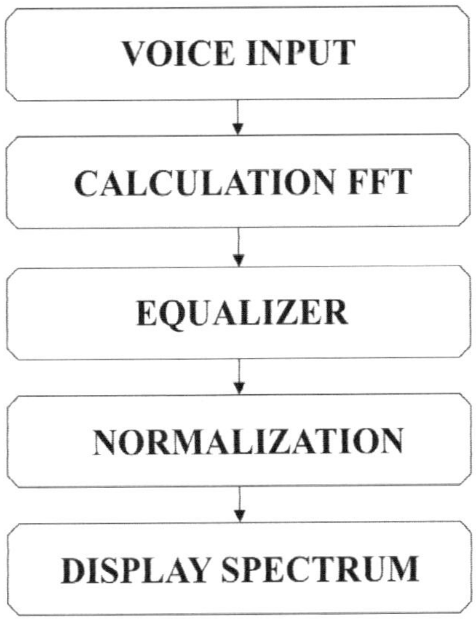

Figure 2. Diagram of the Audio Spectrum Analyzer

Process of FFT calculation implemented in the development of Spectrum Analyzer Audio used method of FFT Butterfly Radix-2 [10]. FFT calculation process was divided into several parts, as follow (1) sorting data, and (2) calculating complex multiplication, in which the results of the multiplication processed were complex number data in the frequency domain [2]. Furthermore, the calculation process in the FFT algorithm would always produce a frequency that has a different amplitude level power, even in certain conditions there can be a large-amplitude between frequencies. As a result, changes occurred at each change in frequency would be very difficult to detect [11]. Steps to overcome this problem the author would apply the equalizer method, the equalizer method was used to create a balance between the frequency levels that occur with existing frequencies [2]. The equalizer can be formulated as follow:

$$X'[n] = F[n] * X[n] \quad (1)$$

In which X[N[as data source of frequency and F[n] is response from Equalizer.

In each snippet of real-time sound caught would be FFT calculation. spectrum will be obtained with different amplitude [2]. The thing that needed to be conducted was to equalize every amplitude that occurs on each spectrum. Equalization of amplitude values on each spectrum used normalization procedures [11]. The process of stages of normalization can be seen in Figure 3.

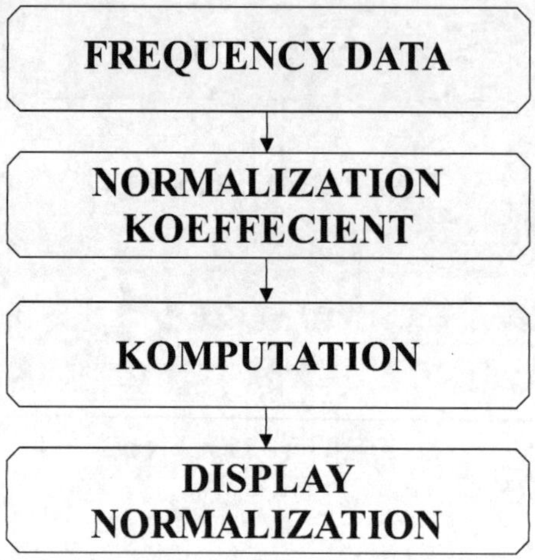

Figure 3. Diagram of Normalization Process

Process of normalization stages conducted was to implementation of algorithm from software of spectrum analyzer audio. Those processes were conducted some stages [2] suh as (1) conducting the highest frequency searching from as spectrum; and (2) conducting the research of normalization coefficient.

$$K_n = 255/\, X_{highest} \quad (2)$$

In which K_n is value from normalization coefficient and $X_{highest}$ is the highest data value from normalization coefficient.

After conducting the previous stage, it then conducted transfer of all data from the spectrum taken with normalization coefficients. Search for normalization values is found with the following formula.

$$X_{norm}[n] = K_n \,.X[n] \quad (3)$$

In which X_{norm} is spectrum data which has been normalized and X is spectrum data.

Each calculation process conducted, the result of spectrum data obtained from the normalization process was then configured as an array. The last step after all the data processes were stored, all the data would be displayed in graphical form on the audio spectrum analyzer software.

2.2 System Implementation
System implementation of the entire assembly and configuration of the tools used was to realize an audio spectrum analyzer. The figure shows the overall implementation of the hardware after being assembled. After all devices could work, the next step was to create a program which could have the performance of audio spectrum analyzer software using Matlab by utilizing the Guide User Interface (GUI) feature. Whereas the hardware components were external soundcard, Real-Time Analyzer (RTA) microphone and loudspeaker.

Figure 4. Implementation of the Whole System

Figure 4 shows program implementation of spectrum analyzer audio which had been designed. The main interface of the program interface designed would display several frequency spectrum with additional features. The features contained in the audio spectrum analyzer developed by the diagram are as follow:

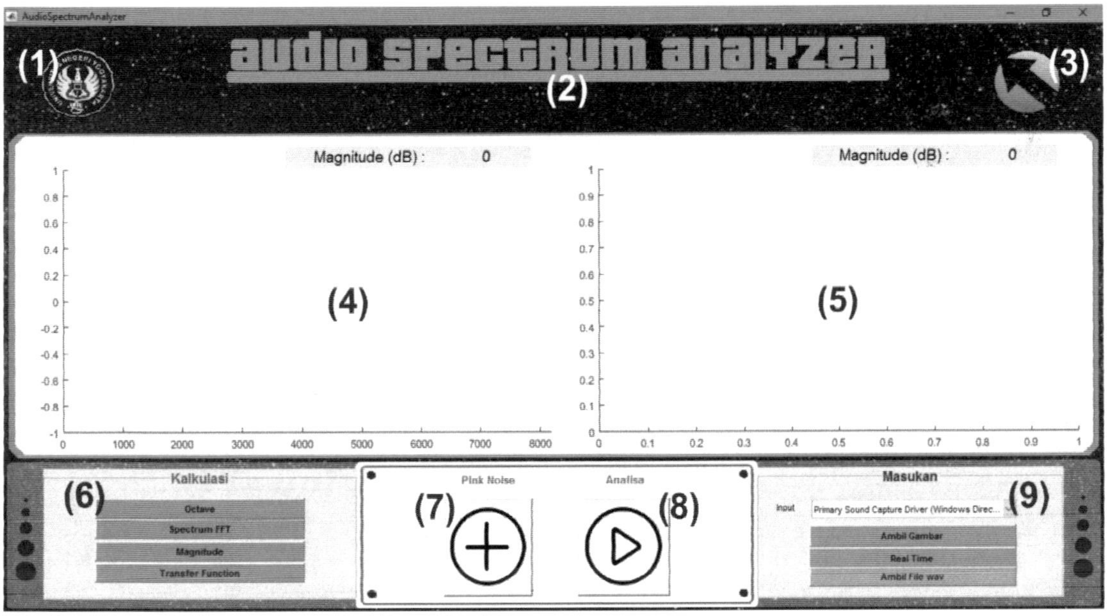

Figure 5. Main/Home Page Display

Table 1. Features of Audio Spectrum Analyzer

No	Information
1	Logo of Yogyakarta State University
2	Software Name
3	Software Logo
4	Display of one channel spectrum
5	Display of two channel spectrum
6	Calculation: a. Octave b. Magnitude c. Transfer Function
7	Pinknoise button
8	Start Analysis button
9	Input: a. Display of External Device Interface b. Real-Time c. Open file from .Wav

3. Results and Discussion

One of the function of spectrum analyzer audio is to be able to communicate with external soundcard. Therefore, to test ability of audio spectrum analyzer software which had been created, it was necessary to test to communicate with hardware. The output obtained from this test was that there was an audio spectrum analyzer software which could communicate with hardware with the appearance of the name of the hardware in the audio spectrum analyzer software.

Table 2. Hardware External Soundcard Communication Test

No	Device	Test
1	EDIROL UA 25	Detected
2	Behringer U-Phoria um2	Detected
3	Focusrite Scarlett 2i2	Detected
4	Edirol FA-66	Detected

Test result in table proves that software audio of spectrum analyzer built can demonstrate being able to communicate with external soundcard hardware from various brands in the market. Further test was conducted by comparing spectrum resulted by application. This would be compared to the results of the smaart audio spectrum analyzer spectrum from Rational Acoustic. The purpose of the measurement was to determine the level of sensitivity of the system built on the audio signals captured by the RTA microphone. The type of audio spectrum used to compare was smaart from Rational Acoustic.The objective of this test was to compare the spectrum form produced by software developed with paid applications.

One method of testing used pink noise.In general, method used was Laudspeaker given signal of Pink Noise input.Pink Noise input is signal sound source which was generally used to analyze loudspeakers/retrieval of loudspeaker technical data. RTA is a device used to analyze the sound energy captured by the microphone used. If it analyzed a loudspeaker by using a music sound source, the results obtained were the sound energy contained in the sound of the music.In analyzing/taking data of laudspeaker, we want to give an input signal in the form of flat/ flat energy at all frequencies. Thus, we can see how influential the change in the loudspeaker is.

Pink noise has the same energy content in each octave and this is closely related to how our ears map frequencies[9]. In terms of retrieving loudspeaker data, pink noise has more important roles and many methods are commonly used [5]. Pink noise is a type of noise with a spectrum which is inversely proportional to the frequency component. Tests were carried out on the monitor speaker Recording Tech RT-5. Tests are compared at frequencies in octave form. An example of a comparison of measurements from an octave spectrum from the results of the development and smaart application is shown in the figure.

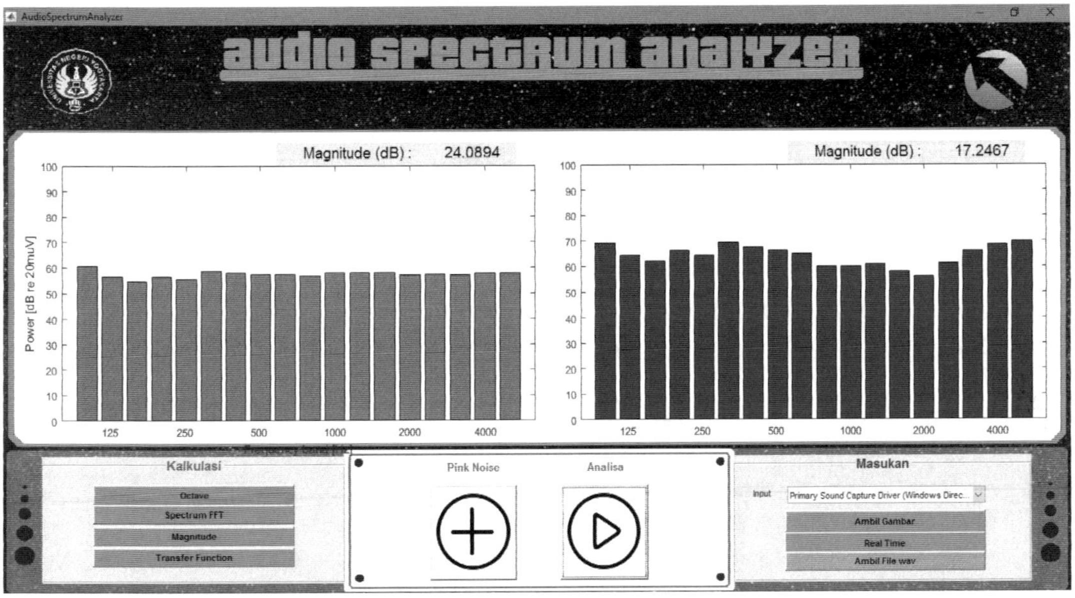

Figure 6. Testing of Audio Spectrum Analyzer

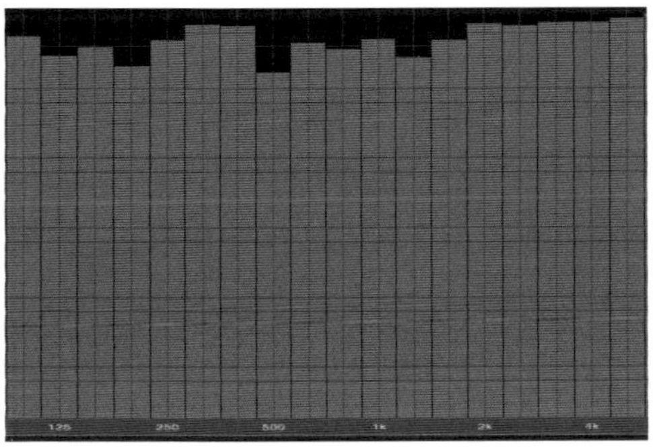

(1) (2)

Figure 7. Comparison of Smaart v.7 (1) and Development Results (2)

Figure 7 shows visualization of spectrum from audio signal of pink noise caught from microphone from a loudspeaker. The form of the spectrum produced by the software developed has similarities, it can be said that the results of the spectrum produced by audio spectrum analyzer software approach the test results with smaart devices. The form of the spectrum produced by having a similar resemblance to the frequency response displayed, thus this result can be categorized as close to the spectrum results of the test equipment with the paid application smaart.v7.

4. Conclusion

Based on the result and the discussion conducted, it can be concluded that the results of the concoction produced are audio spectrum analyzer software combined by hardware in the form of external soundcards and Real-Time Analyzer (RTA) microphones. The results of the testing show that software can communicate with hardware with various brands in the market. Furthermore, the results of the comparison of measurements using the smart v7 audio spectrum analyzer indicate that the system can show the exact location of the audio frequency spectrum.

5. References

[1] C.-P. Chuang, M. Jou, Y.-T. Lin dan C. T. Lu, "Development of a situated spectrum analyzer learning platform for enhancing student technical skills," *Interactive Learning Environments,* 2013.

[2] L. W. Santoso dan R. Sulistio, "Application of Spectrum Analyzer to Analyze Loudspeakers," dalam *Proceedings of the National Seminar on Informatics Engineering*, Surabaya, 2012.

[3] A. N. A. Thohari dan A. E. Putro, "Build a Spectrum Analyzer Using Fast Fourier Transform on the Sigle Computer Board," *IJEIS,* pp. 71-82, 2017.

[4] J. W. Cooley dan J. W. Tukey, "An Algoritm for The Machine Calculation of Complex Fourier Series," *Mathematics Of Computation Journal,* pp. 297-301, 2000.

[5] H. S. Kristianto, Practical Way of Collecting and Interpreting Technical Data of a Loudspeaker, Purwokerto: Hadi Sumoro, 2012.

[6] A. Saxena, A. K. Sinha, S. Chakrawarti dan S. Charu, "Speech Recognition Using MATLAB," *International Journal In Computer Science and Cloud Computing,* pp. 26-30, 2013.

[7] N. Sharma dan S. Sardana, "Designing a Real Time Speech Recognition System using MATLAB," *International Journal of Computer Applications,* pp. 0975-8887, 2016.

[8] K. Sieczkowski dan T. Sondej, "A Method for Real-Time Data Acquisition Using Matlab Software," *International Journal, Mixed Design of Integrated Circuit and Systems,* 2016.

[9] R. G. Lyons, Understanding Digital Signal Processing, United States: Prentice Hall, 2012.

[10] J. G. Prolakis, Pemrosesan Signal Digital, PT Prenhallindo, 1995.

[11] C. L. Phillips, J. Parr dan E. Riskin, Signal, Systems and Transforms 4th ED, Prentice HALL, 2007.

Mini CNC Design to Increase Students' Programming and Control CNC Competencies

E Prianto, T H T Maryadi, Sunomo, C A J Malik, M W Purnomo, and O M Caecaria
Electrical Engineering Education Department, Universitas Negeri Yogyakarta, Yogyakarta, Indonesia

Email : eko_prianto@uny.ac.id

Abstract. Practical tools used in learning must support and adapt to the needs of the Industrial Revolution 4.0 era with the characteristics of working automatically, 3D printing, internet-based work of things (IoT) and Data of Things. One practice tool that works with these principles is a CNC machine. Mini CNC design that has been made has been equipped with modern computing technology. Mini CNC learning media aims to make students not only skilled in terms of operating the machine, but students can develop their skills in the design and programming of a CNC machine. This research was conducted based on the ADDIE model, namely Analysis, Design, Development, Implementation and Evaluation. The results of the design and manufacture of IoT-based Mini CNC 3D printers for learning media consist of designing hardware and software. Linearity test results showed there was no significant difference between the setting value and the measured value of the results of the work process of the tool and being able to do the cutting process from wood, fiber or aluminum, so that the Mini CNC was suitable for use as a learning media.

1. Introduction

Graduates competency of Vocational High School is currently faced with meeting the needs of the workforce in the Industrial Revolution Era 4.0. The process of producing goods and services using digital technology is better known as "Industry 4.0". This term was first known in Germany in 2011. German Chancellor, Angela Merkel, introduced the term at the WEF 2015 annual meeting. Industry 4.0, which is integrating the online world with industrial production. The current industry 4.0 applications include fields such as: digital printing 3D, routers, Internet of Things (IoT), smart factories, robotics to innovative systems of production of goods and services integrated with information and communication technology (ICT) [1].

The availability of instructional media is still limited and there is no development of suitable learning media in Vocational High Schools, especially in schools that are located far from urban areas, so that basic competency goals cannot be achieved. This has an impact on learning outcomes that are still low on average with minimal completeness criteria that have not been achieved [10]. The competency of Vocational High Schools graduates who supported the era of the Industrial Revolution 4.0 can be achieved one of them by increasing psychomotor learning which is reflected in the learning practices in school. In the success of learning practice in schools, there are several supporting factors that should be met by a school. One of these factors is the facilities related to school facilities and infrastructure, especially in the field of media for learning or practice equipment. Practical tools used in learning should support and adapt to the needs of this era with the characteristics of working automatically, 3D printing, work based on the Internet of Things (IoT) and Data of Things. Vocational High Schools try to improve the quality of graduates by improving facilities and infrastructure in the form of laboratory learning practice and CNC machining. The aim is to produce graduates who have the skills to operate CNC, especially for industrial processes [9].

One of the practical equipment that works with the principle as above is a CNC machine. The Mini CNC design that was made has been equipped with modern computing technology. CNC learning media aims to make students not only skilled in operating the machine, but students will also develop their skills in designing and programming to make an object. Design using relatively inexpensive materials so that this learning media can be used in schools that need to improve student skills.

1.1. Mini CNC

CNC is one where the function and movement of machine tools is controlled through a program that is prepared which contains alphanumeric-coded data. CNC can control the movement of the workpiece or tool, input parameters such as feed, depth of cut, speed, and functions such as turning on / off the spindle, turning on / off the cooler. There are many types of CNC machines. Common CNC machines are two-axis and three-axis. The two-axis machine can move only vertically and horizontally which are the X and Y axes. The three-axis machine can make movements starting with the three main axes namely the X, Y and Z axes. The Z axis is paralleled with a shaft [5].

CNC machine operations begin by gathering data from programming then extracting it using Computer-Aided Design (CAD) and Computer-Aided Manufacturing (CAM) programs. The programs generate computer files and then extract commands to operate the machine. The program will be transferred through the post-processor and then loaded into the CNC machine to start machining. The flow of CNC machine operations is shown in figure 1.

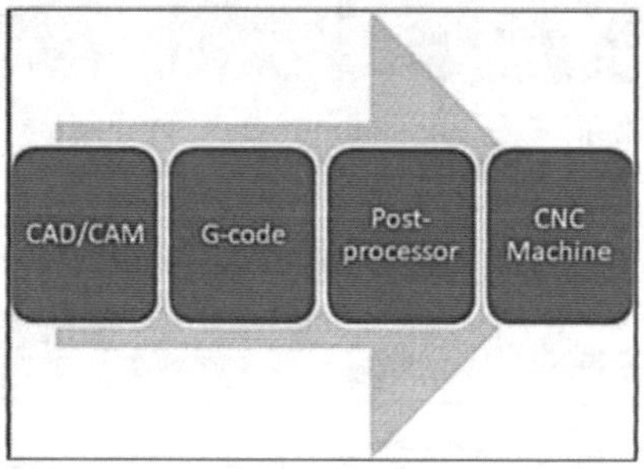

Figure 1. CNC Machine Operation Process [4]

CNC is a system. In a CNC machine system, there are 4 main components namely mechanical design, drive modules, system software and Automatically Programming Tool (APT) postprocessor. Mechanical design system, this part is the hardware part of the machine which is part of the frame. Drive system, the command signal is received from the microprocessor. Microprocessors consist of motors, amplifier units and power supplies. For software systems, the software generates a program to the CNC machine to start the movement of tools and workpieces. For APT post-processors, it was developed to produce G-codes and M-codes that can be used by CNC machines.

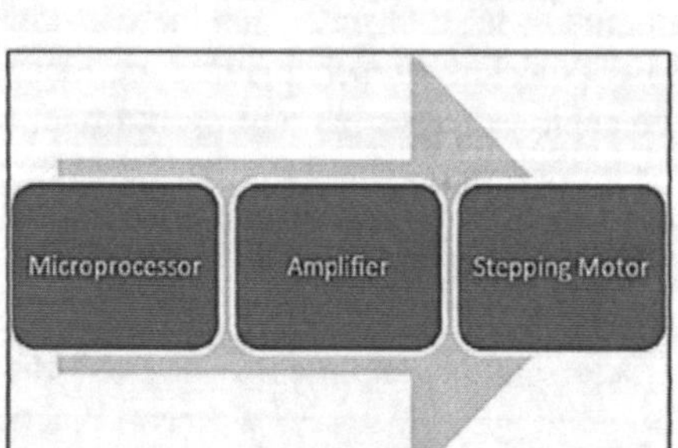

Figure 2. CNC machine workflow [4]

CNC applications cover both machine-tool and non-machine-tool areas. In the machine-tool category, CNC is widely used for lathes, drill presses, milling machines, milling units, lasers, sheet metal processing machines, tube bending machines and so on. Highly automated machine tools such as turning centers and machining centers that automatically change cutting tools under CNC control have been developed. In the non-machine-tool category, CNC applications include welding machines (arcs and resistances), coordinate measuring machines, electronic assembly, laying tape and filament machines for composites and so on [2].

The benefits of CNC machines are (1) high accuracy in manufacturing, (2) short production time, (3) greater manufacturing flexibility, (4) contour machining (2 to 5-axis machining), (5) reduction in human error. The disadvantages include high costs, maintenance, and the need for skilled programmers [6]. This research includes the design, manufacture and application of three-axis mini CNC machine tools to meet the needs of practical equipment in Vocational High School.

Design of Frame Construction in Prototype 3 Axis CNC Milling Machine, which uses electrical components used in the form of power supply, DC servo drive, DC stepper motor, Break of Board produces a prototype of a small CNC Axis 3 milling (mini) as a medium of teaching, training and research [8]. Transfer of Project Development Training To Vocational High School Students To Three Axis Mini CNC Router Design: An Applied Study, has the objective to determine the success of project development and training provided to students [4]. In his research, theoretical and practical training is given to students on the theoretical calculations of machine construction, machine design and manufacturing, the stage of system assembly and analysis and system studies. As a result of the training, students were asked to design and make a three axis CNC mini router. During the design and manufacturing process, the required technical support is provided to students who have problems in making the project. The results of this study, it has been seen that students have successfully transferred the theoretical knowledge they received to the design and manufacture of CNC routers. After the manufacturing process is complete, the sample is processed and the functionality of the mini CNC router machine is tested and the results obtained [4].

1.2. Era Revolusi Indsutri 4.0

The Industrial Revolution Era 4.0 produced an intelligent factory where all used the Internet of Think (IoT). IoT is a concept such as industrial machines, electricity generators, vehicles, household appliances, to wearable devices that are interconnected through networks to exchange in real time [3]. The special characteristic of the Industrial Revolution Era 4.0 is that it occurs through communication between humans, machines and their resources. The fourth industrial revolution was marked by a paradigm shift from being centrally controlled to a decentralized production process. The world of education must also be able to evolve in accordance with the development of existing technology and must be able to predict the needs of labor and technology in the industry. Education must be relevant to current and future employment. In the sense that education must be able to adapt and be flexible both teachers, students, as well as existing facilities and infrastructure. As Dagli and Durmus stated that "Vocational and Technical Education should provide students with appropriate information equipment in accordance with the requirements of the advanced age in their general skills ...... provides a practical training in laboratories equipped with new technology and cooperates with all common working partners "[5].

Major changes are needed in the world of Education to support the industrial revolution 4.0, to realize this, it is necessary to have facilities including infrastructure in learning, namely practical tools as a support for the achievement of education in the industrial revolution era 4.0. Attention to school facilities such as practical tools used in learning are factors that have a strong influence on output. As Ekundayo and Haastrup said in their research journal, "Availability of school physical facilities and the conduciveness of the school learning environment have been said to be potential factors influencing students' achievement in the affective and the psychomotor domains" [6]. The results of this study are expected to produce learning media in the form of mini CNC that can be used in improving the programming and control competence of CNC machines in Vocational High Schools. Competencies that can be supported are competencies in the subject of Mechatronic System Control Engineering. The mini CNC made is expected to meet learning standards in the Industrial Revolution 4.0 era, especially in the IoT section.

2. Research ethod

This research uses the Research and Development method. The study was conducted based on the ADDIE model namely Analysis, Design, Development, Implementation and Evaluation. The process carried out at the analysis stage is to identify what competencies are needed by SMK students in terms of CNC programming and analyzing material requirements, surveying the type of material / equipment, and selecting the appropriate material quality. The design phase is carried out by designing a Mini CNC in accordance with the requirements at the analysis stage. This design is done in the form of hardware and software. The development phase is carried out by making the Mini CNC starting from making and assembling its hardware consisting of mechanical and electronic parts, then installing and connecting the software to the hardware that has been made. The setting is also done on the software so that it can communicate with the hardware that has been prepared, then test the mini CNC performance that has been made. Implementation phase, in this stage the implementation of learning the CNC programming using 3D Printed Mini CNC is implemented. Evaluation is done to get improvements from the implementation that has been done. Improvements in hardware, software, lesson plan and user manuals are carried out at this stage. Data retrieval is done in two ways, namely observation used to get data about the competencies required by vocational students in programming and controlling CNC, and testing the performance of tools that have been made is also carried out in the process of taking data by taking the parameters of the tool that will produce specifications and manuals.

3. Research Results and Discussion

The analysis phase is carried out by identifying what competencies are needed by SMK students in terms of CNC programming, generating data based on the Vocational Curriculum of the Mechatronics Engineering Program. Mini CNC can support the achievement of competency skills in the Mechatronic System Control Engineering Subjects, namely related to the scope of control techniques and regulatory systems, open and close loop control systems, drawing method in control techniques, analog to digital converters and digital to analog converters, logic circuit control and the workings of control circuits with electronic media using analog and digital electronic components, as well as brushed and brushless DC motor controller circuits. All components in the mini CNC support this competency.

The implementation of Mini CNC design is in accordance with the requirements at the analysis stage. This design is done in the form of hardware and software. The design of work flow chart is as follows:

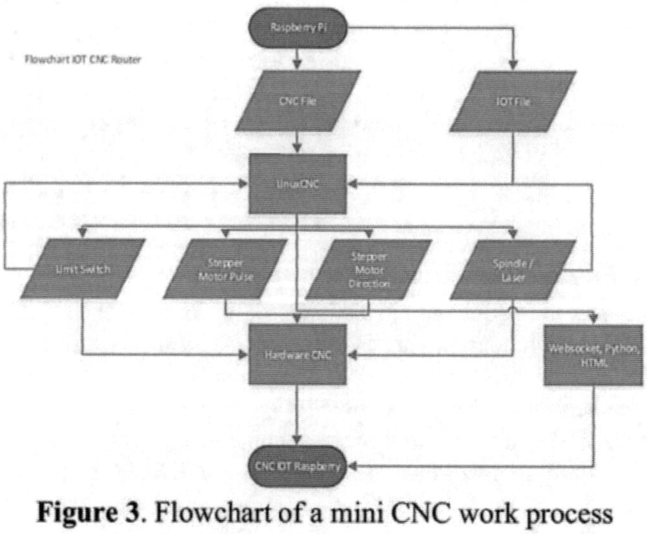

Figure 3. Flowchart of a mini CNC work process

Development of Mini CNC uses Raspberry PI Mini PC with Linux Operating System to run previously created .gcode files. Raspberry PI Mini PCs along with LinuxCNC can control the spindle or laser on or off to spindle or laser speed settings. Mini PCs along with LinuxCNC can control the motion and speed of each Stepper Motor until knowing the maximum working distance of the machine using the Limit Switch sensor. In addition, Mini PCs along with LinuxCNC can be controlled through the IoT (Internet of Things) program that uses python programming language, HTML5, and websocket connections so that this machine is called the IoT-based CNC Mini Low Cost Machine. The process flowchart is used as a basis for making software. The software used is based on the Linux operating system using Machinekit.

Figure 4. 3D Printed Mini CNC design

The development stage produces a system of connections between hardware components as shown in figure 5.

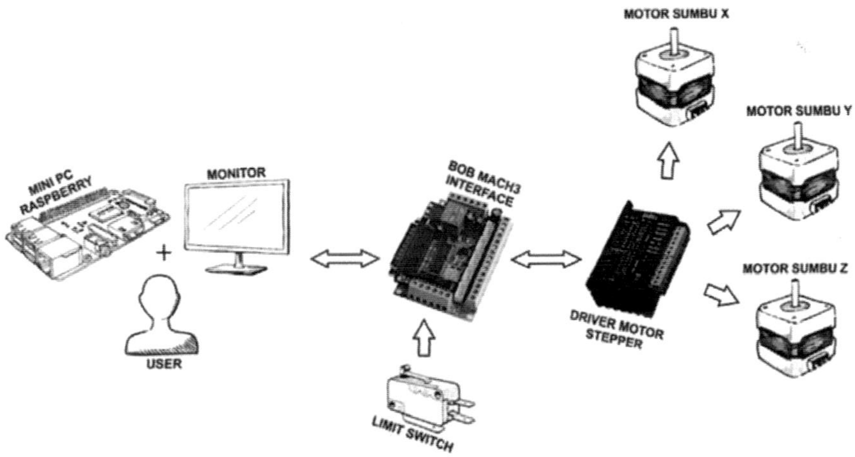

Figure 5. Connection between components on a mini CNC

The Raspberry Mini PC is the center of control and programming of the Mini CNC. Raspberry Mini PCs have input facilities from the keyboard and output on the monitor screen, so that the CNC machine program is done like operating a Personal Computer (PC). Mini CNC programming has been installed in the Mini PC and because of its ability to run programs specifically it can use this equipment. The Raspberry Mini PC makes it possible to explore computing, and learn programming languages such as Scratch and Python. Raspberry Mini PCs are also able to do everything expected from a desktop computer, from using the internet and processing high-resolution videos, to creating spreadsheets, word processing, and playing games. The Raspberry Mini PC has the ability to interact with the outside world with its easily accessible GPIO, this device has been used in various digital projects. One of the outputs of the Raspberry Mini PC is connected to the Break of Board (BOB) Interface where the BOB functions as a terminal in controlling the components contained in the CNC module. BOB is connected to each stepper motor driver, so that the stepper motor rotation can be set based on commands from the Mini PC. BOB also has a terminal that is connected to the limit switch where the limit switch is installed at each axis limit. The limit switch will limit the working area of the CNC machine. If the limit switch is active / pressed, BOB will instruct the motor on that axis to stop rotating because the position is already at the maximum point of the axis.

The result of software development used in developing mini CNC is Machinekit where the software is in the Linux operating system. The Graphical User Interface (GUI) display of the LinuxCNC Machinekit is shown in figure 6.

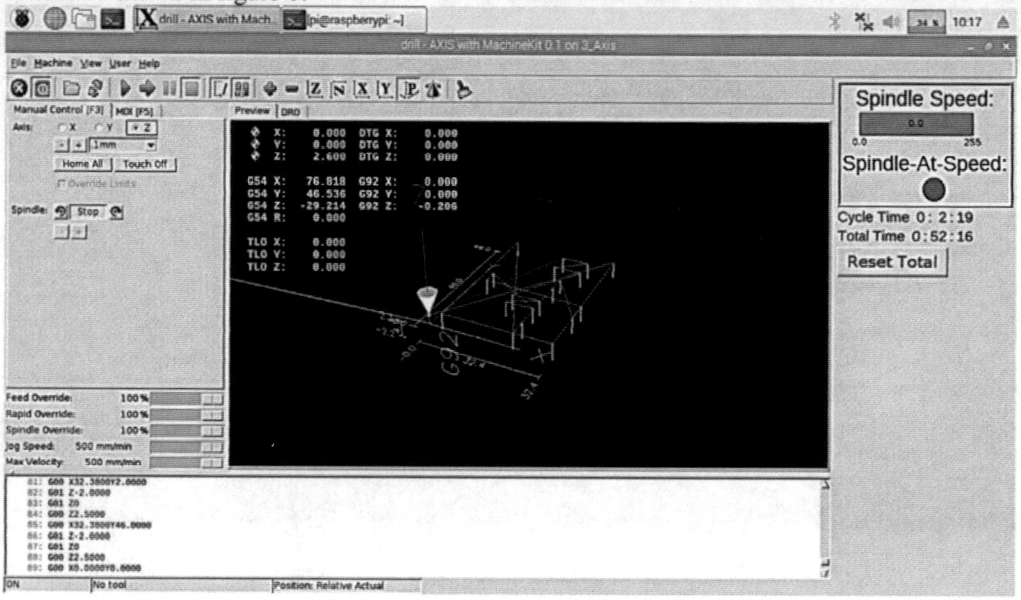

Figure 6. Display of the Machinekit GUI

Table 1. Buttons in the Toolbar group

Buttons	Function	Buttons	Function
	Shutting down the system in an emergency		Zoom in
	Turn on the machine		Zoom out
	Open G-Code file		Display top view
	Reload current file		Rotate the top view
	Run the program file to be executed		Display side view
	Run the next line file program		Display front view
	Pause the execution of program files		Isometric display
	Stop executing program files		Toggle between Drag and Rotate Mode
	Toggle Skip lines with "/"		Clean the application display
	Toggle Optional Pause		

Table 2. Buttons in the Control group

Buttons	Function	Buttons	Function
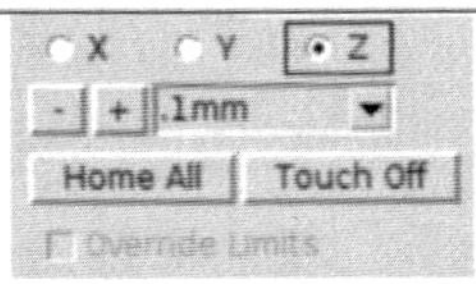	Adjust the sledding motion of each axis		Adjust the direction and speed of the spindle rotation.
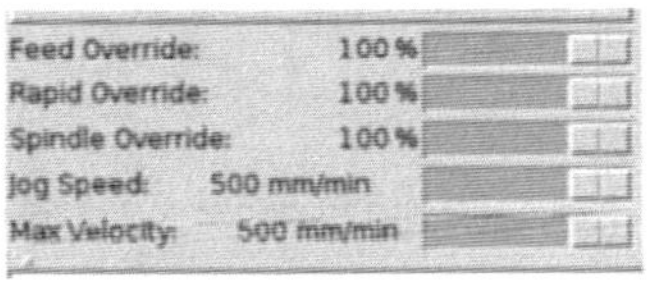	Set the work speed of the machine		Spindle rotation indicator
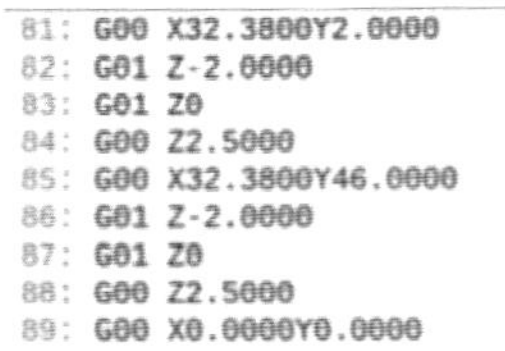	Display program code	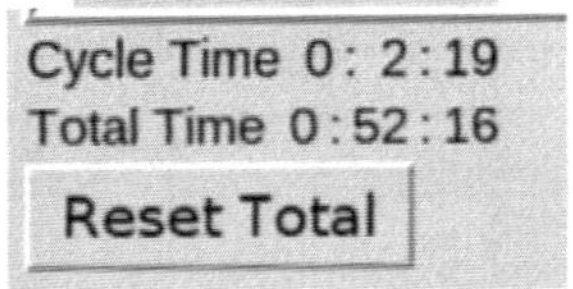	Time indicator when working

The implementation is carried out with the application of CNC programming learning using 3D Printed Mini CNC. The implementation test is carried out in the course of Maintenance and Repair of CNC Mechatronics Engineering Education Study Program where students are asked to identify material requirements, arrange materials, identify work processes, test by making a product in the form of a Print Circuit Board (PCB).

Students in the learning process are asked to produce a product in the form of a Print Circuit Board (PCB). This PCB is obtained from a design created using PCB design software that produces a PCB image file. The image file is then converted into numerical codes used in CNC machine processing. The product results from manufacturing PCBs using mini CNC are shown in Figure 7.

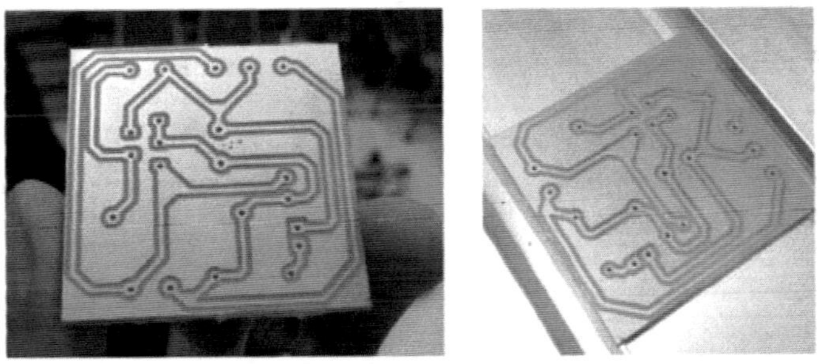

Figure 7. The results of making PCB using mini CNC

The evaluation phase is carried out to get improvements from the implementation that has been done. Linearity test is carried out in this stage to test whether there is a difference between the value of the amount of input (setting) on each axis equal to the results of the work (measured value) mini CNC.

Table 3. Data on the results of the mini CNC linearity test

Positive X-axis (mm)		**Negative X-axis (mm)**		**Positive Y-axis (mm)**		**Negative Y-axis (mm)**		**Positive Z-axis (mm)**		**Negative Z-axis (mm)**	
Setting	Measured Value	Setting	Measured Value	Setting	Measured Value	Setting	Measured Value	Setting	Measured Value	Setting	Measured Value
30	30	30	30	30	30	30	30	10	10	10	10
60	60	60	60,4	60	60	60	60	20	20	20	20
90	90,4	90	90,4	90	90	90	90	30	30	30	30
120	120,4	120	120,4	120	120	120	120	-	-	-	-
150	150,4	150	150,4	150	150	150	150,5	-	-	-	-

From the results of the linearity testing, it can be seen that in general there is no significant difference between the setting value and the measured value either in the X-axis, Y-axis or Z-axis processing. while for the Z axis values ranging from 30 to 90 mm. The value of this setting is adjusted to the maximum length of each axis which then describes the maximum dimensions of the mini CNC. Based on the linearity value, the dimensions of the tool are in the range of 150 mm x 150 mm x 90 mm and are in the form of beams.

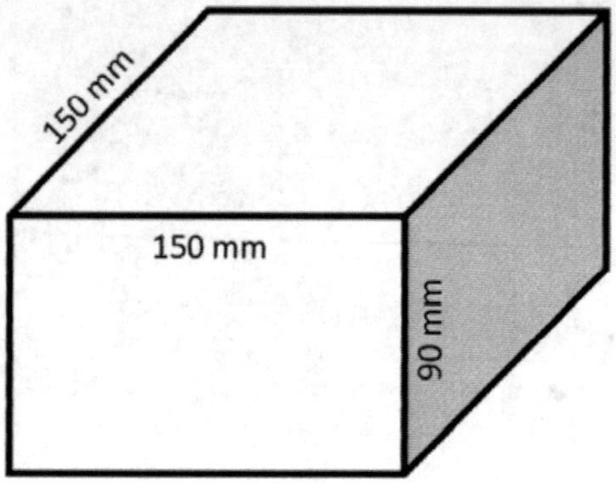

Figure 8. Dimensions of mini CNC work

This can be interpreted that the mini CNC that has been developed can serve the design of the maximum dimensions of 150 mm x 150 mm x 90 mm. The material used in the trial implementation using a cutting tool made from high-speed steel (HSS) can make the cutting process from wood, fiber or aluminum.

4. Conclusion

Competencies that can be supported from a Mini CNC 3D Printer Learning Media are competencies in the subjects of Mechatronic Control System Engineering related to the scope of control engineering and control systems, open and close loop control systems, depiction methods in control techniques, analog to digital converter and digital to analog converter, logic circuit control and the workings of a control circuit with electronic media using analog and digital electronic components, as well as a brushed and brushless DC motor controller circuit where all components in the mini CNC support this competency. The design and manufacture of IoT-based Mini CNC 3D printers for Learning Media consists of hardware and software design.

The performance of the 3D IoT-based Mini CNC printer can be seen from the results of the linearity test where there is no significant difference between the setting value and the measured value of the tool performance test results and is able to do the cutting process from wood, fiber or aluminum, so that the Mini CNC is feasible to be used as a means in learning. Implementation in learning from the Mini CNC 3D printer is carried out in the course of maintenance and repair of CNC on Mechatronics Engineering Education study programs where students are asked to identify material requirements, arrange materials, identify work processes, test by making a product in the form of a Print Circuit Board (PCB).

5. Reference

[1] Anonim. *Revitalisasi SMK untuk Mendukung Suksesnya Industri 4.0.* diakses melalui halaman http://eksis.ditpsmk.net/artikel/revitalisasi-smk-untuk-mendukung-suksesnya-industri-4-0 pada tanggal 14 Januari 2019

[2] Asif Hussain Ansar Md., dkk. (2016). *Features and applications of CNC machines and systems*. International Journal of Science, Engineering and Technology Research (IJSETR),5 (3), March.

[3] Asnan Furinto. (2017). *Menelusuri Inovasi: Esensi Berbagai Studi untuk Inspirasi Inovasi.* Jakarta. Penerbit PT Gramedia Pustaka Utama.

[4] Ferdi YILDIRIM, dkk. (2017). *Transfer Of Project Development Training To Vocational High School Students To Three Axis Mini CNC Router Design: An Applied Study*. International Journal of Education Science and Technology, Aralık/ December 2017, 3 (3), 125-134 e-ISSN:2458-8628.

[5] Ali Durmus dan Abdulkadir Dağlı. (2017). *Integration of Vocational Schools to Industry 4.0 by Updating Curriculum and Programs*. International Journal of Multidisciplinary Studies and Innovative Technologies, Volume: 1, Number: 1 tahun 2017.

[6] Ekundayo, Haastrup Timilehin. (2012). *School Facilities As Correlates Of Students' Achievement In The Affective And Psychomotor Domains Of Learning*. European Scientific Journal, March edition vol. 8, No.6, e - ISSN 1857- 7431.

[7] Nur Suraini Binti Hashim. (2012). *Design of Mini CNC Machine*. Faculty Of Manufacturing Engineering, Unıversıtı Malaysıa Pahang.

[8] Sigit Widiyanto. (2017). *Rancang Bangun Training Kit CNC Mill 3 Axis Sebagai Materi Praktek CADCAM Di STT Duta Bangsa*. Jurnal Sains & Teknologi (JUS TEKNO) Vol. 01 No. 01 Mei 2017 ISSN 2580-2801.

[9] Martono dan Wagiran. (2016). *Developing A Learning Module Of Computer Numerically Control GSK 983 Machines To Enhance Students' Learning Outcomes*. Jurnal Pendidikan Teknologi dan Kejuruan, Volume 23, Nomor 2, Oktober 2016, 184-190 e-ISSN: 2477-2410.

[10] Suyitno. (2016). *Pengembangan Multimedia Interaktif Pengukuran Teknik Untuk Meningkatkan Hasil Belajar Siswa SMK*. Jurnal Pendidikan Teknologi dan Kejuruan, Volume 23, Nomor 1, Mei 2016, 101-109 e-ISSN: 2477-2410.

Development and Testing of Microcontroller-Based Learning Media for the Internet of Things Lab Work

R Asnawi, A C Nugraha, D B Hertanto, and F Surwi
Electrical Engineering Education Department, Universitas Negeri Yogyakarta, Yogyakarta, Indonesia

Email: rustam@uny.ac.id

Abstract. These days, the Internet of things (commonly abbreviated as IoT) has been applied in various industrial sectors such as manufacturing, health, restaurant, electricity generation, transmission, and distribution, etc. Therefore, IoT learning materials need to be included in the learning process both at the college and high school level. The most common obstacle is the lack of learning media about IoT. This paper reports on the process of developing an IoT laboratory work (lab work) learning media based on the Arduino UNO and NodeMCU microcontroller. Furthermore, this paper also discusses the results of the feasibility test on learning media. The stages of the research are need assessment, planning, development and implementation, and also evaluation. Outcomes of this development process are the Arduino and NodeMCU-based IoT lab work modules, short manuals and lab sheets. Product testing with student respondents shows that the Arduino and NodeMCU-based IoT lab work modules, manuals and lab sheets are considered to be very feasible for use in learning. The test results show students considered this module can increase learning interest and make it easier for them to understand IoT learning materials.

1. Introduction

The era of the industrial revolution 4.0 has 4 main characteristics namely cyber physical systems, internet of things, cloud computing and cognitive computing. Cyber physical system can be interpreted as the connection of physical systems in cyberspace. Connections between physical systems will require technology and the concept of the internet of things (abbreviated as IoT), namely the connection of "things" through the internet network [1]. Internet of Things can be interpreted as a connection of everything (things) namely objects or objects around us that can communicate with each other through the internet network [2] [3]. The initial idea of the Internet of Things was first presented by Kevin Ashton in 1999 in one of his presentations. Now many large companies, such as LG and Samsung, are starting to explore the Internet of Things and begin to apply the concept of IoT in its innovative products. Many predict that the influence of the Internet of Things is "the next big thing" in the world of information technology, because IoT offers a lot of potential that can be explored [4] [5].

IoT architecture according to Shivangi Vashi et al. [6], is divided into 5 main layers namely: (1) Perception layer; (2) Network layer; (3) Middleware layer; (4) Application layer; (5) Business layer. Some characteristics of a system that uses the concept of IoT are Intelligence, Connectivity, Sensing, Safety and Energy. To establish connections and communication between things, sensors controlled by the microcontroller will send data to the microcontroller and then the microcontroller will send the data to the internet so that it can be accessed by anyone in anywhere [7] [8] [9]. The basic concept of IoT can be illustrated as in Figure 1. A "thing" can be interconnected with both "person" and other "things" using the internet network. Data communication between "things" can occur anywhere, anytime, and under any conditions. More broadly, the connectivity and data communication between these "things" will be able to be used to support services and any business.

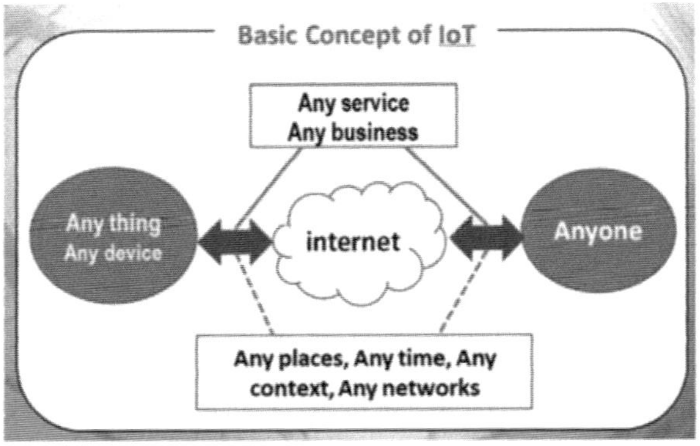

Figure 1. Basic Concept of IoT

The concept of IoT has been used in various sectors of life, for example: smart health, energy with smart grids, management and urban planning with smart cities, housing with smart homes, education with smart classes, and so on. Therefore, IoT must start to be included in the curriculum of primary, secondary and higher education. However, due to the wide scope of this IoT, IoT learning needs a strategy and a level of material depth that is appropriate to the level of education. Specifically at the level of higher education, IoT learning can be a variety of ways and strategies including: (1) IoT as part of the course, (2) IoT as a course; (3) IoT can be a scientific concentration; (4) IoT as a study program.

Meanwhile, in the Department of Electrical Engineering Yogyakarta State University, IoT learning is still limited as part of a course. IoT is given in 2 or 3 meetings to complement material such as in Interfacing, Data Control and Acquisition, and Technology Innovation. IoT learning in JPTE has actually been implemented but the learning modules related to IoT, especially on internet connection material via Wi-Fi (hotspot) are still very minimal. Therefore it is necessary to develop and also test the feasibility of two types of learning media for IoT laboratory work (lab work); one developed using an Arduino microcontroller equipped with ESP8266 Wi-Fi module and the other using NodeMCU.

Because this IoT will be part of a course, Lite-IoT is used, i.e. several layers in its implementation are combined into one as in the Blynk application. The Blynk application is actually a middleware application, but Blynk also combines the middleware layer with the application layer (and also some network layers). The use of the Blynk application to support the development of a product / system with an IoT concept will be more efficient in the manufacturing process.

Blynk is suitable for beginners who are learning IoT, but for expert users, the development of IoT systems that are more complex and require further data analysis will be better if you use the ThingSpeak middleware application. Unlike Blynk, ThingSpeak is more transparent and pure in that it only supports middleware layers. If you use ThingSpeak to develop a product or system with an IoT concept then you need to make a supporting application so that the data on ThingSpeak can be processed into information and can be accessed by anyone and anywhere. In developing this learning module, middleware applications will be used namely Blynk and ThingSpeak.

Blynk is a mobile application that can be installed both on Android and IOS. The Blynk application is specifically designed to support the development of IoT technology-based projects. For Android users, the Blynk application can be downloaded via Google play. While for IOS users, Blynk can be downloaded at the App Store. This special application to support IoT simplifies the project development process because it includes applications in the middleware layer. Blynk supports a variety of hardware that can be used for IoT projects. Blynk is a digital dashboard with graphical interface facilities in making its projects. Adding components to Blynk Apps by means of Drag and Drop making it easier to add Input / output components without the need for Android or iOS programming capabilities. Blynk was created with the aim of remote hardware control and monitoring using internet data communication or intranet (LAN network). The ability to store data and display data visually using numbers, colors or graphics makes it easier to create projects in the field of the Internet of Things.

ThingSpeak is an open source platform Internet of Things (IoT) application that provides API features for storing and retrieving data using HTTP protocol over the Internet or through the Local Area Network. ThingSpeak allows the creation of sensor logging applications, location tracking applications, and social networks with status updates. ThingSpeak was originally launched by ioBridge in 2010 as a service to support the IoT application. ThingSpeak has been integrated with numerical computing support from MATLAB MathWorks. This allows ThingSpeak users to analyze and visualize data uploaded using MATLAB without requiring the purchase of a MATLAB license from MathWorks. The use of ThingSpeak can be free (but there is a delay of about 15 seconds between the read and write data) or even paid.

App Inventor will be used to build an application that will access data on ThingSpeak. App Inventor is a software system for creating applications on android devices online, making applications created with graphical visual interaction instead of using lines of program code. The programmer's interaction with MIT App Inventor is almost completely through a visual interface with drag and drop operations. The programmer only needs to specify the behavior of the application you want to create by arranging the appropriate blocks. One of the advantages of MIT App Inventor is that it can be accessed via the web on an online browser (http://ai2.appinventor.mit.edu), thus there is no need to install the MIT App Inventor software on a computer.

The hardware used for the development of this learning module is Arduino Uno, Wi-Fi module ESP8266-01 (version 1) and NodeMCU. Wi-Fi Module ESP8266-01 is a Wi-Fi module that is commonly used by microcontrollers (usually Arduino) to communicate (make connections) with the internet. Physically, ESP8266 differs from other components, because its power supply needs exactly voltage +3.3V (not more or less). There are many versions of the ESP8266 module from version 01 to version 12. Using the ESP8266-01 Wi-Fi module on the IoT system is quite complex because it requires several wiring techniques and sometimes resetting the firmware and baud rate speed is needed

NodeMCU is a development board (development board) in which the ESP8266 Wi-Fi device is also integrated to make connections with the internet via Wi-Fi hotspot. NodeMCU is a System on Chip (SoC) hardware system based on ESP-12 modules. This device features a 4MByte flash memory capacity, 80MHz system clock, around 50K for usable RAM and an on-chip Wi-Fi Transceiver. NodeMCU can be analogous to the ESP8266 Arduino board. The NodeMCU programmer only requires the same USB data cable extension as an Android smartphone data cable. Because NodeMCU is Arduino-like, for NodeMCU programming it can be used by Arduino-IDE.

2. Module Development Method

This IoT learning module was developed using the steps of the Analyze, Design, Develop, Implement and Evaluation steps [11] [12]. The philosophy of developing this learning module must be student centered, innovative, authentic and inspiring. Analysis is related to the analysis of the situation and the environment, so that it can be determined what products will be developed. Design is a product design activity in accordance with what is needed. Development is a product manufacturing and testing activity. Implementation is an activity using a product, and Evaluation activity evaluates whether every step that has been made is according to specifications.

3. Results and Discussion

3.1. Results of Analysis Phase

Includes the results of need assessment or needs analysis, literature study, syllabus observation or RPS. Need assessment is done by observing and evaluating the learning process of the Full Practice and Data Acquisition courses. The following are the results of evaluations and observations on the course of the practice of Control and Data Acquisition.

a. In the curriculum of S1 Mechatronics Engineering Education Study Program, this lab work course is a new subject in the Study Program. Mechatronics Engineering (PBM has only been carried out twice).
b. The content of this course is closely related to technological developments in the industrial revolution era 4.0, especially about the Internet of Things (IoT).
c. The enthusiasm of students when attending the introductory course is very enthusiastic, especially when given a brief theory about IoT and its application.
d. There needs to be learning about IoT and simple IoT practices in this course.
e. IoT learning is given not at the beginning of the lecture but after the students have completed their competence in control and data acquisition using a single board microcontroller.
f. Learning media for IoT lab work is still minimal and needs to be adjusted to suit the competency demands of the course.
g. This practical learning media designed as simple and as easy as possible for practice for students.

3.2. Results of the Design Stage

The design stage is divided into hardware design and software design. The hardware design prepares the hardware components that will be utilized such as NodeMCU, Wi-Fi module ESP8266-01, sensors and actuators etc. Software design is in the form of designing programs (codes) for Arduino and designing Android application which are all for developing the learning media in the classroom. Next, design a lab sheet and a brief handbook for supporting the IoT Lab work.

For the hardware module that based on NodeMCU, if using ThingSpeak then the design of the hardware module will realize the circuit with sensors, actuators, microcontroller (included the Wi-Fi module ESP8266), which can be depicted in Figure 2.

Figure 2. Hardware Module Design Based on NodeMCU with ThingSpeak

Data from the sensor will be acquired by NodeMCU, and then will be sent to the ThingSpeak database in the cloud. Everyone can access and view the data from these sensors, anywhere at any time as long as there is internet access using an Android smartphone. Android users can control actuators by sending a control data to ThingSpeak using their smartphone. Furthermore, the data will be read by NodeMCU via a hotspot Wi-Fi network, which will then be used to control the actuator. The Android application for reading from and writing to ThingSpeak is built using MIT App Inventor. Meanwhile, using the Blynk middleware application, the schema of the NodeMCU-based hardware module can be drawn as shown in Figure 3 below. Blynk will merge the programming in the middleware, application and partly at the network layer.

Figure 3. Hardware Module Design Based on NodeMCU with Blynk application

Meanwhile, for the design of the hardware module based on Arduino and ESP8266-01 with the ThingSpeak is as described in Figure 4. In principle it is same as the hardware module based on NodeMCU (Figure 2), only in this configuration there are a microcontroller and a separated Wi-Fi module.

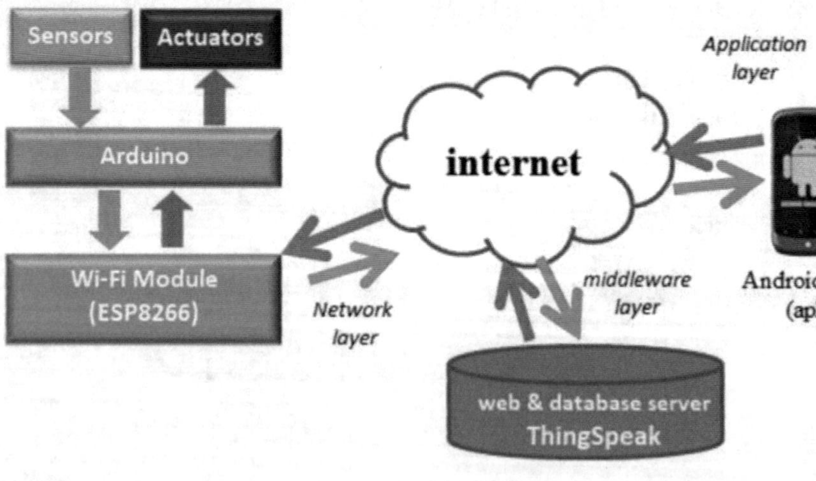

Figure 4. Hardware Module Design based on Arduino and ESP8266 with ThingSpeak

The scheme for hardware modules based on Arduino and ESP8266 with the Blynk application is the same as the scheme in Figure 4, only for the NodeMCU component is replaced with the Arduino and ESP8266 Wi-Fi modules.

3.3. Results of the Development Stage

The development stage is a process to realize the design of both hardware and software that has been made in the previous stage. Along with the development of hardware modules, lab sheet modules and brief handbook were also compiled and developed.

This stage ends with testing the functionality of the developed product. Functionality testing is done by entering various kinds of input data and treatment variations in the process, and observing the output. For the hardware modules that are tested include: data acquisition testing from sensors by Arduino and

NodeMCU, actuator control test with Arduino and NodeMCU, data communication test between Arduino and Wi-Fi module ESP8266, data communication test between microcontroller and ThingSpeak, test data communication between ThingSpeak and Android, and test data communication between Android and a microcontroller via ThingSpeak, test connection and data communication between NodeMCU and ThingSpeak, test data communication between Android and NodeMCU via ThingSpeak, test the connection between Android and the Arduino using Blynk and test the connection between Android and NodeMCU using Blynk.

3.4. Results of the Implementation Stage

In this stage the results of the previous stage (development) that is the learning media products both hardware modules and the lab sheets and the brief handbook are used as teaching aids for teaching and learning process in an IoT lab work class. For this research we used the Data Control and Acquisition Lab work class in the Mechatronics Engineering Education Study Program, Yogyakarta State University. This class has participant 35 students. Next, there were 2 implementations: first implementing the Arduino and ESP8266-01 module and second implementing the NodeMCU module. Both are using ThingSpeak and Blynk application. For the first implementation (using Arduino and ESP8266-01 module with ThingSpeak), all students need more than 3 hours to accomplish all the jobs in the lab sheet.

For the second implementation, using NodeMCU and Blynk Application, the step by step of implementation is as follow. Beginning with distribute the lab sheets and the brief handbook modules to all students. The students are given 10 minutes to read and study the job sheets and the brief handbooks. During in such 10 minutes, the hardware components are distributed to each student. After all is ready, all the students start immediately to do one-by-one instructions in the lab sheet and the stopwatch starts. Once a student finishes all the instructions in the lab sheet, record the time spent of such student. Repeat recording the time spent for all students.

The result shows that from 35 students who took part in the lab work using this learning media, they had all passed. This means they can understand and practice 2-way communication between Blynk application and NodeMCU (sensor-actuators). Only 7 students out of 35 students took more than 1 hour to complete the lab sheet. So it can be resumed that in general students can complete the lab sheet in less than 1 hour. There are even students who only need 40 minutes to complete the entire lab sheet. When compared with Arduino and ESP8266-01, practicing the IoT using NodeMCU (especially with Blynk app) can be faster than using Arduino and ESP8266-01 with both ThingSpeak and Blynk.

3.5. Results of Evaluation Phase

After carrying out a series of research stages ranging from Analysis to Implementation, there are Evaluation results. For the general evaluation based on the questionnaire sheets written directly by all students, the following is a summary of the evaluation.

The material (meaning the basic theory of IoT) in the lab sheet is still lacking, it needs to be added related material in more detail. The codes in the lab sheet are still too difficult to understand and practice. For this problem, there has been a follow up by reviewing programs for both Arduino and NodeMCU. The code lines in the lab sheet need to be given an explanation in each line. The pictures and text in the handbook are still unclear. It needs to be attractive and more interesting. Explanation related to the MIT App Inventor programming need more detail. The IoT learning media in the form of hardware, software and learning modules, there are still some shortcomings, including the ThingSpeak middleware that is used which is still a free account so that there is around 15 second delay in each writing data to the field. The Blynk application still uses the standard (free) version so the features of the components are very limited. The output of this research, those are the hardware module for IoT lab work with ESP8266 and NodeMCU, a brief handbook and lab sheets still need to be updated and improved.

4. Feasibility Test and Discussion

At the implementation stage, after students finishing on the lab sheets and see the results, then students are given a questionnaire to assess this learning media. From the results of the questionnaire it can be analyzed for the feasibility of learning media in the form of the lab sheets, the brief handbook and the IoT lab work hardware module is as follows.

In this study used quantitative data analysis techniques. The assessment data obtained from students were analyzed descriptive qualitatively and used as a reference for revising the learning media products. Eventually it will produce a learning media that is feasible to use. The instructional media products are assessed by students using an assessment sheet. For the lab sheet module and the brief handbook, aspects of the assessment include: the appropriateness, completeness and suitability of the material in supporting IoT competencies, ease in understanding the module, appropriateness of the lab work questions. Then for the IoT lab work hardware module the evaluation aspects include: suitability, appropriateness and completeness of the material with the competencies required in the syllabus especially those related to IoT, ease of compiling and programming, and its attractiveness.

The results of the assessment of these aspects are measured on a Likert scale. Likert scale is a number of statements both positive and negative about an attitude object. The main principle of the Likert scale is to determine the position of a person's attitude in a continuum of attitudes towards the object of attitudes ranging from very negative to very positive [13].

In this study, the answer items for each instrument statement were classified into 5 choices. Each indicator measured is given a score with a scale range from 1 up to 5. A score of 5 means very agree / very appropriate / very appropriate / very good. A score of 4 means agree / feasible / appropriate / good. A score of 3 means quite agree / quite feasible / Quite appropriate / good enough. A score of 2 means less agree / less suitable / less appropriate / less good, and a score of 1 means disagree / not appropriate / not appropriate / not good.

The next step is to assess or test the feasibility of the learning media. The way to do this is by implementing the learning media in a class attended by 35 students. After implementing the learning media, all students fulfill the questionnaire (instruments of the feasibility test). Next the data is obtained, and to determine the weight of each response and calculate the average score using the following formula:

$$\overline{x} = \frac{\sum x}{n} \quad (1)$$

Where:

$\overline{x}$: the average score

N : number of respondents

$\sum x$: each total score

The percentages were calculated with the following formula:

$$Result = \frac{Score}{Maximum\ Score} x100\% \quad (2)$$

According to Arikunto S. [14], as quoted by Ernawati [15], the feasibility category is based on the following criteria:

Tabel 1. Feasibility Criteria of Media

No.	Score in percent (%)	Eligibility Category
1	<21%	Very Poor (VP)
2	21-40%	Poor (P)
3	41-60%	Acceptable (A)
4	61-80%	Good (G)
5	81-100%	Very Good (VG)

4.1. Feasibility Testing of the IoT Hardware Lab Work Module

From the total of 21 statements in the questionnaire provided for the respondents (students), the statements that were subsequently identified were related to the feasibility test for the lab sheet and the manual book. The following statements are related to this: P6, P8, P9, P10, P11, P12, P13, P14, P18, P19, P20 and P21, there are 12 statements in total. The distribution of the answers from 35 respondents is presented in the following table.

Table 2. Results Assessment Questionnaire of the Hardware Module

No	Statement	VP	P	A	G	VG	Sum
1	P6			2	22	11	35
2	P8			1	22	12	35
3	P9			3	14	18	35
4	P10			4	16	15	35
5	P11			1	19	15	35
6	P12			5	15	15	35
7	P13			1	19	15	35
8	P14			4	17	14	35
9	P18			2	13	20	35
10	P19			3	18	14	35
11	P20			1	15	19	35
12	P21			1	18	16	35
	Sum	0	0	28	208	184	420
	Total Score	0	0	84	832	920	1836

Table 1 show that all statement items are dominated by the answers of "Good" and "Very Good". There is not a statement item responded with "Very Poor" or "Poor". Then, if it was analyzed using the statistical formula, the calculation is as follows:

Number of statement items = 12 items

Maximum score = 5 x number of statements x number of respondents
= 5 x 12 x 35
= 2100

Results (percentage of feasibility) $= \frac{1836}{2100} x100\%$
= 87.4%

Based on the criteria table, the results were categorized as very feasible.

From the feasibility testing process, several points can be obtained as follows:

a. The materials in the manual book and the lab sheet are in accordance with the Semester Lesson Plan of the subject of KAD.
b. The manual book and the lab sheet were practical and easy to understand
c. The module for the IoT hardware lab work was particularly in accordance with the competency demands in the subject of KAD lab works
d. The module is user friendly because the manual book and the lab sheet are provided
e. The learning media are generally interesting and highly motivating students' learning
f. The learning media is appropriate to support learning process, especially in the KAD lab works.

4.2. Feasibility testing of the Lab sheets and the Manual Book

From a total of 21 statements in the questionnaire addressed to the respondents (students), the statements that were subsequently identified were related to the feasibility test of the lab sheet and the manual book. The following statements are related to this matter: P1, P2, P3, P4, P5, P7, P15, P16, P17 and P21. The distribution of the answers from 35 respondents is presented in the table 3.

Table 3. Results Assessment Questionnaire of the Labsheet Module and the Manual Book

No	Statements	Disagree	Not Agree	Quite Agree	Agree	Strongly Agree	Total Number
1	P1			2	17	16	35
2	P2			1	22	12	35
3	P3			1	25	9	35
4	P4			2	17	16	35
5	P5			2	20	13	35
6	P7			1	24	10	35
7	P15			2	28	5	35
8	P16			2	16	17	35
9	P17		1		14	20	35
10	P21			1	18	16	35
	Total	0	1	14	201	134	**350**
	total score	0	2	42	804	670	**1518**

Table 3 indicates that all of statement items were dominated by the answers "Agree" and "Strongly Agree". There is only one statement item that was responded by a student with "Disagrees", namely in the statement of "Examples of programs in the lab sheets are easy to understand and practical". However, in general students answered agree and strongly agree. Then, if it was analyzed using the statistical formula, the calculation is as follows:

Number of statement items = 10 items

Maximum score = 5 x number of statements x number of respondents
= 5 x 10 x 35
= 1750

Results (percentage of feasibility) $= \frac{1518}{1750} x 100\%$
= 86.7%

Based on the criteria Table 3, the results were categorized as very feasible.

5. References

[1] Somayya M.,et al. 2015. *Internet of Things (IoT): A Literature Review*. Journal of Computer and Communications, 3,164-173

[2] Tutorialspoint. 2016. *Internet of Things, Simple Easy Learning*. Tutorials Point (I) Pvt.Ltd.

[3] Huansheng Ning. 2013. *Unit and Ubiquitous Internet of Things*. CRC Press Taylor & Francis Group

[4] Zeinab Kamal, et.al. 2017. *Internet of Things Applications*, Challenges and Related Future Technologies. World Scientific News.

[5] Shivangi Vashi et.al.2017. *Internet of Things: A Vision, Architectural Elements and Security Issues*. International Conference on I-SMAC, 2017.

[6] M. Athar & R. Asnawi. 2017. *Monitoring Kecepatan dan Arah Angin Pada Pembangkit Listrik Tenaga Bayu dengan Konsep Internet of Things*. Seminar Nasional Pendidikan Teknik Elektro X.

[7] M. Athar & R. Asnawi. 2017. *Monitoring Kecepatan dan Arah Angin Pada Pembangkit Listrik Tenaga Bayu dengan Konsep Internet of Things*. Seminar Nasional Pendidikan Teknik Elektro X.

[8] Salsabillan Ulfa Tian & R. Asnawi. 2017. *Prototipe Sistem Monitoring Parameter Pembangkit Listrik Tenaga Surya Berbasis Internet of Things*. Proyek Akhir Program Studi Teknik Elektro, Jurusan Pendidikan Teknik Elektro, FT UNY

[9] Ardis Bany Sutrisno & R.Asnawi, 2017. *Monitoring Parameter Pembangkit Listrik Tenaga Surya Dengan Pendekatan Konsep Internet Of Things*. Proyek Akhir Prodi Teknik Elektro, JPTE FT UNY.

[10] ---------. What is Arduino?. Diakses dari https://store.arduino.cc/usa/arduino-uno-rev3 pada 15 Januari 2019

[11] Sugiyono 2010 *Metode Penelitian Kuantitatif, Kualitatif dan R&D*. Bandung: CV. Alfabeta

[12] Wagiran 2013 *Metodologi Penelitian Pendidikan (Teori dan Implementasi)* Deepublish Yogyakarta

[13] Arikunto S, Safrudin AJ 2009 *Evaluasi Program Pendidikan* Bumi Aksara Jakarta

[14] Ernawati I, Sukardiyono T 2017 *Uji Kelayakan Media Pembelajaran Interaktif Pada Mata Pelajaran Administrasi Server* Elinvo (Electronics, Informatics and Vocational Education, Volume 2, Nomor 2, November 2017, Yogyakarta

Health Care Monitoring System Based-on Internet of Things

Taryudi, I Prasetyo, A W Nugraha, and R S Ammar
Faculty of Engineering, Universitas Negeri Jakarta, Indonesia

Email: taryudi@unj.ac.id

Abstract. The development of technology in the era of the industrial revolution 4.0 have led to the emergence of the Internet of Things (IoT). IoT allows an electronic devices to send and capture data through the internet and provides more methods of data interoperability. Nowadays, IoT plays an important role not only in communication, but also in monitoring, recording, storage and display. Currently, the application of IoT in health monitoring system to monitor conditions in the vital signs of the human body becomes interesting issues as an effort to prevent disease. In this article, a system that can monitor vital sign parameters and transmits data through wireless communication was developed, then those data transferred to the network via Wi-Fi modules. Data can be accessed at any time and represent the patient's current condition. So that a nurse or doctor can monitor the patient's data in real time through an application software on smartphone. Overall the system developed consists of DFRobot Firebeetle ESP32 as a controller that has been equipped with Wi-Fi modules, Sensor MAX30100 as a pulse detector and MLX90614 sensor as a body temperature detector. Also added is a GPS (Global Positioning System) module to detect patient locations in real time and an OLED (Organic light-emitting diode) module as data viewer. The test results show that the overall system was worked properly with a 6.5% error for pulse detection, 4% for oxygen detection and 1.7% for body temperature. As for the accuracy of the position generated by GPS that is 7 meters.

1. Introduction

Development in society in somehow have a negative impact on the health condition. For example, currently, more people consume fast food, smoke, drink alcohol, and lack of physical activity, that become the major reason of increased heart disease. Heart disease is one of the main health problems and the leading cause of death in the world. Data from the World Health Organization (WHO) in 2018 stated that more than 17 million people worldwide died from heart and blood vessel disease [1]. More than 75% of deaths from heart and blood vessel disease occur in developing countries with low to moderate income [1]. It's interesting that the current trend of heart disease is not only suffered by the elderly population, but also has been found at a young age. In Indonesia, according to the National Basic Health survey in 2018 showed that 1.5% or 15 out of 1,000 Indonesians suffer from coronary heart disease [2].

Prevention of heart disease can be done by detecting vital signs on the body. It is urgently needing a manufacture and development of tools that can monitor vital sign includes heart rate and body temperature in more effective ways, real time, anytime without complex equipment. Internet of Thing (IoT) become major interesting issues to be applied for patients monitoring in real-time without complex equipment and can be documented well-in the mobile phone.

Previous study conducted by Agung and Achmad, design human body temperature using a Bluetooth communication [3]. The system of sending health condition data information to the user uses the Bluetooth module, but the device does not have a wide range of health condition data sending information to the user using Bluetooth, so there is limited distance for sending data information. In a study by Marti and Setia that development of heartbeat monitoring applications through finger test using Arduino [4]. Application monitoring on this system is accessed using a computer, so it will be difficult to be carried anywhere, where in the era of industry 4.0 almost everything can be done anywhere. Previous studies have shown that remote patient monitoring has a positive impact on the prevention of infectious diseases such as tuberculosis [5] and monitoring the temperature and toxic gas conditions in the room [6]. Therefore, this study aimed to develop an IoT-based health monitoring system that can be accessed anywhere online using a smartphone.

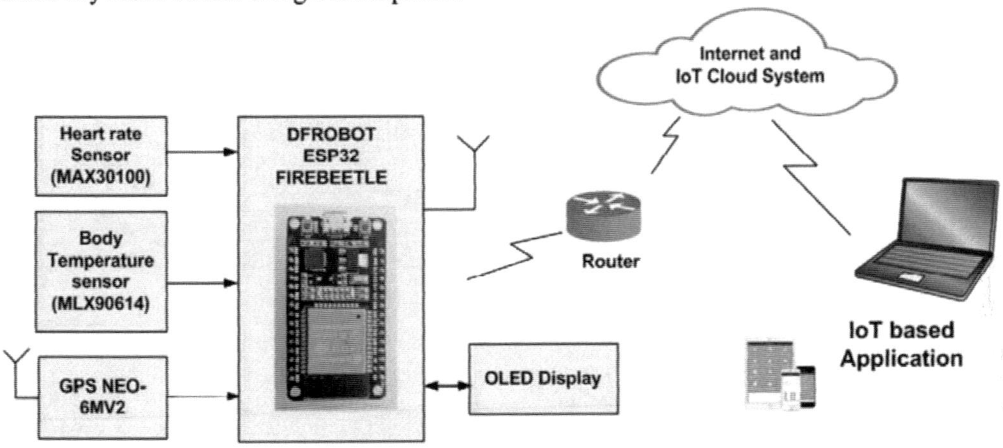

Figure 1. System architecture

2. Materials and Method

2.1. System Architecture

This study was conducted using research and development with four steps to design a hardware and software system. The first stage was searching for definitions, specifications, theories, datasheets, and application notes to determine the specifications of prototype. The results of the analysis in phase 1 were used to determine the specifications of each sub-system of hardware and software. Step three was designed a prototype, then in the final stage this prototype was tested using the specific criteria. The architectural system for monitoring patient health is shown in **Error! Reference source not found.** which consists of two sensors to detect body temperature, heart rate and oxygen levels in the body, GPS module to detect patient location, and OLED to display sensor data. Finally, the controller system was used a microcontroller that has been integrated with the wifi communication system.

2.2. Hardware System Design

This health monitoring hardware system consists of MAX30100 sensor to detect heart rate, MLX90614 sensor to detect body temperature, and Neo-6M GPS module to detect the location. Besides these sensors, an OLED SSD1306 data viewer is also used to display the pulse and body temperature data directly in mobile application. While, the microcontroller was used DFRobot Firebeetle ESP32 to receive data from sensors using serial communication, then process and display it, and send the data wirelessly to the IoT cloud system. In detail the diagram of the system is shown in Figure.

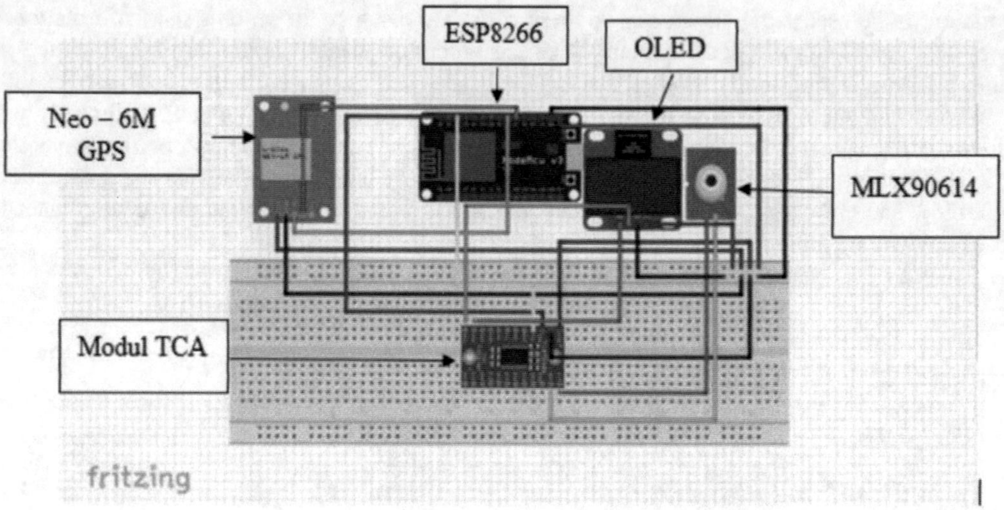

Figure 2. Wiring diagram of the hardware system

2.3. *Software System Design*

Software development in the health monitoring system was carried out in two parts, namely on the microcontroller and smartphone. In the microcontroller, Arduino IDE software is used to read sensor data and process it, then display the data and perform communication with the IoT cloud system. In the smartphone application, a user interface is developed to monitor the location of the system used in accordance with the data generated from the GPS module as shown in Figure and displays the pulse and body temperature data. In detail the program algorithm is explained in the next section and the user interface design of the blink application is shown in Figure .

Figure 3. Location data from GPS

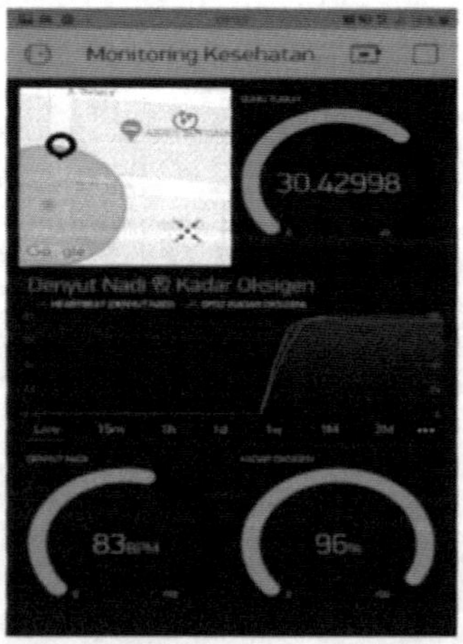

Figure 4. User interface system design

- *Microcontroller System algorithm*
  1. Initialization system
  2. Connect to internet access point until connected.
  3. Read data from all sensors then save to buffer data sensor
  4. Read data from GPS, then save to buffer Location data
  5. Formatting all data sensors and location
  6. Send all data to IoT cloud system
  7. In Blynk application system, read all data package from IoT cloud then display it to the GUI.

3. Results and Discussion

The results of the design and development of the system were prototype hardware controller and the hand-band system as shown in Figure and Figure. While the results of the development of the user interface on an IoT-based application using the Blynk application are shown in Figure. Overall, the system has been integrated and tested on each sub-system and calibrated. The system showed good performance with a 6.5% error for pulse detection, 4% for oxygen detection and 1.7% for body temperature detection. As for the accuracy of the position generated by GPS that is 7 meters.

Figure 5. Controller hardware system

Figure 6. Hand-band sensor system.

However, each sub-system of sensors and GPS as well as the controller still not integrated in a single package, the life time used of power supply was not taken into consideration yet, and the accuracy of the sensor is less accurate. Therefore, further development needs to be done in order to be used at clinical setting.

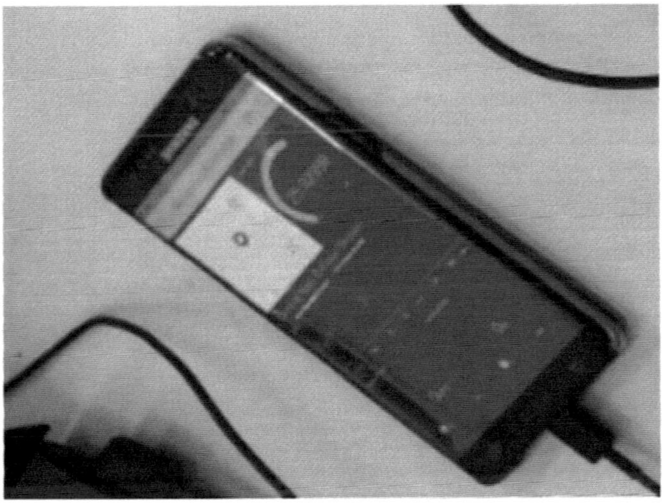

Figure 7. IoT-based application software.

While the results of software development, still using IoT-based applications that are already on the market, so that data security and system reliability are still dependent on the application. For this reason, it is necessary to develop software independently with better system reliability.

4. Conclusion
In conclusion, this study showed that the prototype of monitoring vital sign using IoT had a proper function with a 6.5% error for pulse detection, 4% for oxygen detection and 1.7% for body temperature detection. As for the accuracy of the position generated by GPS that is 7 meters. However, the prototype system still uses sub-system modules separately. Further research need to integrate system in one device such as in a smart watch.

5. References
[1] WHO. *Heart Disease and Stroke Statistics — 2018 Update A Report From the American Heart Association.* 2018. Epub ahead of print 2018. DOI: 10.1161/CIR.0000000000000558.
[2] Kementerian Kesehatan. HASIL UTAMA RISKESDAS 2018.
[3] Wijaya AB. Rancang bangun alat pengukur detak jantung dan suhu tubuh manusia berbasis komunikasi bluetooth. 2010; 1–4.
[4] Sari MW. RANCANG BANGUN APLIKASI MONITORING DETAK JANTUNG MELALUI FINGER TEST BERBASIS ARDUINO.
[5] Taryudi, Daryanto, Irma Darmawati, Heni Purnama LL. Peningkatan Kapasitas Tenaga Kesehatan dalam Monitoring Penyakit Menular Berbasis Internet of Things. *ABDIMAS BSI* 2019; 2: 332–339.
[6] Taryudi ADB, Wahyu A, Ciptoning B. Iot-based Integrated Home Security and Monitoring System. Epub ahead of print 2018. DOI: 10.1088/1742-6596/1140/1/012006

Acknowledgements
The authors would like to express their appreciation and extremely grateful to the student in running this study for their gracious participation.

The Design of Inverted Pendulums as Learning Sources of Adaptive Control

A A Baiti, Mashoedah, M Zakarijah, Suprapto and N Ahwan
Universitas Negeri Yogyakarta, Yogyakarta, Indonesia

Email: aawaluddin@uny.ac.id

Abstract. Inverted Pendulum is a control application which is the basis of science in today's technologies such as missiles, aircraft, satellite control, etc. This inverted pendulum receives input then the parameters are determined and the system responds. This research aims to design an inverted pendulum as a source of learning adaptive control. The design process uses the RnD method with phases including requirements analysis, design, implementation, and evaluation. Inverted pendulum design includes mechanical, electronic and source code design.

1. Introduction

The demand of industrial revolution 4.0 has become an unavoidable challenge **[1]**. Minister of Research, Technology and Higher Education (MoRTHE), Mohamad Nasir, explains that based on the initial evaluation of the country's readiness in facing this current era 4.0, Indonesia has been predicted to have high potential **[2]**. Regarding to these circumstances, it is necessary to reorient the curriculum to be more relevant for a 21st-century world, especially with the massive technological developments [3]. The current technology is not just about turning on/off the machine power but more on controls that can adapt to the surrounding environment. For instance, [4] development of self-driving car or often called as smart car. This kind of vehicle has various automatic sensors that can detect the surrounding environment to operate as a human driver. It urges academicians and researchers to master up-to-date technology to have global competitiveness. To boost technological literacy, a proper model is truly used to represent this requirement.

Inverted Pendulum is one model that represents a simple adaptive control. To learn control in this model, it requires some innovation that involves mathematical functions. This control is commonly used in industries, such as Proportional Integral Derivative (PID) controller, and Sliding Mode Controller (SMC). This control system has been the basis of applications in present technology, like rockets, airplanes, unicycles (single-wheeled vehicles), modern motor vehicles, wheelchairs, satellite position control, games (Cubic Li) and others.

This study aims at designing an inverted pendulum as a learning source of adaptive control. The control in this Inverted Pendulum employes PID (Proportional Integral Derivative). PID was chosen because it became the basis of adaptive control so that students can learn the basic concepts of adaptive control in the form of simple modeling.

Journal of Physics: Conference Series **1413** (2019) 012009 doi:10.1088/1742-6596/1413/1/012009

2. Literature Review

2.1. Inverted Pendulum

The Inverted Pendulum System is a non-linear control system that is unstable as a training module of theoretical control system and various control system experiments [5]. In the control system experiment of the Linear Inverted Pendulum, a pendulum with a center of mass above the pivot point is attached to a linear actuator and is moved to be in balanced condition. Figure 1 below is a general example in basic control system experiments with feedback especially in the design of control systems with PID (Proportional Integral Derivative) algorithm.

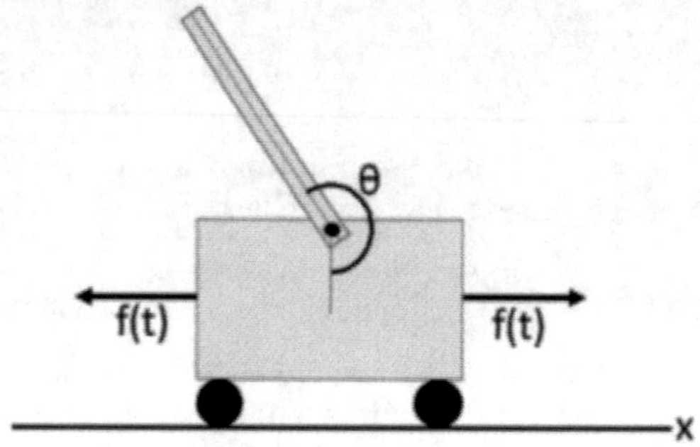

Figure 1. Linear Inverted Pendulum System

Inverted pendulum conditions that are unstable can be an interesting laboratory experimental objects and can be easily connected to real-world applications, such as rocket and missile control systems or electric vehicles with automatic balance systems [6].

There are several studies related to an inverted pendulum that has been done, such as Goegoes Dwi [7] by designing an inverted pendulum with a circular path. In this study, the Inverted Pendulum was designed as a circular or rotary inverted pendulum to eliminate the length limit of the path contained in the pendulum inverted with the transverse path. Another study is from Sudipta Chakraborty thesis [8] that examines the systems and hardware in the Inverted pendulum, system dynamics, state-space models, derivatives in the transfer of functions, and the strategy of making controls on the inverted pendulum. This thesis implemented 2 PID loops, PID + PI and LQR control that showed a satisfying response.

2.2. Mathematics Model

Inverted pendulum systems include classical control problems with the non-linear processes and unstable [5]. Inverted pendulum systems generally have one input signal with several output signals. The main purpose of a linear system of inverted pendulums is to balance a pendulum stick vertically on a motor-driven cart.

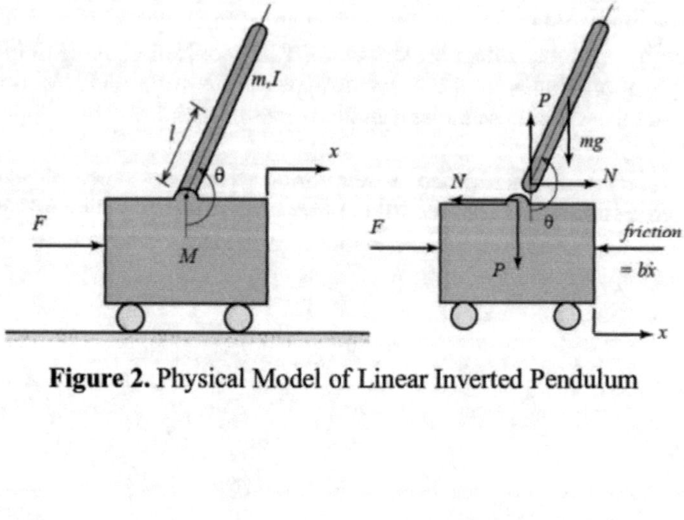

Figure 2. Physical Model of Linear Inverted Pendulum

Image 2 above is a Physical model of linear inverted pendulum system.
M : cart mass
m : pendulum mass
b : coefficient of friction cart
L : pendulum length
I : moment of pendulum inertia
F : given force on cart
x : cart position
θ : pendulum angle

2.3. PID Controller

The PID controller is a classic control that has been used in a variety of control system applications. The PID controller has a simple control structure that can be easily understood. The PID controller has the ability to optimize the performance of a control system by tuning the parameter values that involve Proportional (P), Integral (I) and Derivative (D). In a linear inverted pendulum control system, the use of PID controller can be described in the block diagram as follows.

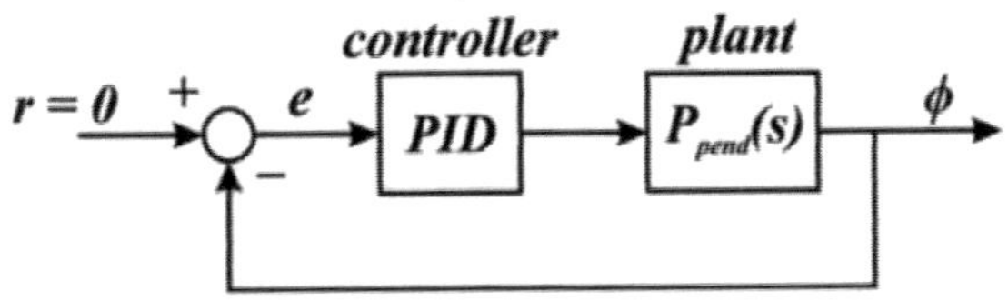

Figure 3. Block Diagram of PID Controller on Linear Inverted Pendulum

In the block diagram above, the inverted pendulum system has one input signal and one output signal. From the block diagram, a transfer function must be made to determine the characteristics of the control system contained in the linear inverted pendulum.

2.4. Learning Sources

Learning according to Oemar Hamalik [9] is a modification or reinforce behaviour through experience. According to this definition, learning is seen as a process or activity and it is not just as an outcome or a goal. It means learning can not be simply defined as remembering, but it is more providing meaningful experience. The outcomes of learning do not only in the form of content mastery but also behaviour change.

To make learning run well, it requires an object as a source of learning. Fatah Syukur NC [10], explains that learning resources are all things (environment, and experience) that can be used to support the learning process more effectively and efficiently. It should be available directly or indirectly whether concrete or abstract. According to Januszewski and Molenda [11] learning resources refer to all sources including messages, people, materials, tools, techniques, and settings that can be used by students both individually and in a group to facilitate their learning activities and improve their learning performance. Similarly, Seels and Richey explain that learning resources are all supporting sources for learning activities, such as supporting systems, materials and environment. Based on the various statements above, it can be inferred that a learning source is a learning object that can make students easier to understand the materials to achieve the learning objectives.

Learning resources are not only tools and materials used in learning, but also people, budgets, and facilities. It can include anything available to help someone to learn. According to Putu Sudira [12] how big the impact of education or learning is strongly influenced by how many students obtain their learning experience and it is influenced by several factors, i.e. teacher/lecturer performance, learning facilities and infrastructure, learning environment, and ICT support. Some research on learning resources, especially for the development of instructional media in engineering field by Eko Marpanaji et al. [13]

creates the PID Controller trainer using dual brushes controllers. Meanwhile, in this study PID Controller employed inverted pendulum.

2.5. Adaptive Control
Adaptive control refers to a system control that is able to adaptively change the way it works according to environmental conditions and these changes are based on units of time [14]. This system can be basically applied to plants with a broader class, in this case if the system is applied then the controller will automatically adjust the system to ensure the plant operation in accordance with the designed parameters. The adaptive control in the inverted pendulum indicates that the system will make adjustments so that the pendulum shaft remains upright. It can be done by reading the rotary encoder sensor if more than 0 degrees then the system will move the slider to the right and left. This inverted pendulum stabilization is also needed as presented by Sekiguchi et al study [15] that the scope of the pendulum's movement in the form of horizontal and vertical should be taken into consideration. The stabilization of vertical and horizontal movements requires a dynamic addition.

3. Research Method
This research employed to research and development (R & D) models. Basically, this procedure was the stage in designing the inverted pendulum as a learning resource for adaptive control. This learning resource will be developed through validation both closed-model (in a small group) and experts judgments or called alpha testing. Beta testing was done afterward and revisions were made based on the suggestions before the real implementation. This development model refers to the Lee & Owens model below.

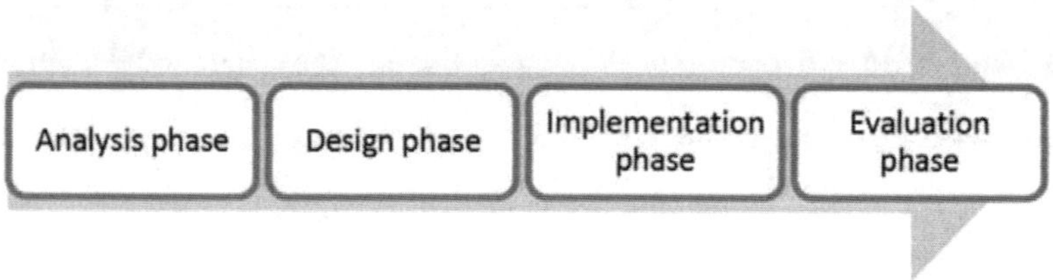

Figure 4. Lee & Bowyer Development Model

Descriptions of the pictures above were: 1) assessment phase, as the reinforcement of the objectives of this study. This phase consisted of need analysis regarding needs in the field and front-end analysis which aimed to bridge the gap between the existing goals and the objectives to be achieved 2) Design Phase, as the project planning so that it can run well 3) Implementation phase, as the realization of the previous phases. This phase involved the alpha and beta test process, 4) The evaluation phase as the stage to answer the core problem. In this phase, the level of accuracy, function, material content and instructions in using the module adaptive control were presented.

4. Finding and Discussion
4.1. Analysis Phase
The research aims at designing a learning model of adaptive control in the form of Linear Inverted Pendulum based on the following findings:

(1) The aadaptive control modelling so far had only been through the MATLAB software and can not be easily understood by students. It raised the importance of media as a source of learning
(2) The students who have studied control materials need a real practice model that had been undertaken, especially on adaptive control
(3) There is no learning model of Linear Inverted Pendulum as the basis for adaptive control learning.
(4) It can be the right momentum to introduce, understand and implement an up-to-date technology, such as control, instrumentation, and manufacturing.

4.2. Design Analysis
This phase consisted of designing both administrative and product design as presented below.

4.2.1. Administrative Design
This design contained the schedule of development activities, team, and provided mapping information.

4.2.2. Product Planning
The initial step before designing a tool was knowing the structure and the system model of the tool. It is important to make it easier to do testing, analysis, and troubleshooting if there are problems in the manufacturing process and tool operation. The overall block diagram of the linear inverted pendulum system is presented below.

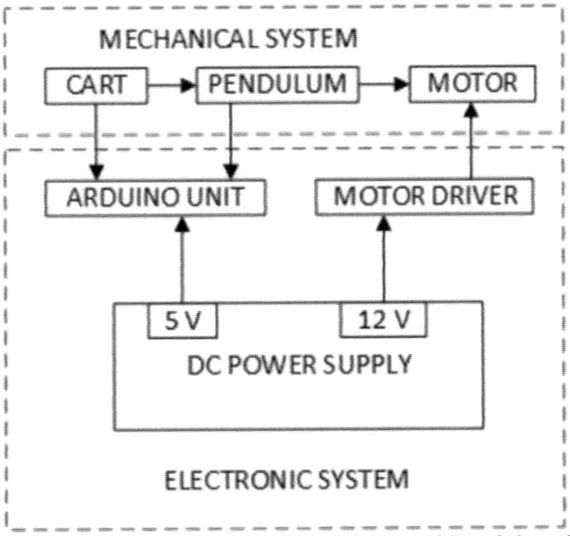

Figure 5. Block Diagram of the Linear Inverted Pendulum System

Based on the block diagram above, it can be seen that the linear inverted pendulum system consisted of two main parts, namely the mechanical and the electronic system. These two parts will be related to each other when the pendulum balancing process was in a vertical position.

4.2.2.1. Design
4.2.2.1.1. Mechanic System Design
In general, the mechanical system of a linear inverted pendulum consisted of three main parts, i.e. cart, pendulum, and motor. Besides these three main parts, there were other important components that make the mechanical system of the linear inverted pendulum can work properly. The following is a mechanical system construction design of a linear inverted pendulum.

(1) Pendulum shaft
(2) Rotary Encoder Sensor
(3) Cart or Slider line
(4) Supporting Legs
(5) Cart or Slider
(6) Belt and Pulley System
(7) Stepper Motor

4.2.2.1.2. Electronic System Design

The electronic system design of the linear inverted pendulum as a whole consisted of three main parts namely the Arduino Uno microcontroller, driver motor, and power supply. The microcontroller functioned as the brain which regulated the operation of the linear inverted pendulum. This microcontroller can be programmed with the PID algorithm to get a programmable control system that was reliable and easy to be operated. The input from this microcontroller was the data from the pendulum bar angle which was read by the sensor. The data from the angle reading will be then processed using PID algorithm in order to obtain the appropriate data to adjust the position of the cart so that the pendulum shaft remains balanced.

4.2.2.1.3. Coding

Software design was a program created to manage the workflow of the Arduino Uno microcontroller. This program will control the working system of the linear inverted pendulum according to the desired instructions. The important step in a program creation was making a flowchart of the program. The flowchart can show the explanation of the workflow of the running program within the linear inverted pendulum system. The following picture is a flowchart of the created program.

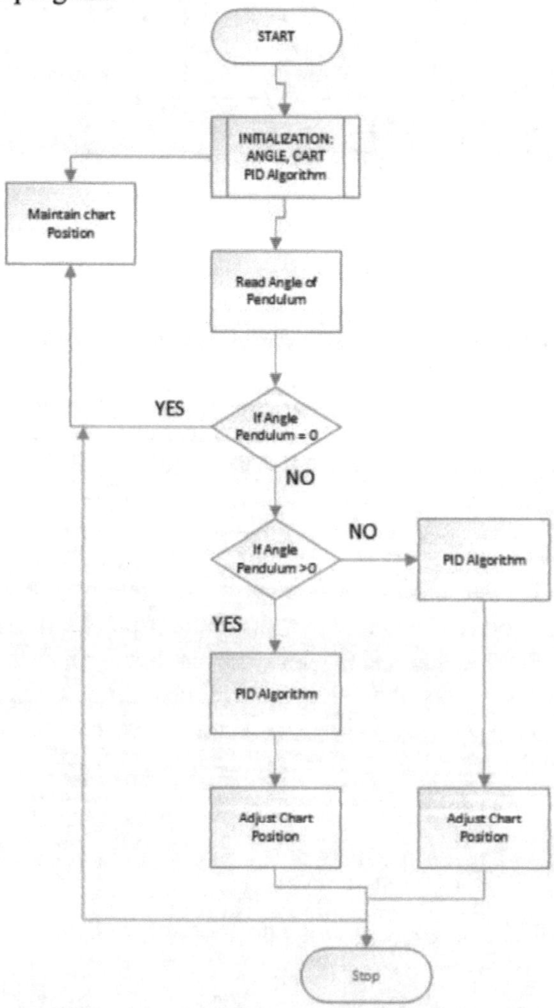

Figure 6. Flowchart of Linear Inverted Pendulum System

4.2.2.2. Planning Results
The results of the planning design of Linear Inverted Pendulum

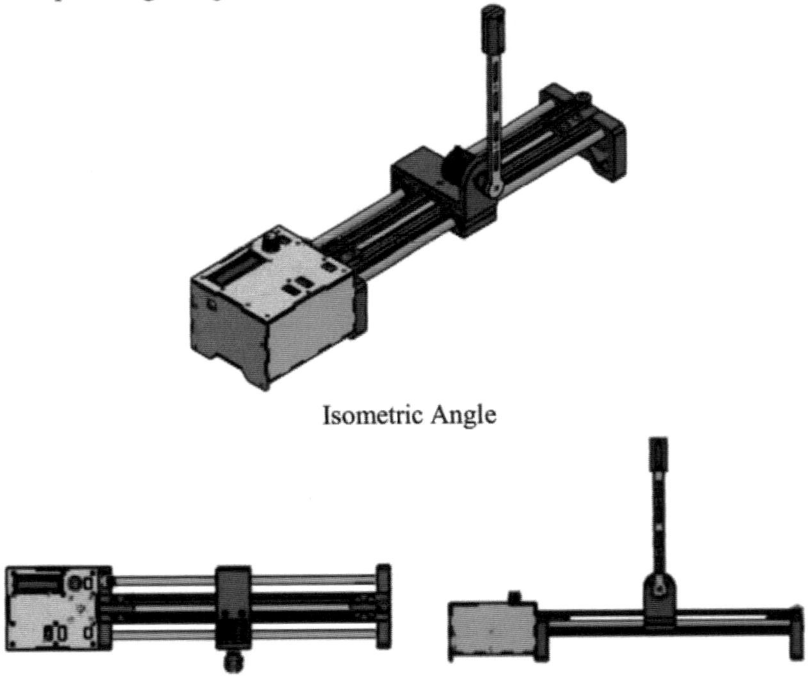

Figure 7. Design of Linear Inverted Pendulum

4.3. Implementation Phase
This phase is realizing the design in the tangible form including pre-production, production, post-production and quality review. These stages began with the preparation of manufacturing, mechanical manufacture, electronics, and coding in order to match the tool purpose. Moreover, additional practice guidelines were needed to support the learning process. Then, it was necessary to perform a performance test to determine overall performance. After the performance was fulfilled the desired target, a quality review was held.

4.4. Evaluation Phase
In this phase, the alpha and beta tests were done. The alpha testing was done by testing the products through expert judgment from the material and media experts. After the trial, there were some revisions and improvements. After going through an alpha test, then it was through the beta test where the product was tested on two students who have received adaptive control courses. The beta test was then ended by trying out a group of students to reveal the practicality level of the learning resources.

5. Conclusions and Suggestion
Based on the research findings, some conclusions and suggestions can be drawn as follow.
(1) It is necessary to provide the tangible forms of the learning sources for adaptive control material
(2) The design of an Inverted pendulum as a learning source of adaptive control had been through several stages, i.e. Needs analysis, design, implementation, and evaluation.
(3) Inverted Pendulum design consists of mechanical, electronic and program design.

6. References

[1] Rizky Prabowo Rahino, Artikel published in tribunpontianak.co.id entitled Firdaus: Revolusi Industri 4.0 Tidak Bisa Dihindari, https://pontianak.tribunnews.com/2019/02/14/firdaus-revolusi-industri-40-tidak-bisa-dihindari., Penulis: Rizky Prabowo Rahino, and Editor: madrosid, "FiFirdaus: Revolusi Industri 4.0 Tidak Bisa Dihindari published in tribunpontianak.co.id entitled Firdaus: Revolusi Industri 4.0 Tidak Bisa Dihindari, https://pontianak.tribunnews.com/2019/02/14/firdaus-revolusi-industri-40-tidak-bisa-dihindari. Penulis: Rizky Prabowo Rahino Editor: madrosid rdaus: Revolusi Industri 4.0 Tidak Bisa Dihindari," 14-Feb-2019.

[2] Nada Marsudi, "Menristekdikti Nasir: Indonesia Siap Menyambut Globalisasi Pendidikan dan Revolusi Industri ke-4 Read more at https://www.ristekdikti.go.id/siaran-pers/menristekdikti-nasir-indonesia-siap-menyambut-globalisasi-pendidikan-dan-revolusi-industri-ke-4/#ooZTf4jkPtcBPVKg.99," *RIstekdikti*, 29-Jan-2018.

[3] Biro Kerjasama dan Komunikasi Publik, Biro Kerjasama dan Komunikasi Publik, and Kemenristekdikti, "Menristekdikti: Perlu Reorientasi Kurikulum untuk Meningkatkan inovasi Perguruan Tinggi di era Revolusi Industri 4.0."

[4] "Merasakan Canggihnya 'Sentuhan' Teknologi Mobil Tanpa Sopir." [Online]. Available: https://oto.detik.com/ototest-mobil/d-3645606/merasakan-canggihnya-sentuhan-teknologi-mobil-tanpa-sopir. [Accessed: 21-Aug-2019].

[5] KHALIL SULTAN and ASHAB MIRZA, *Inverted Pendulum Analysis, Design and Implementation*. Pakistan: IIEE Visionaries.

[6] K. H. Lundberg and T. W. Barton, "History of Inverted-Pendulum Systems," *8th IFAC Symp. Adv. Control Educ.*, vol. 42, no. 24, pp. 131–135, Jan. 2010.

[7] G. D. Nusantoro, M. A. Muslim, P. Purwanto, and R. I. C., "Rancang Bangun Rotary Inverted Pendulum (RIP) dengan Menggunakan Kontrol PID," *J. EECCIS Vol 6 No 2 2012*, 2013.

[8] S. Chakraborty, "An Experimental study for Stabilization of Inverted Pendulum," NATIONAL INSTITUTE OF TECHNOLOGY ROURKELA, 2014.

[9] O. Hamalik, *Media Pendidikan*. Bandung: Citra Aditya Bakri N, 1994.

[10] F. Syukur, *Teknologi Pendidikan*. 2005.

[11] A. Januszewski and M. Molenda, *Educational Technology*. NewYork & London: Routledge Taylor & Francis Group, 2008.

[12] P. Sudira, "KURIKULUM DAN PEMBELAJARAN PENDIDIKAN DAN PELATIHAN VOKASI MENYONGSONG SKILL MASA DEPAN," presented at the Pengembangan Kurikulum, Politeknik Negeri Bali.

[13] E. Marpanaji, "TRAINER PID CONTROLLER SEBAGAI MEDIA PEMBELAJARAN PRAKTIK SISTEM KENDALI," *Elinvo Electron. Inform. Vocat. Educ.*, vol. 2, p. 27, Apr. 2017.

[14] Iven Mareels and Jan Willem Polderman, *Adaptive System An Introduction*. Springer Science+Business Media, LLC, 1996.

[15] K. Sekiguchi, K. Tahara, and M. Sampei, "Experimental study of stabilization of the inverted pendulum with horizontal and vertical movement via exact linearization based on the dynamic extension," in *2010 IEEE International Conference on Control Applications*, 2010, pp. 1433–1438.

Hybrid Power Plant System Analysis in Seruni Beach, Bantaeng District, South Sulawesi

H Mubarok and M Farid
Electrical Engineering Department, Universitas Islam Indonesia, Yogyakarta, Indonesia

Email: mubarok.husein@uii.ac.id, 14524056@students.uii.ac.id

Abstract. Seruni Beach is one of the tourist attractions in Bantaeng Regency, South Sulawesi. As an area that has potential as a tourist place, it must be supported by adequate electricity supply. The electricity source in Seruni Beach is only sup-plied by the State or PLN power plants. To reduce the electricity supply from the National Electric Company, a hybrid power plant is designed that utilizes the potential of alternative energy sources in the form of wind and solar energy. The purpose of this study is to produce efficient solar and wind hybrid plants as a backup generator in sup-plying the loads that are around the coast. The study uses HOMER as an optimal generator determinant. The indicators are used as economic valuations, namely payback period (PP), net present value (NPV), net present cost (NPC), and cost of energy (COE). The study uses 2 schemes to calculate the indicators used. Scheme 1 only uses the grid as a supplier of electricity, and in scheme 2 uses wind turbines, solar panels, and grids. The results of this research are so-lar and wind power hybrids producing 390,620 kWh / year. Scheme 1 has a higher NPC value of Rp 4,632,183,000 while scheme 2 has an NPC value of Rp 2,555,496,645. The costs incurred by scheme 2 to produce electricity per kWh are smaller, namely Rp. 304,709 compared to the scheme 1 in the amount of Rp. 1467.28. in scheme 2, the NPV is positive in the amount of Rp. 1,037,919,493 and from the results of the calculation of the payback period, the time needed to recover the investment costs of solar and wind power plants is 5 years 10 months.

1. Introduction

Bantaeng is a district located in the province of South Sulawesi which has an area of 395.83 km^2. Bantaeng Regency is located in an area with a coast that has potential as an industrial and tourism area. As an area that has potential as a tourist place, it must be supported by a good supply of electricity. Electricity sources in Bantaeng regency are only supplied by the State or PLN power plants in the area even though the area has several beaches that can be used to build solar and wind power plants. Wind energy is an abundant resource available in nature as an alternative energy source that can be renewed (renewable energy) which can be used as a wind power plant. Likewise with solar power plants that utilize solar thermal energy, but the use of renewable natural resources is still very minimal in its use. PT PLN (Persero) stated that in 2010-2019 it was estimated that electricity needs reached 55,000 MW, so the average increase was 5,500 MW. Of the total 32,000 MW of power, (57%) is built by PLN, the remaining 23,500 MW will be built by the private sector [1]. From these data it can be concluded that electricity generation, especially in Indonesia, still relies on PLN which uses non-renewable energy as its energy source and still lacks utilization of renewable energy sources.

Bantaeng Regency, which is located on the coast, is very suitable to develop hybrid power plants because it has the potential of natural resources in the form of solar and wind heat energy. With the availability of these hybrid power plants, it is expected to become a backup power plant in the area. The design is done using Homer software as software that designs, tests, and simulates the results of the design.

2. Basic Theory

2.1. *Solar Power Plant*

Hybrid power plants are power generation systems that have two or more integrated ones. Hybrid plants consist of two plants, namely renewable and conventional power plants or consist of a combination of renewable power plants. The purpose of a hybrid system is to get efficiency or use alternative energy as a source of electrical energy. Alternative energy sources commonly used in designing hybrid systems use solar energy using photovoltaic (PV) and wind energy using wind turbines as generators. [2]

2.2. *Grid Tie Inverter*

Grid tie inverter or commonly known as synchronous inverter is an inverter that functions to convert DC electric cur-rent into AC current [3]. The inverter converts the DC current to a component such as a solar cell, the battery becomes an AC. This grid inverter can not stand alone and must be connected to the PLN. The advantages of kWh obtained from PLTS from the presence of this inverter can be channelled to the PLN and instead PLN pays to the PLTS provider. There are two systems that are used on the Grid Tie System, namely the off grid and on grid systems that have their respective differences. The difference between off grid and on grid is as follows:

- Off Grid System

Off grid system is a generating system that can only produce electricity by relying on energy sources from the plant that is owned as its main energy. Off grid system is used in areas that have not been reached by PLN electricity. Examples of off grid systems are SHS (Solar Home System).

- On Grid System

On grid system is a generating system that generates electricity from a generator owned, the difference from an off grid sys-tem is an on grid system connected to the PLN electricity net-work. The purpose of this system is to reduce electricity consumption from PLN and reduce electricity bills.

2.3. *Payback Period*

Payback period is the time needed to return the investment costs of the project. To get the payback period value by dividing the investment costs with the income generated by the generator in a year [4]. The payback period can be calculated using the following equation (1):

$$Payback\ Period\ (t) = \frac{\text{investment cost}}{income\ per\ year} \qquad (1)$$

Payback period can be said to be feasible if the payback time for investment costs is smaller than the time the project is done.

2.4. *Net Present Value*

Net Present Value is income earned in the future whose interest has been paid at the beginning or discount. The data required to find the value of NPV in the form of investment costs, maintenance and repairs, and the income derived from plants are designed. The Net Present Value can be calculated using equation (2):

$$NPV = \frac{R_t}{(1+i)^t} \qquad (2)$$

Where:

i = bank interest rate
t = year period
Rt = net income in time t

2.5. *Cost of Energy*

Cost of Energy is the expenditure cost per kWh of electrical energy produced by the generator used. COE can be calculated by dividing the total plant cost per year by the total energy load per year (kWh). COE can be calculated using the follow-ing equation (3):

$$COE = \frac{TAC}{E_{tot.served}} \quad (3)$$

Where:
E (tot. served) = total annual energy used to serve the load (kWh)
TAC = total annualized cost incurred for hybrid generation.

3. Research Method

3.1. *Research Flow*

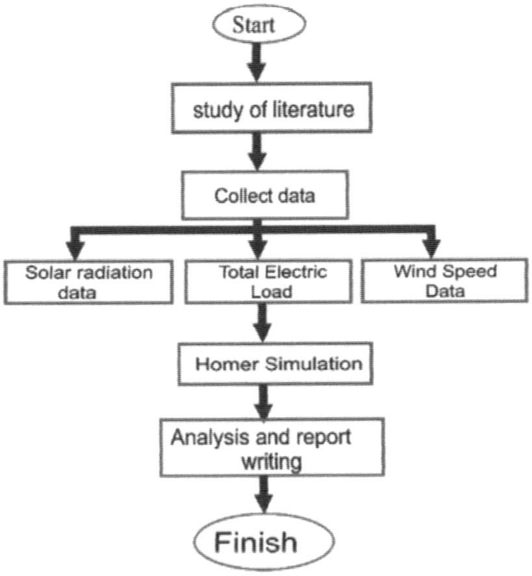

Figure 1. Flowchart Diagram

The flow chart as shown in Figure 1 is the stage carried out during the study. The steps taken are:

- The literature study is looking for and studying various literatures that discuss topics around the research conducted.
- Data retrieval is to collect data to be used during research, in the form of solar radiation data, wind speed data, and electrical load data on Seruni Beach.
- Simulate data that has been taken in the HOMER software to get optimal results. Then do the preparation of the report.
- Research completed.

3.2. *Design of Solar and Wind Hybrid Power Generation Systems Using HOMER*

The simulation that will be carried out uses HOMER as a software, while the components used are photovoltaic, batteries, converters, and wind turbines. The simulation is divided into 2 parts as shown in Figure 2 and Figure 3, first only using the grid as electricity supply then in the second part using a grid accompanied by wind and solar power plants.

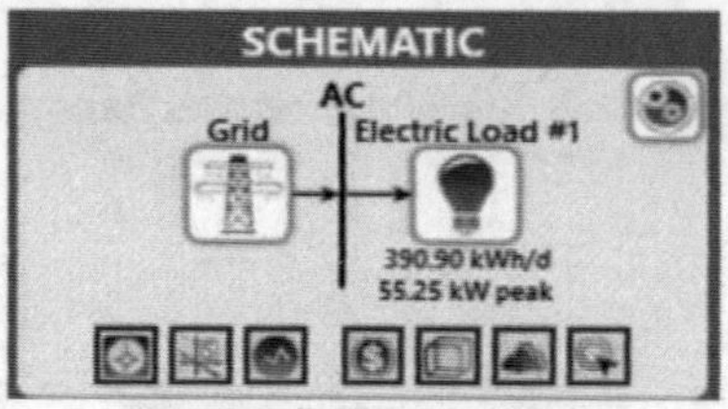

Figure 2. Schematic 1

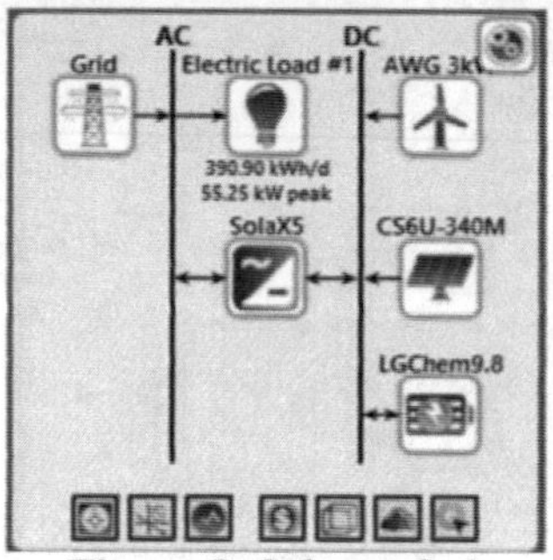

Figure 2. Schematic 2

4. Results and Discussion

The design of solar and wind power hybrid plants in this study uses homer as software. In this study, grids sourced from PLN will be compared to hybrid plants that are designed, with the aim of looking at the benefits of the generator designed, namely from the economic side, the energy produced, and the influence on the environment. Timeline of projects designed at HOMER for 25 operational years.

4.1. Analysis of the Scheme Plant 1

Table 1. Total Energy Used in a Year and Its Parameter

Parameter	Amounts
Net Present Cost	4.632.183.000
Cost Of Energy	1.467,28
Energy consumption (kWh/tahun)	142.668
Renewable Penetration	0%

The Table 1 shows the total use of electrical energy used in a year. The use of electrical energy comes from the load used which is equal to 142,668 kWh / year. The cost of energy calculated at HOMER in scheme 1 is 1,427.28.

4.2. Result Calculation of Net Present Cost Scheme 1

Figure 4. Net Present Cost Scheme 1

The net present cost value obtained from the results of calculations performed by HOMER as shown in Figure 4 is 4,632,183,000. In addition to HOMER calculations, NPC values can also be calculated using equations

$$NPC = 4{,}632{,}183{,}177 + 0 + 0 + 0 + 0$$
$$NPC = Rp\ 4{,}632{,}183{,}177$$

4.3. Result Calculation of Cost of Energy Scheme 1

Table 2. Energy Consumption and Total Annual Cost

Parameter	Amounts
Energy Consumption	142.668 kWh/tahun
Total Annualized Cost	Rp 209.333.946

The Table 2 shows data on electricity consumption and total annualized costs. COE values can also be calculated using equations (3).

$$COE = \frac{\text{Rp } 209{,}333{,}946}{142{,}668 \text{ kWh}}$$
$$COE = \text{Rp } 1{,}467.28/\text{kWh}$$

4.4. Result Calculation of Payback Period Scheme 2

The electricity selling price for the region in South Sulawesi is based on the 2017 Minister of Energy and Mineral Resources regulation of Rp 974.00 [14], the total power generated by the plant is 438,073 kWh / year. So that the annual income generated by the generator is:

$$Rp\ 974 \times 438{,}073\ kWh = 426{,}683{,}102.00, -year$$

The investment cost for designing a hybrid power plant is Rp 2,555,496,645, - so the payback period is:

$$Payback\ Period\ (t) = \frac{\text{Rp}2.555.496.645}{426.683.102}$$
$$Payback\ Period\ (t) = 5.98\ years = 5\ years\ 10\ months$$

From the calculation above, to restore the capital cost for the generation of hybrid electricity of solar and wind power it takes 5 years and 10 months.

4.5. Result Calculation of Net Present Value Scheme 2

Net present value can be calculated by calculating income and costs discounted by 11% (interest rate) for 25 years [5]. Net present value can be calculated using equation (2):

$$NPV = \frac{R_t}{(1+i)^t}$$

So that the NPV value can be seen at the Table 3.

Table 3. Net Present Value of Hybrid Solar Wind Plant

Year		Cash in	Bank interest	Cash value
0	Rp 2.555.496.645		1,00	Rp 2.555.496.645
1		Rp 426.683.102	1,11	Rp 384.399.191
2		Rp 426.683.102	1,23	Rp 346.305.577,5
3		Rp 426.683.102	1,37	Rp 311.987.006,7
4		Rp 426.683.102	1,52	Rp 281.069.375,4
5		Rp 426.683.102	1,69	Rp 253.215.653,5
6		Rp 426.683.102	1,87	Rp 228.122.210,4
7		Rp 426.683.102	2,08	Rp 205.515.504,9
8		Rp 426.683.102	2,30	Rp 185.149.103,5

9	Rp 426.683.102	2,56	Rp 166.800.994,1
10	Rp 426.683.102	2,84	Rp 150.271.165,9
11	Rp 426.683.102	3,15	Rp 135.379.428,7
12	Rp 426.683.102	3,50	Rp 121.963.449,3
13	Rp 426.683.102	3,88	Rp 109.876.981,3
14	Rp 426.683.102	4,31	Rp 98.988.271.49
15	Rp 426.683.102	4,78	Rp 89.178.622,96
16	Rp 426.683.102	5,31	Rp 80.341.101,77
17	Rp 426.683.102	5,90	Rp 72.379.370,96
18	Rp 426.683.102	6,54	Rp 65.206.640,5
19	Rp 426.683.102	7,26	Rp 58.744.721,18
20	Rp 426.683.102	8,06	Rp 52.923.172,23
21	Rp 426.683.102	8,95	Rp 47.678.533,54
22	Rp 426.683.102	9,93	Rp 42.953.633,82
23	Rp 426.683.102	11,03	Rp 38.696.967,41
24	Rp 426.683.102	12,24	Rp 34.862.132,8
25	Rp 426.683.102	13,59	Rp 31.407.326,85
	Total		Rp 3.593.416.138
	NPV		Rp 1.037.919.493

Table 3 above shows the net present value of Rp. 1,037,919,493 (positive value), it can be concluded that the capital of hybrid power plants is accepted and can be done.

5. Conclusion

Wind and solar power hybrid power plants have the potential to be built because natural and wind resources have the potential to be used. In addition, solar and wind power hybrids that have a capacity of 24.8 kW can produce electrical energy of 438,073kWh/year.

The cost of energy in scheme 2 has a smaller value that is equal to 79.27% of the value of scheme 1 because in the scheme 2 the energy produced mostly comes from renewable energy generation.

In scheme 2, to return the investment value used it takes 5 years 10 months, but the NPV value of the generator is positive, which means that the value of the investment has bene-fits and the construction of the plant can be carried out.

6. References

[1] K. Daljeet and P. S. Cheema, "Software Tools for Analyzing the Hybrid Renewable Energy Sources:-A Review," *Int. Conf. Inven. Syst. Control*, pp. 1–4, 2016.

[2] H. Mubarok, "Optimal specification analysis of hybrid-PV-battery-diesel power generation based on electrical outage cost as an industrial reserve power, " IEEE, vol. pp. 253 – 257, October 2017, International Seminar on Application for Technology of Information and Communication (isemantic) Semarang, 2017.

[3] M. H. Nehrir *et al.*, "A review of hybrid renewable/alternative energy systems for electric power generation: Configurations, control, and applications," *IEEE Trans. Sustain. Energy*, vol. 2, no. 4, pp. 392–403, 2011.

[4] A. K. Pradhan, S. K. Kar, and M. K. Mohanty, "Off-Grid Renewable Hybrid Power Generation System for a Public Health Centre in Rural Village," *Int. J. Renew. Energy Res.*, vol. 6, no. 1, pp. 282–288, 2016.

[5] H. Mubarok, "Economic Studies of Wind Turbine Diesel Hybrid Power Generation System (Case study at Queen of the South Beach Resort Hotel, Yogyakarta, Indones." IEEE, September 2018, International Seminar on Application for Technology of Information and Communication

(iSemantic) Semarang, 2018..

[6] Aziz Saleh Ali dkk. "*Feasibility Analysis Of PV/Wind/Battery Hybrid Power Generation*", *A Case* Study, *International Journal of Renewable Energy Research*,Vol.8.No.2.2018.

[7] Betha Divya dkk. "*Design and Control of Grid Connected PV/Wind Hybrid System Using* 3 Level VSC", *Publisher : IEEE, International Advance Computing Conference* (IACC), India.2017.

[8] Foster R., Ghassemi M., Cota A."*Solar Energy : Renewable Energy and The Environment*", Boca Raton, FL : CRC Press, Taylor & Francis Group, New York. 2010.

[9] Tan D., Seng Ang K. "*Handbook For Solar Photovoltaic* (PV) *System*", *Published by Energy Market Authority, Building and Construction Authority*, Singapura, ISBN: 978-981-08-4462-2, Hal. 4-7. 2011.

[10] Wei Tong. "*Wind Power Generation and Wind Turbine Design*", Kollmorgen Corp., USA, Published *by* WIT Press Southampton, Boston, Hal. 15-16. 2010.

[11] Lambert T., Gilman P., Lilienthal P. "*Micropower* Sistem *Modeling With Homer*", *Integration Of Alternative Sources Of* Energi, *by* Felix A., Farret *and* M. Godoy Simoes., John Wiley *and Sons*, *Inc*, Hal. 379-397. 2006.

Performance of A Programmable Logic Controller Based Electrical Machine Trainer Kit

Sukir, Soeharto and A S J Wardhana
Electronics Education Department, Universitas Negeri Yogyakarta, Yogyakarta, Indonesia

E-mail: sukir@uny.ac.id

Abstract. The objectives of this study were: (1) to develop Programmable Logic Controller (PLC) Based Electrical Machine Trainer Kit; and (2) to examine the performance of the developed trainer kit to be applied in the Electrical Engineering Practicesin the Department of Electrical Engineering Education at Faculty of Engineering, UniversitasNegeri Yogyakarta. The method used in this study was research and development by referring to the ADDIE model from Branch. The results of the study showed: (1) the size of the developed PLC-based Electrical Machine trainer kit for the bottom side and the back side were 44.1 cm x 100 cm and 92.7 cm x 100 cm respectively. It had a front tilt angle of 80 °, a panel board made of acrylic, a body made of aluminum plate, and an installed PLC with the brand name of Zelio SR2.201FU; and (2) the PLC-based Electrical Machine trainer kit shows satisfactory performance, indicated by all of job descriptions can function properly as planned.

1. Introduction

One of the courses in the curriculum of the Study Program of Electrical Engineering Education is Electrical Machines Practice. This course has a very strategic role in the curriculum since it underlies courses on the concentration of Industrial Control and the Industrial Electricity in the following semester, and it supports the Industrial Practices and Final Projects in the eighth semester. In addition, through this course, it is expected that students will gain the provision of practical electrical machine skills in the industry required for vocational teachers who will educate their students to enter the industry.

In the implementation of electrical engineering practice in the last five years, it is still less than ideal in increasing the achievement of the students. This was indicated by the results of individual practice tests that did not pass an average of 35% in each class and must be remedied to pass the minimum grade of B. In addition, only 10% of the students achieved the A grade. One of the reasons for this problem was inadequate types, number and quality of available equipment for electrical machine practice in the laboratory. The available equipment at the electrical machines laboratory was first procured by the World Bank Project in 1979. In that year, the main practical equipment were 3 units of Terco electric mechanical practice. Along with the age of its use, in 1999, one of them suffered a total damage so it could no longer be used. Since 1999, there were only 2 units of Terco practice machines with low quality. Moreover, with the increasing number of classes and students where each class should accomodate approximately 20 students, electrical machines practice was conducted in large groups and was not effective. In addition, the topics practiced in the Electric Machine Practice course were also limited, namely DC motors, DC generators, three-phase AC induction Motors with wound rotor, three-phase AC induction Motors with squirrel-cage rotor, synchronous generators and synchronous motors and transformers. Other topics such as 1 phase electric motor and its application, permanent magnet DC motor, starting, rotation settings, braking in induction motor, etc., have not been practiced. On the other hand, the control of electrical machines has evolved using Programmable Logic Controllers (PLC).

Referring to these problems, it was necessary to conduct a study on the development of a PLC-based electrical engineering practice trainer kit in the study program of Electrical Engineering Education, Faculty of Engineering, Universitas Negeri Yogyakarta. In this study, the trainer kit is limited to the starting trainer kit, rotation settings, and DC motor braking, fan and mixer motor rotation settings, and 3-phase induction motor braking. The objectives of this study included: (1) producing a PLC-based electrical machine practice trainer kit; and (2) examining the performance of the trainer kit to be used for electrical machines practice.

Rauner, Maclean, Pabst, & Zimmer [1] suggested several types of learning media in terms of their forms that include real objects, model objects or artificial objects. The media of real objects or artificial objects can be realized in the form of a trainer kit. A trainer kit as a learning medium consists of tools or tangible or artificial objects, and an explanation of the use of the tools in the form of jobsheets or learning modules.

Furthermore, Rauner, Maclean, Pabst, & Zimmer explained that the use of media in learning, can influence the process of students to acquire learning experiences, as illustrated in the cone of Edgar Dale's experience. In the cone of Edgar Dale's experience, the learning experience stretches from the top to the bottom, which shows the learning experience from abstract to concrete. The learning experiences include: verbal, symbolic, visual, radio, film, television, exhibitions, field trips, demonstrations, experiences through drama, experiences through artificial objects, and direct experience. The use of learning media which is increasingly concrete will further facilitate the students' learning. Conveyor trainer kit with a monitoring system developed in this study included the Edgar Dale cone experience, which is the experience of learning through artificial objects. This means that the conveyor kit trainer with this monitoring system can deliver a more concrete learning experience to facilitate students in mastering the determined competencies.

The existing studies related to the trainer kit, among others, were carried out by Gill, Kumar & Kumar [2] on "Design and fabrication of electro-pneumatic trainer kits", which produced electropneatic trainer kits for learning in the Department of Mechanical Engineering, Gulzar Group of Institutes, India. Jamaluddin, Ghani, Rahman, & Deros [3] examined the "Development of training kit for learning Taguchi methods and design of experiments", which showed that the Taguchi experiment was successfully carried out through the trainer kit and the factorial design tested was in principle in accordance with information obtained from the Taguchi method. Samanol, Hamid, &Ramli [4] studied the "Development of a pneumatic trainer kit for polytechnic students", which produced pneumatic trainers for the learning of pneumatic basics for students of the Department of Engineering, Seberang Perai Polytechnic, Malaysia. Sean [5] conducted a study on "BeRobot-the robotic scientific education development kits", which produced a human-like robot trainer kit called BeRobot, to be used as a training tool for people working or interested in the robot field.

Another study was conducted by Akparibo, Appiah, & Antwi [6] entitled "Development of a programmable logic controller training platform for the industrial control of processes". The type of study used was research and development. The results showed that an interactive, cheaper and more portable PLC trainer kit was obtained for industrial process control simulations. The parts of the PLC trainer kit consisted of: PLC Delta DVP14SS2 PLC, WPLSoft software, programming devices, switches as inputs and pilot lights as outputs. The PLC trainer kit assisted the students of the Department of Electrical and Electronic Engineering in Ghana to install and program all types of inputs and outputs according to their choices. The PLC trainer kit was easily carried using a carrying case to the desired location, which accomodated the students to study the PLC comfortably anywhere. The application of the PLC trainer kit in learning could improve the students' achievement compared to the achievement before applying the PLC trainer kit. It was still limited since the developed product was a PLC trainer kit for simulation purposes only which was still abstract.

Other studies related to the trainer kit was conducted by Bhise & Amte [7] with the topic of "Embedded PLC trainer kit with industry application". This study was research and development. It showed that the embedded PLC trainer kit was obtained for application simulations in the industry, which had the characteristics of: safe, low cost, it had an interface with a personal computer, it met the standard, the programming language used IEC 6-1131, and it had an interface with electrical components. In the application of a trainer kit for students at MTech National Institute of Electronics and Information Technology, Auranagabad, Maharashtra, India, it revealed that the students' learning achievement can be improved. The study is different from this study in terms of the objects and the use of the PLC trainer kit in tertiary institutions.

Studies related to other trainer kits was conducted by Eko Nugroho Nur Rohman & Sunomo [8] entitled "PLC trainer kit module with load relay safety as learning media for Wonosari Vocational School". The method used in was research and development with the ADDIE model whose research steps including: analysis, design, development, implementation, and evaluation. The results of the study showed that: (1) the performance of the PLC trainer kit module with the load relay safety operated effectively based on the results of three trials; (2) the feasibility level of the PLC trainer kit module according to the user as a whole achieved a score of 76.31%, which was categorized as good; and (3) based on the students' learning outcomes in the trial of the PLC trainer kit in the learning of the Electronic Control System at SMK YAPPI Wonosari, the PLC trainer kit can effectively improve students' learning outcomes. The difference between the existing study and this study lies in the developed trainer kit, namely PLC trainer kit with load relay safety, and the product of the trainer kit was used in the subject of the Electronic Control System for Vocational High School competencies in Electronic Electronics.

2. Method

This study is research and development which produce the PLC-based Electrical Machine Practice trainer kit. The development model referred to the ADDIE development model according to Robert Maribe Branch. The research steps included: Analyze, Design, Develop, Implement, and Evaluate [9]. Data collection techniques and instruments used in this study included: (1) in the needs analysis, the data were collected through observation and focus group discussions (FGD), while the instruments used were observation sheets and FGD sheets; (2) in the evaluation of the results of needs analysis and design, the data collection technique was observation, while the instruments were observation sheets; and (3) in the performance testing, the data were obtained through observation with instruments in the form of observation sheets.

To obtain instrument validity, a validity test was performed. In this study, the validity of the instruments was determined with face validity and content validity through expert judgment. To obtain the reliability of the instruments, a reliability test was performed on the instruments. The Likert scale questionnaire with the students as the respondents, , was determined through the calculation of the reliability coefficient using the Croanbach alpha formula. The instrument reliability was in the form of an observation sheet used by two observers to observe the performance of the developed product. It was determined by using the reliability of the Scot. Data obtained in the study were further analyzed descriptively.

3. Results and Discussion

During the analysis stage, observation and focus group discussion (FGD) were conducted. The results indicated that the course of Electrical Machine Practice really requires a PLC-based Electrical Machine Practice trainer kit that can be used for several practical activities, namely: (1) rotation control of DC motors; (2) rotation control of three phase induction motor; (3) rotation control of one phase induction motor; (4) starting DC motor; (5) starting 3 phase induction motor using autotransformator; (6) dynamic

DC motor braking; (7) DC motor braking by plugging; and (8) braking 3-phase induction motor by DC current injection. At the design stage, the PLC-based Electrical Machine Practice trainer kit has been produced as illustrated in Figure 1 and Figure 2.

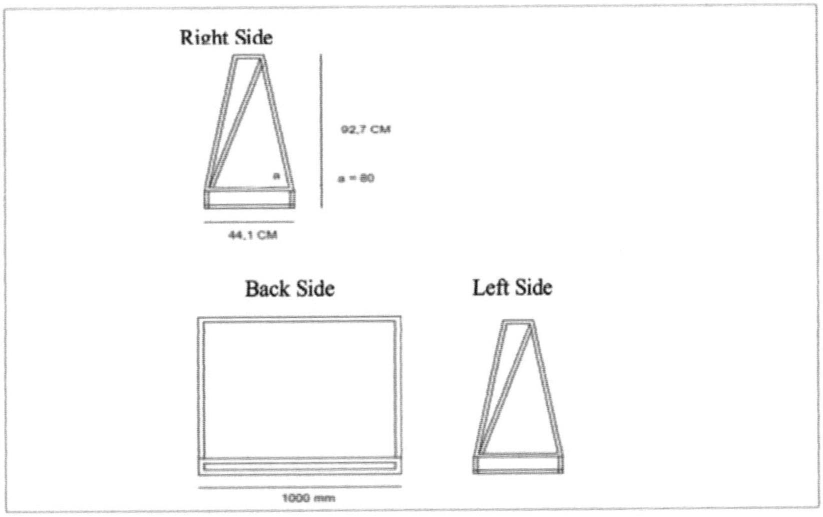

Figure 1. Dimensions of PLC-based Electrical Machine Practice Trainer Kit

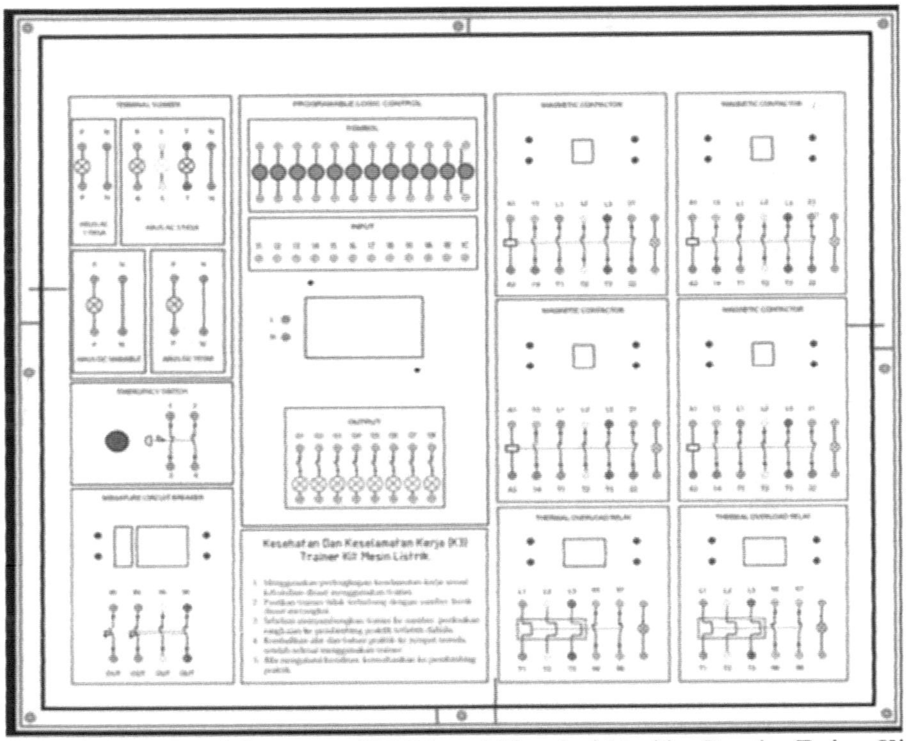

Figure 2. Front Panel Structure of PLC-based Electrical Machine Practice Trainer Kit

In the development stage, the PLC-based Electrical Machine Practice trainer kit was produced. The trainer kit had the following specifications: the size of the developed PLC-based Electrical Machine trainer kit for the bottom side and the back side were 44.1 cm x 100 cm and 92.7 cm x 100 cm respectively. It had a front tilt angle of 80 °, a panel board made of acrylic, a body made of aluminum plate, and an installed PLC with the brand name of Zelio SR2.201FU. The results of developing the electrical machines practice trainer kit are presented in Figure 3 and Figure 4.

Figure 3. The Side View of the PLC-Based Electrical Machine Practice Trainer Kit

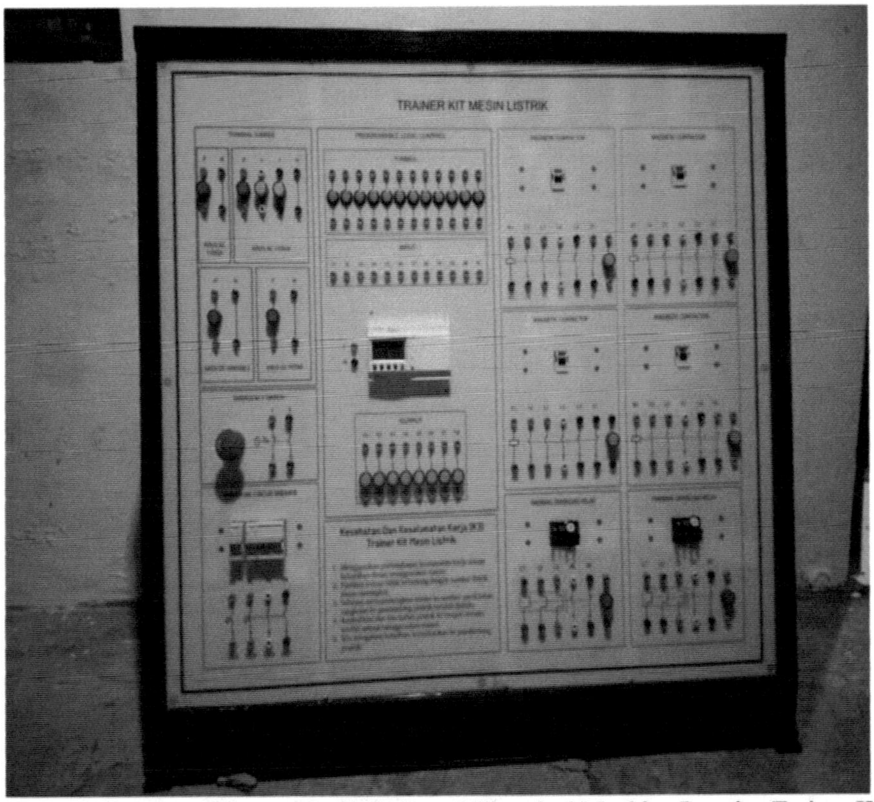

Figure 4. The Front View of the PLC-Based Electrical Machine Practice Trainer Kit

Performance testing of the product development in the form of PLC-based Electrical Machine Practice trainer kit obtained results described in Table 1 and Table 2.

Table 1. Performance Testing for the Components of the PLC-Based Electrical Machine Practice Trainer Kit

No.	Part	Performance Description	Function Yes	Function No	Explanation
1.	AC Power supply	Voltage: Neutral Phase = 220 Volt; Phases = 380 Volt.	√		Good
2.	DC Power supply	Voltage: 220 Volt DC. Voltage: 0 - 220 Volt DC	√		Good
3.	PLC	Power supply 220 Volt AC, On if given a power supply.	√		Good
4.	Input I1 - I12 Output Q1 - Q8	Running a simple Ladder diagram I1 activates Q1 dst.	√		Good
5.	Output Indicator light (green)	On, if the power is being supplied to the PLC and Off, if the power is not being supplied to the PLC.	√		Good
6.	TOR Indicator light (red)	On, if there is an overload on the load and TOR trip and Off, if the load is normal	√		Good

7.	MC Indicator light (green)	On, if the MC coil has electric current and Off, if the MC coil has no electric current	√	Good
8.	Source indicator light (m/k/k))	On, if there is a connected source and Off, if the source is not connected	√	Good
9.	Push button	If it is pressed the contact is connected, if not the contact opens.	√	Good
10.	Emergency Switch	NC is connected when it is normal, NC is disconnected when it is pressed, NO is disconnected when it is normal, and NO is closed when it is pressed	√	Good
11.	Magnetic Contactor	If the coil has a working voltage, NO is closed and NC is open, if the coil does not have a working voltage, NO is open and NC is closed. Main contact performs like NO	√	Good
12.	TOR	When in a normal state, the main contact, the auxiliary NC contact are closed and the NO contact is open. When in a overload state, the main contact and NC auxiliary contact are open, and the NO contact is closed	√	Good

Table 2. Test Results of the PLC-Based Electrical Machine Practice Trainer Kit

No.	Use of Practice	Performing its function		Explanation
		Yes	No	
1.	DC motor rotation control.	√		Good
2.	3-phase induction motor rotation control.	√		Good
3.	1 phase induction motor rotation control.	√		Good
4.	DC Starting motor	√		Good
5.	Starting 3 phase induction motor using autotransformator.	√		Good
6.	Dynamic DC braking.	√		Good
7.	DC motor braking by plugging.	√		Good
8.	DC injection Braking of 3-phase induction motor	√		Good

Based on the performance testing results of the PLC-based electrical machine practice trainer in Table 1, all electrical components of the trainer kit are in good condition and can perform their functions. This means that 100% of the components can work appropriately. In addition, Table 2 shows that PLC-based Electrical Machine Practice trainer kit can be used for 8 kinds of practical experiment where the traner kit perform its function in each experiment. These points demonstrate that the developed PLC-based electrical machine practice trainer kit has a good performance.

4. Conclusion

The PLC-based electrical machine practice trainer kit has been produced with the following specification: the size of the developed PLC-based Electrical Machine trainer kit for the bottom side and the back side were 44.1 cm x 100 cm and 92.7 cm x 100 cm respectively. It had a front tilt angle of 80°, a panel board made of acrylic, a body made of aluminum plate, and an installed PLC with the brand name of Zelio SR2.201FU. The PLC-based electrical machine practice trainer kit has a good performance, indicated by all of the electrical components and the practical work description can function appropriatedly as planned.

5. References

[1] Rauner, F., et. al. 2008. *Handbook of technical and vocational education and training research.* Bremen: Springer Science Business Media.
[2] Gill, K.J.S., Kumar, R., & Kumar, S. 2015. Designing and fabrication of electro-pneumatic trainer kit. *International Conference of Advance Research and Innovation (ICARI-2015), India, p.59-66*
[3] Jamaluddin, H., et.al. 2014. Development of Training Kit for Learning Taguchi Method and Design of Experiments. *Jurnal Teknologi (Sciences & Engineering) 68:4 (2014), 19–25.Penerbit UTM Press, Malaysia.*
[4] Samanol, S.B., Hamid, S.B.A, & Ramli, M.N.B. (2014). Development of Pneumatic Trainer Kit For Polytechnic Student. *Journal of Mechanical Manufacturing (J-Mfac). Vol. I. 2014. p. 178-183.*
[5] Sean, C. 2015. BeRobot-The Robotic Scientific Education Development Kits. *Lovotics, an open access journal, Volume 3, Issue 1, 2015. p. 1-4. ISSN: 2090-9888.*
[6] Akparibo, A.R., Appiah, A., & Antwi, O.F. 2016. Development of a Programmable Logic Controller Training Platform for the Industrial Control of Processes. *American Scientific Research Journal for Engineering, Technology, and Sciences (ASRJETS, (2016), Volume 15, No 1, pp 186-196.*
[7] Bhise, K., & Amte, S. 2015. Embedded PLC Trainer Kit with Industry Application. *International Journal of Engineering Science and Innovative Technology (IJESIT). Volume 4, Issue 3, May 2015, ISSN. 2319-5967. P. 58-65.*
[8] Nugroho, E., Rohman, N. & Sunomo. 2016. Modul Trainer Kit PLC dengan Pengaman Relay Beban Sebagai Media Pembelajaran SMK YAPPI Wonosari. Prodi Pendidikan Teknik Elektro, Vol.6, No.3, October 2016 : 27-35. E-Journal Universitas Negeri Yogyakarta http://journal.student.uny.ac.id/ojs/index.php/elektro.
[9] Branch, R.M. 2009. *Instruction design: the ADDIE approach.* New York: Springer Science + Business Media.

Automatic Tsunami Early Warning System Based on Open Data of Indonesia Agency for Meteorological, Climatological, and Geophysics

R Atika, A E Raditya, R N Marjianto and H S Pramono
Electrical Engineering Education Department, Universitas Negeri Yogyakarta, Yogyakarta, Indonesia

Email: rizaatika.2018@students.uny.ac.id

Abstract. Indonesia frequently experiences earthquakes and tsunami since Indonesia is geologically located in the ring of fire and sits along three major tectonic plates of the world thus this natural phenomena have the potential to continuously occur for example a notable tsunami that hit Aceh in 2004 and killed 160,000 people. This condition is not yet supported by adequate means of disaster mitigation and a tsunami early warning system. The objective of this study was to create an automatic tsunami early warning system that utilizes tsunami potential data from the Indonesian Agency for Meteorological, Climatological and Geophysics' website to be conveyed through loudspeakers in places of worship whose area will be affected by the tsunami. This study used the Research and Development (RnD) method with the ADDIE (Analyze, Design, Development, Implementation and Evaluate) procedure. This study developed a tsunami early warning system that works effectively with a fast average response time of 13.48 seconds to download tsunami potential information on the BMKG website and proceed it to trigger the loudspeaker at places of worship.

1. Introduction

Indonesia is located at the confluence of three major tectonic plates of the world, namely Eurasian, Indo-Australia and Pacific. Indonesia sits along a line of active volcanoes or ring of fire, where several tectonic plates collide and many volcanic eruptions, earthquakes, and tsunami occur. The tsunami that struck Palu and Donggala on September 28, 2018 left significant loss of lives, environmental damages and socio-economic losses which became a national disaster by killing 2,101 people. In this incident, 1,373 people were also missing and 206,219 people were evacuated (Source: www.bbc.com/2018). At the end of 2018, on December 22, 281 people died, 1,016 people were injured and 11,687 people were evacuated due to the tsunami occurred at the Sunda Strait in Banten and Lampung (Source: news.detik.com).

Fifteen years ago, an earthquake that measures 9.1 to 9.3 on the Richter scale caused a tsunami with waves reaching up to 35 meters. It devastated the Veranda of Mecca, Aceh, killing 160,000 people (Source: news.detik.com/ 2019). Various efforts have been made by the government to minimize death and damages due to tsunami, including by installing disaster mitigation and tsunami early warning system such as the InaTEWS (Indonesian Tsunami Early Warning System) conducted by BMKG (Indonesian Agency for Meteorological, Climatological and Geophysics), EWS (Early Warning System) by the BPBD (Regional Disaster Management Agency) Kulonprogo and BISS (Bantul Integrated Siren System) by BPBD (Regional Disaster Management Agency) Bantul. However, currently there are only 52 units of InaTEWS throughout Indonesia whereas ideally according to the Indonesian National Disaster Management Agency (BNPB), it should be 1,000 units. In addition to the limited units, not all of coastal area in Indonesia have InaTEWS towers because the components still have to be imported from abroad, thus the price is very high. On the other hand, the EWS uses a 140-450 megahertz HT (Handy Talky) radio frequency thus it is prone to other radio signal interventions and requires more power.

Based on these problem, this paper aims at proposing an innovation of the Automatic Tsunami Early Warning System designed for early informing the occurrence of earthquakes that potentially lead tsunami in certain regions or area to minimize the casualties. This automatic tsunami early warning system utilizes earthquake and tsunami potential data available in xml format that can be reused and downloaded through BMKG's official website, http://data.bmkg.go.id which is updated in real-time using a mini-PC Raspberry Pi 3 to be processed and sent to targeted locations via SMS (Short Message Service) to automatically trigger the loudspeakers in places of worship.

Loudspeakers at places of worship were selected because its accessibility in every place not only on the coastal area. It is considered a close object that is always available in the public and used as a means of delivering information in the community. This system can work quickly with a minimum cost because it was installed as an additional device of the loudspeaker thus it does not require tsunami early warning towers required by the pre-existing tsunami early warning devices. In addition to its easy and simple maintenance, the system can be produced in large quantities due to its small size and domestic components.

2. Research Method

This automatic tsunami early warning system was developed with the Research and Development (RnD) method that uses the ADDIE (Analyze, Design, Development, Implement, Evaluate) procedure adopted from Robert Maribe Branch (2009) presented in Figure 1.

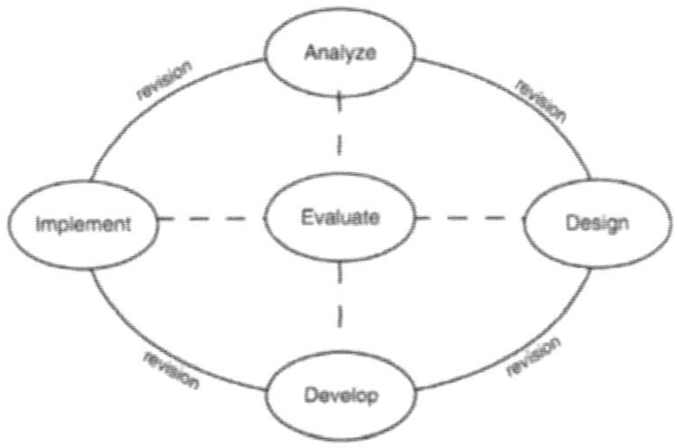

Figure 1. ADDIE Development Procedure
(Robert Maribe Branch, 2009: 2)

2.1. Analysis Stage

At the analysis stage, the concept of the was prepared based on data obtained from field observations in several related agencies such as BMKG Yogyakarta Geophysics Station, BPBD Bantul, PUSDALOPS Bantul, BPBD Kulonprogo and PUSDALOPS Kulonprogo which analyze the weaknesses of the existing tsunami early warning system.

The observations revealed that currently, Indonesia used tsunami early warning systems called InaTEWS (Indonesia Tsunami Early Warning System) whose quantity is only 52 units in total throughout Indonesia, BPBD Bantul has BISS (Bantul Integrated Siren System) and Kulonprogo BPBD has EWS (EWS) Early Warning System). BISS and EWS have a collaboration system and the chain process from BMKG. BPBD broadcasts using radio waves such as HT (Handy Talky) at a frequency of 140-450 megahertz thus it requires large transmitter power, has a weakness against the weather, and transmits data that were vulnerable to other nearby radio signal interventions.

2.2. Design Stage

At this stage, based on the results of the analysis, an automatic tsunami early warning system was designed. It was generally divided into two main parts, namely the server and the client, as shown in Figure 2. The server will be located in the centre of the city that has an internet network to take the earthquake and tsunami data from the BMKG server using the Raspberry Pi 3 mini-PC and process it into the required information then send it to the location of the potentially tsunami-affected client via SMS (Short Message Service). The client receives information sent from the server using the GSM SIM 900SA module to trigger the loudspeaker at the places of worship to produce sound automatically. If an earthquake occurs but it is not potential to cause a tsunami, it will be informed in the form of a voice instructing the public to remain calm and displaying running text on the dot matrix that informs the magnitude and depth of the epicentre.

Whereas if there is a potential for a tsunami, then the voice will inform the community to immediately evacuate to a higher and safer place and then be followed by a long siren. The dot matrix will display a running text that displays potential tsunami hazards and instructs the community to immediately evacuate.

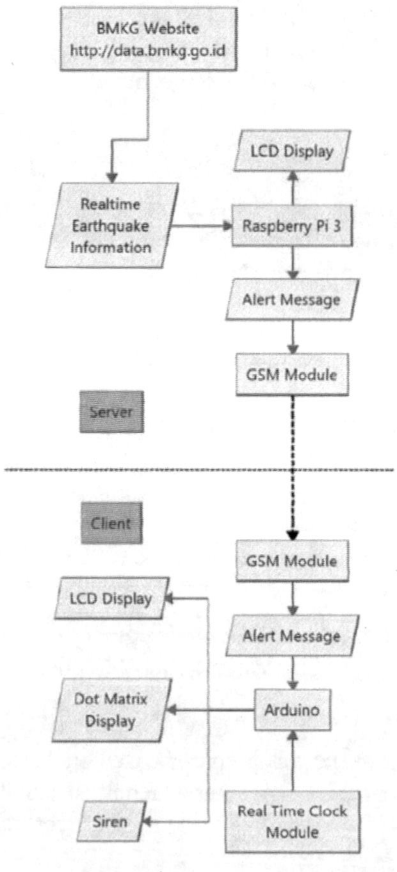

Figure 2. System Block Diagram

2.3. Development Stage

At this stage, the automatic tsunami early warning system was developed according to the results of the previous draft.

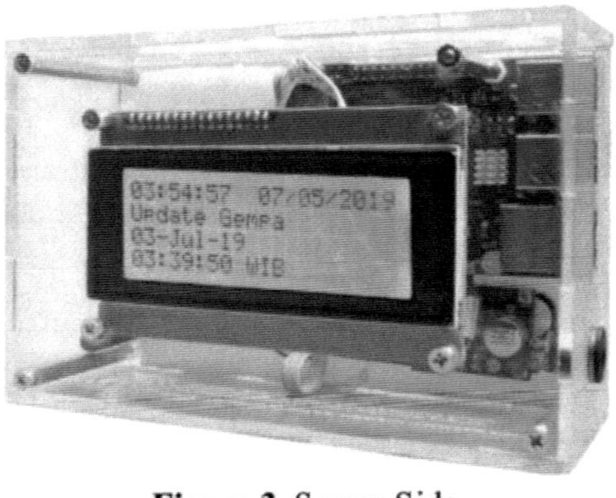

Figure 3. Server-Side

Figure 4. Client-Side

The limitation of the problem in this study was in testing the system to broadcast tsunami warning sirens. It was determined for a limited area of Parangtritis coastal area, Yogyakarta. This test was to examine how the performance and the accuracy of the broadcast warning messages according to the predicted location being affected. The server as the main controller in processing tsunami potential data from BMKG has an important role in determining which area need to be sent tsunami alerts, thus it will not lead unnecessary panic in the community.

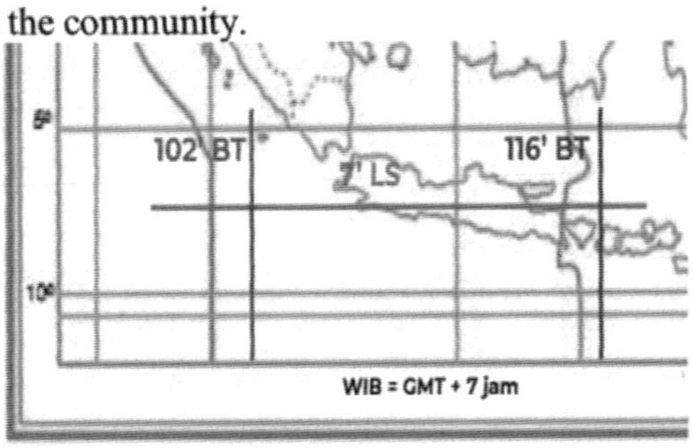

Figure 5. Regional Experiment Mapping

The method used for this problem is to determine the coordinates of the location where the earthquake occurred. If an earthquake that has tsunami potential is between the coordinates of 102° BT to 116° BT and more than 7° LS, it will send a tsunami warning message to the Parangtritis coastal area, Yogyakarta. In the other side, if the earthquake location is outside the determined coordinates, the server will not send a tsunami warning message. This also applies if there is an earthquake that does not have the potential for a tsunami. If an earthquake arrives at the location specified above, the system will only send information messages that the earthquake has no tsunami potential. Table 1 presents the test results.

Table 1. Regional Mapping

No.	Location	System Reaction
1.	8° LS 99° BT	It did not send the message
2.	2° LU 110° BT	It did not send the message
3.	9° LS 120° BT	It did not send the message
4.	9° LS 117° BT	It did not send the message
5.	6° LS 115° BT	It did not send the message
6.	8° LS 110° BT	It sent the message
7.	8° LS 103° BT	It sent the message
8.	9° LS 112° BT	It sent the message
9.	10° LS 115° BT	It sent the message
10.	10° LS 115° BT	It sent the message

2.4. Implementation Stage
After the system is completed and operates as expected, the next step is to implement it in Assalam mosque, Parangtritis, Bantul, Yogyakarta.

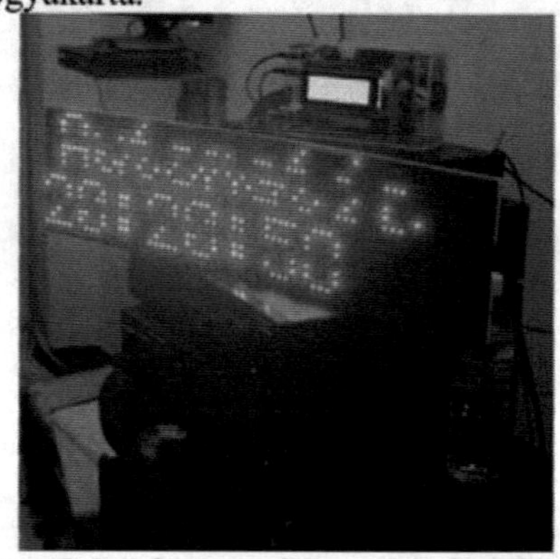

Figure 6. Implementation at Assalam Mosque

2.5. Evaluation Stage
The next stage is to evaluate based on the performance, by testing the system at the Yogyakarta Geophysics station and obtaining some input for evaluation and improvement.

3. Results and Discussion
The results of the automatic tsunami early warning system development showed that on the server-side located in the center of the city with adequate internet access, the system can retrieve data from the official BMKG website using Raspberry Pi as a server computer and process it to be sent to the client-side. This data collection is carried out every 10 seconds to ensure the system gets the latest information from the BMKG, so that if an earthquake occurs, a potential tsunami warning can be sent faster. The client-side has been able to receive messages sent by the server using the SIM 900A Module and process them using the Arduino Nano microcontroller as the main controller in activating the siren and display earthquake and tsunami information in the form of running text on the dot matrix display to be able to distinguish whether the warning sirens are turned on tsunami or just an information viewer on the LCD screen and dot-matrix.

This tsunami early warning system has been tested at the Assalam mosque, located in the coastal area of Parangtritis, Bantul, DIY and has been tested at the Yogyakarta Geophysical Station to analyze its performance and to identify its weaknesses for evaluation and improvement. The testing method was to simulate a post-earthquake tsunami, by creating a dummy website or a replica website from the official BMKG website because it is not possible to wait for a tsunami in real-time for the trials. The data contained in the website page can be changed according to the need for the trials in the form of magnitude, coordinates and the potential for a tsunami to trigger the server computer to send tsunami information immediately to the client-side.

The trials were designed to examine the level of accuracy of the server computer and information transmission to the predicted locations which are affected and not affected by the tsunami. The test result data can be seen in Table 2.

Table 2. Test Results

Experiment	Response Time (second)
I	13.8
II	12.2
III	14.5
IV	13.4
V	13.5
Average	**13.48**

From the test results, it was obtained the average response time of 13.48 seconds.

4. Conclusion

The developed automatic tsunami early warning system is divided into two main parts, namely the server and the client. The server is a data reader program of earthquakes which have tsunami potential from the BMKG website, the information processor and the information sender section to the location of the tsunami. Whereas the client is a recipient program sending information from the server and the driver section that turns on the loudspeaker automatically. The developed automatic tsunami early warning system showed high performance and can work effectively as expected. Based on the test results, it was obtained the average response time of 13.48 to proceed the tsunami information on the BMKG website to trigger loudspeakers at places of worship.

The main advantages of this automatic tsunami early warning system consist of low production cost, direct synchronization with the official BMKG website, efficient energy use with only a 12 Volt DC power source, fast and precise with an average response time of 13.48 seconds and easy maintenance.

5. References

[1] Badan Meteorologi, Klimatologi dan Geofisika. *Pedoman Pelayanan Peringatan Tzunami.* 2012.

[2] BBC News Indonesia. *Deretan Bencana Alam Mematikan yang Menerjang Indonesia Sepanjang 2018.* 2018. Accessed from www.bbc.com/indonesia/ majalah-46691586 on 15 June 2019.

[3] BMKG. *Data Gempa Bumi Terbuka BMKG.* 2019. Accessed from http://data.bmkg.go.id/ gempabumi/ on 2 August 2019.

[4] Branch, Robert M. *Instructional Design: The ADDIE Approach.* 2009.

[5] detikNews. *Ini Gempa Magnitudo di Atas 7 yang Pernah Guncang Indonesia.* 2019. Accessed from https://news.detik.com/ berita/d-4650883/ini-gempa-magnitudo-di-atas-7-yang-pernah-guncang-indonesia on 2 August 2019.

[6] MaulidaIffani. *Bantul Integrated Sirine System(BISS) as a Bantul Regency Preparedness in Dealing with Tsunami Hazard.* Sumatra Journal of Disaster, Geography and Geography Education. 2017.

[7] USA:Springer. CNN Indonesia. *BNPB: Indonesia Kekurangan Sirine dan Alat Deteksi Tsunami.* 2016. Accessed from https://www.cnnindonesia.com/ nasional/ 20160303195458-20-115215/bnpb-indonesia-kekurangan-sirene-dan-alat-deteksi-tsunami on June 2019.

Experimental Study of Measuring Radiation Patterns for VHF and UHF Antennas

E Marpanaji[1], K T Yuwono[1], M I Mahali[1], P T Aji[1] and N A B Nugraha[2]
[1]Electronics Engineering Education Department Universitas Negeri Yogyakarta, Yogyakarta, Indonesia
[2]Student Electronics Engineering Education Department Universitas Negeri Yogyakarta, Yogyakarta, Indonesia

Email: eko@uny.ac.id, arispra@uny.ac.id, izzudin@uny.ac.id, purno.tri@uny.ac.id and nikko.adji@gmail.com

Abstract. This paper discusses the experiment of measuring antenna radiation patterns using hardware and software that has been made previously. The antenna measured is the Half Wave Open Dipole antenna that works at VHF frequencies as Antenna Under Testing (AUT). The hardware used is Module-1 and Module-2 which have previously been tested for functionality, and the software is the Antenna Radiation Pattern application that is created using the Visual Basic programming language in the Microsoft Visual Studio package. Radiation pattern measurement experiments were carried out under various conditions, namely (1) without RF signals; (2) with RF signal and the position of the transmitting antenna is 0o with AUT; (3) with the RF signal and the position of the transmitting antenna at an angle of 90o to AUT; and (4) simulation of electromagnetic wave interference when making measurements. Based on the results of experiments conducted, it can be concluded that the Antenna Radiation Pattern application has functioned well and can describe the AUT radiation pattern. Future work of this research is the measurement of radiation patterns for other types of antennas and refinement of applications for the measurement of various antenna parameters.

1. Introduction

The current telecommunications system relies heavily on wireless communication technology. Cellular and mobile communication systems, Internet services via WiFi devices, communication services for monitoring aircraft, ships, and trains also use wireless telecommunications systems or telecommunications using radios. The use of wireless telecommunications systems in various types of communication services shows that wireless telecommunications technology or telecommunications using radios is still needed for telecommunications systems in the future.

Antenna is a very important component in building wireless telecommunications systems especially for radio communication systems. Antennas play an important role in carrying out electromagnetic wave radiation to the air at the transmitter, and receiving electromagnetic waves from the air for the receiver. The distance and direction of communication between the transmitter and receiver is largely determined by the characteristics of the antenna used. If the antenna used does not function properly, then the function of the radio telecommunications system also cannot function as expected.

Antenna characteristics are very important to consider because they determine the success of the radio communication process. Antenna characteristics are usually presented in the form of antenna parameter values that can indicate the performance of an antenna. Antenna parameters are needed to obtain an overview of the performance of an antenna, including: radiation patterns, directionality, gain, bandwidth, efficiency, beamwidth, and aparture. Antenna parameter measurements are performed when testing an antenna design. One of the tools used in antenna measurement is the antenna radiation pattern measurement. This gauge can provide an overview of directivity, aperture, beamwidth, and gain through a graph displayed by this gauge.

This study conducts a study in the engineering of a measuring instrument that functions to measure antenna parameters, especially for antennas that work at Very High Frequency (VHF) and Ultra High Frequency (UHF) frequencies. This research includes observing the possibilities or technological opportunities that can be used in engineering measuring instruments as well as hardware and software components that can be used at relatively cheaper prices.

Hendrik, et al (2013) conducted research on the engineering of measuring instruments using the IC LT 5504 component as the front end of a Radio Frequency (RF) signal receiver, the microcontroller functions to convert analog signals to digital signals and transmit the results of the conversion to a computer via USB port, computer perform signal processing and display results using the Visual Basic programming language [1]. The front end component is actually widely available in addition to the components used in the study. Thus, a comparison can be made which components are most effective.

Another researcher, Kirbi, et al (2015) developed the antenna radiation pattern measurement tool but did not display the radiation pattern graph [2].

This research is looking for alternative solutions related to the components used in engineering the antenna parameter measurement tool. Thus, a measuring instrument can be obtained which is cheaper and can meet the needs of the laboratory for practical purposes and also as a learning medium for antenna measurement.

2. Antenna Parameters

Antenna is one of the main components in the process of radio communication or communication via un-guided media (wireless). Antenna can be defined as a structure associated with the transition region between guided waves and waves in free space or vice versa [3]. Waves are electromagnetic waves that are used in the process of radio communication. Guided waves are high-frequency electrical (current or voltage) waves that pass through the transmission line in the form of coaxial cables or other types of cables that are often used in wireless communication. While free space is free air between the transmitter and the receiver to transmit electromagnetic waves.

The main function of an antenna is to convert high-frequency electric waves into an electromagnetic wave with the same frequency to the transmitting antenna. Electromagnetic waves generated by the transmitting antenna will spread towards the receiving antenna. The function of the receiving antenna is to convert electromagnetic waves into electric waves (current or voltage) again, as shown in Figure-1.

The antenna parameters are generally used to explain the specifications or characteristics of an antenna. The characteristics of a transmitting antenna are identical to those of a receiving antenna.

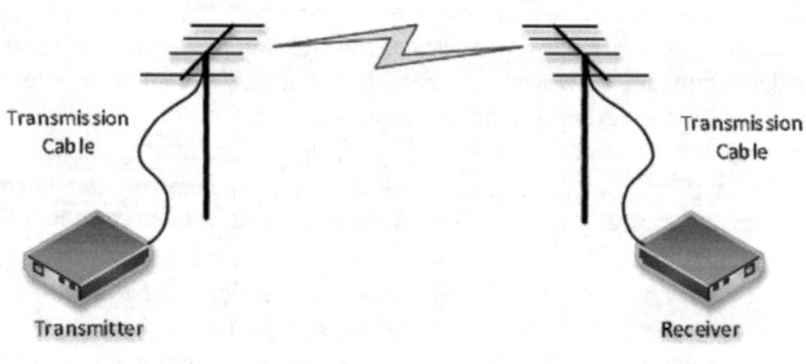

Figure 1. Antennas for radio communication

Before discussing antenna parameters, it is necessary to convey first about the coordinate system that is often used in analyzing antenna parameters, which are spherical coordinates as shown in Figure-2.

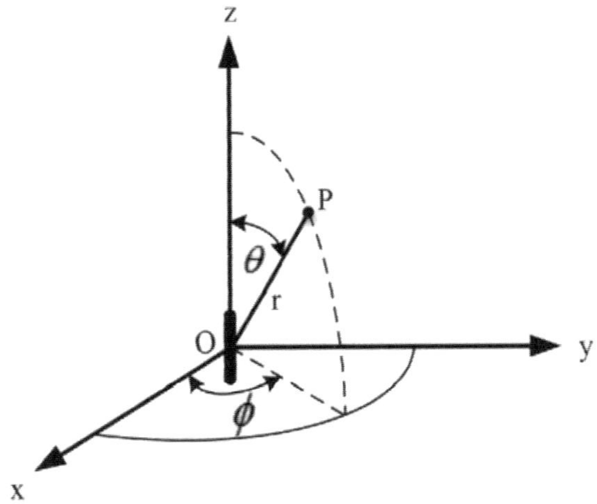

Figure 2. Spherical coordinate.

The point P in spherical coordinates which has a distance r from the center O, forming the angle θ of the z-axis line, and forming the angle Ø of the x-axis can be denoted by P (r, θ, Ø). The line $\overrightarrow{OP}$ can represent an electromagnetic field line issued by an antenna element located on the z-axis (one line with the z-axis). The notation in the form of spherical coordinates is then used in every calculation and analysis of the electromagnetic wave field emitted by an antenna to show the antenna's performance in the form of the parameters of an antenna.

The antenna parameters that are often used to describe the characteristics or performance of an antenna are: radiation pattern, directivity, gain, efficiency, antenna aperture, bandwidth, and beamwidth. The parameters discussed in this paper are focused on measuring the radiation pattern of a VHF antenna, especially the radiation pattern for the horizontal plane (horizontal pattern).

The radiation pattern of an antenna shows the amount of radiation energy released by the antenna at a certain distance according to the intensity of the field in the angular direction θ and Ø. The radiation pattern has a maximum radiation intensity towards the z-axis which is often referred to as the Main Lobe and smaller radiation intensities both towards the front and back of the antenna which is often referred to as the Minor Lobe or Side Lobe. Figure-3 shows the radiation pattern of an antenna [4].

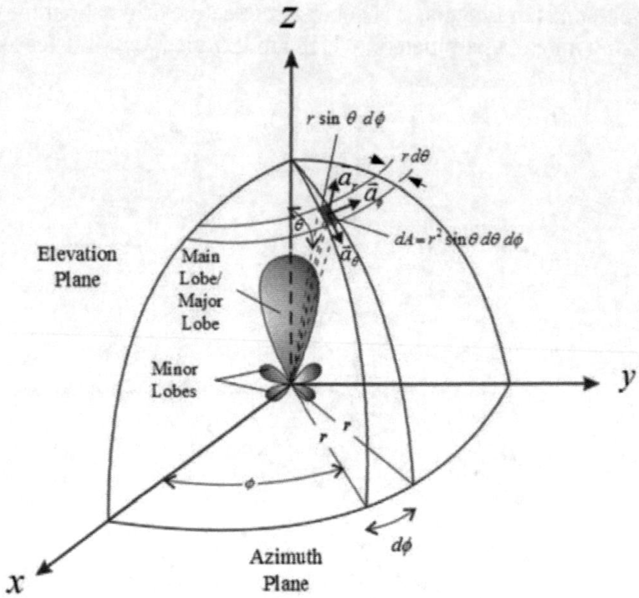

Figure 3. The radiation pattern of an antenna in the z-axis direction

Antenna radiation pattern is a graph of far field radiation from an antenna. More specifically, the radiation pattern is a graph of power emitted from an antenna per unit of solid angle, or radiation intensity *U* [watts per unit of solid angle]. The radiation pattern can be obtained by simply multiplying the power density at a certain distance by the square of the distance *r*, where the density or power density *S* [watts per square meter] is given by the magnitude of the average Poynting vector [5]:

$$U = r^2 S \tag{1}$$

Equation (1) has the nature of the effect of distance on the intensity of U radiation. To eliminate the effect of distance and to ensure that the radiation pattern is not affected by the distance from the antenna, measurements are made with the condition that r is in a far field. The simplest example is an ideal antenna that radiates power in all directions equally or often referred to as an isotropic antenna.

If the total power emitted by the antenna is P, then the power is spread over the plane r radius, and the power density at the distance r for any direction can be stated:

$$S = \frac{P}{luas} = \frac{P}{4\pi} \tag{2}$$

Furthermore the radiation intensity can be determined by the equation:

$$U = r^2 S = P/4\pi \tag{3}$$

Thus, equation (3) is independent of the distance value *r*.

The radiation pattern is usually described in the form of normalized values or described by the value divided by the maximum value of the radiation pattern itself [6]. Thus, the radiation pattern graph has an outer circle with a maximum value equal to 1. These normalized planar pieces are transformed into the two-dimensional domain as shown in Figure-4. The figure shows a radiation pattern of a half-wave open dipole antenna with the antenna element stretching on the 0° – 180° horizontal axis, so that the most radiant energy lies at an angle $\emptyset = 90^o$ dan $\emptyset = 270^o$ in the plane horizontal. The azimuthal angle

Ø (angle moving from 0° to 360° in the horizontal plane) rotates in a direction around the circle of vertical planes (vertical axis). The amplitude of the electric field is plotted along the radius of the horizontal plane circle. Typically, the elevation angle θ (the angle that moves from 0° to 360° in the vertical plane) is determined for the image of the radiation pattern that will be displayed. However, the radiation pattern of the half-wave open dipole antenna rotates symmetrically for the entire elevation angle from 0° to 360° in the vertical plane, so all the pieces are identical [7]. The shape of the radiation pattern of the transmitting antenna will be identical to the shape of the radiation pattern of the receiving antenna.

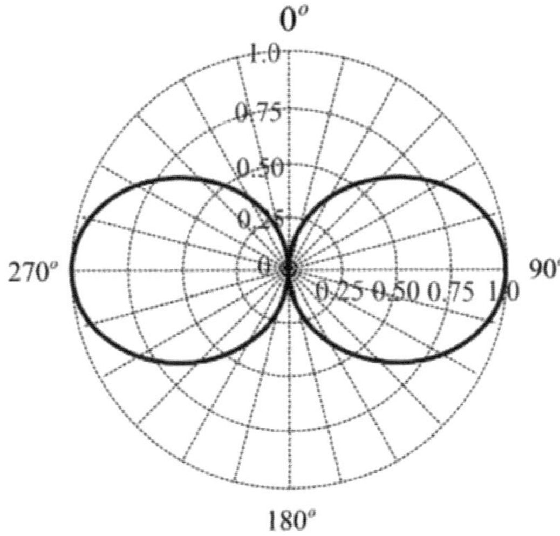

Figure 4. Radiation pattern of a half-wave open dipole antenna in two dimensional shapes in a horizontal plane is common to illustrate the antenna radiation pattern.

3. Area of Antenna Electromagnetic Field

An antenna measurement must pay attention to the electromagnetic field of an antenna, so that the measurement results truly represent the electromagnetic field emitted by the antenna. In principle, the electromagnetic field area of an antenna can be divided into 2 (two) electromagnetic field regions, namely: (1) the area that is close to the antenna or often referred to as the near field region or the Fresnel Zone; and (2) areas with long distances with antennas or often referred to as far field regions or the Fraunhofer Region (Fraunhofer Zone) [3]. The boundary area of this field is determined by the distance (radius) and length of the antenna as shown in Figure-5 below.

The equation used to determine the radius limit of the Fresnel Region with the Fraunhofer Region is:

$$R = \frac{2L^2}{\lambda} \quad (4)$$

Where:
R is the Fresnel area Radius boundary (m)
L is the maximum size of the antenna (m), and
λ is the electromagnetic wavelength (m).

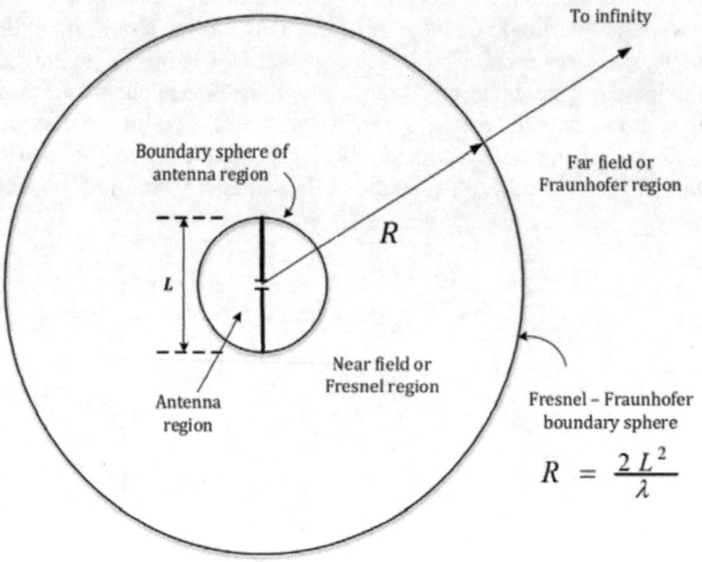

Figure 5. Antenna field area (Fresnel Region and Fraunhofer Region).

Field components that can be measured in the Fraunhofer area will be transversal in the radial direction of the antenna and all the power flow is directed out radially as well. The shape of the terrain pattern in the Fraunhofer area (far field) is independent of distance. Whereas in the Fresnel area, the longitudinal component of the electric field may be more significant and the power flow is not entirely radial. The shape of the field pattern in the Frensnel area is generally very dependent on the measurement distance with the measured antenna. Thus, if we are going to measure the antenna parameters, the location of the antenna to be measured with the measurement point must be located in the Fraunhofer area or the measurement distance meets the equation [3][6]:

$$R \geq \frac{2L^2}{\lambda} \tag{5}$$

Where:
R is the distance of the measurement point in the Fraunhofer area (m)
L is the maximum size of the antenna (m), and
λ is the electromagnetic wavelength (m).

4. Research Method

4.1. Antenna Measurement Configuration

The system configuration for measuring the antenna parameters used in this study is shown in Figure-6. Antenna parameter measurements use a high-frequency electromagnetic wave source (a transmitter or oscillator with sufficient output power) along with a transmitting antenna that functions to radiate electromagnetic waves into the air. The antenna to be tested (AUT = Antenna Under Testing) is placed on a mechanical system that can rotate the AUT at an angle of ϕ. The radio frequency band used is Very High Frequency (VHF), so that the transmitter system, receiver system, and antenna used in this measurement also work on the VHF frequency band.

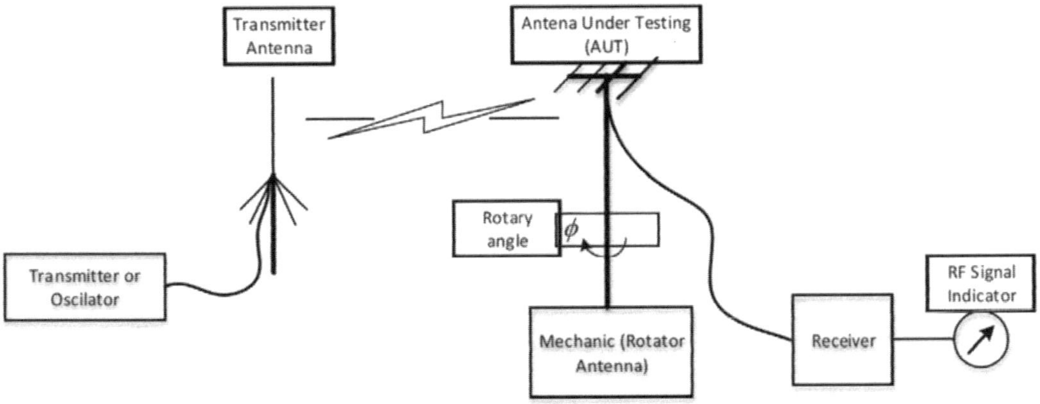

Figure 6. Configuration of antenna parameter measurements.

Mechanical system in principle is an antenna rotator or electric motor with enough torque to rotate the AUT. Furthermore AUT is connected to the receiver system and the receiver system output is connected to an indicator that can show the amount of power received by AUT. The value displayed by the indicator for a certain angle ϕ is the value of the electromagnetic wave power received by AUT at that angle. AUT rotation by rotator is $\phi = 0^0$ s.d. 360^0 will produce an indicator value that can represent the radiation pattern of an antenna or other parameters as explained in the previous section.

4.2. Equipment used

The process of measuring radiation patterns using hardware that has been developed previously. The equipment that has been made is Module-1 in the form of a mechanical rotator antenna and Module-2 which contains a Radio Frequency (RF) to Direct Current (DC) signal modifier and an Arduino Uno microcontroller system that will be connected to a PC or Laptop computer. The hardware function testing of Module-1 and Module-2 had been carried out the previous year.

This experimental study of radiation pattern measurement at the same time testing software created using the Visual Basic programming language in the Microsoft Visual Studio package to display graphs of radiation patterns based on the work functions of Module-1 and Module-2. The Visual Basic application runs on a PC computer that is connected to Module-2 using a USB cable. The software engineering method used is the Water Fall method [8]. In general the block diagram of a system measuring antenna radiation pattern is shown in Figure-7.

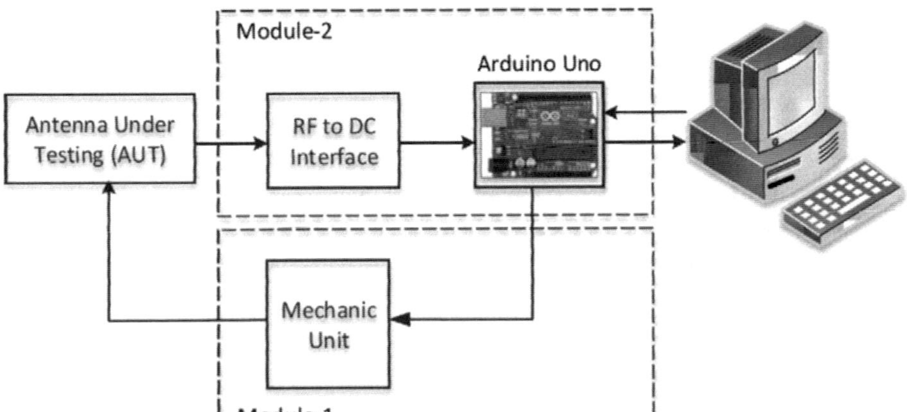

Figure 7. Block diagram of an antenna radiation pattern gauge.

Module-1 is used to put AUT and rotate to the right in the horizontal direction from 0° to 360° and back to its original position (position 0°). Whereas Module-2 functions to start and end rotator motor rotation, adjust the direction of rotation and turn angle of Module-1, convert the RF to DC analog signal voltage into digital RF to DC signal voltage, and then send the results of this conversion to a computer via the port USB The graphic display application is created using the Visual Basic programming language, with a display as shown in Figure-8 below. The application created is titled **Antenna Radiation Pattern**.

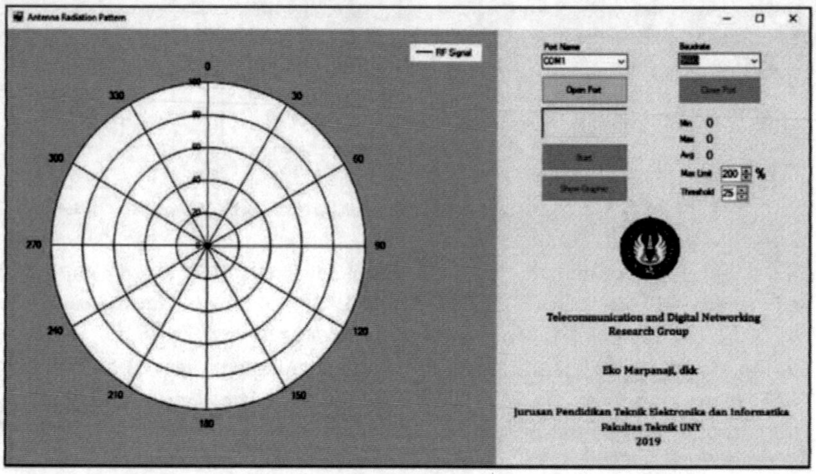

Figure 8. The front page of the Antenna Radiation Pattern application.

5. Result and Discussion

5.1. Specifications Devices used in the measurement of radiation patterns

The specifications of the devices used in the radiation pattern measurement experiment are:

- The transmitter uses a 2 meter band aircraft brand Icom IC-2200H with 5 Watt power at a frequency of 143,250 MHz;
- The transmitter antenna uses a DP-LS2E D quarter-lambda antenna with a ring in the middle;
- The measurement distance between the transmitting antenna and the receiving antenna (AUT) is 4 meters;
- The measurement time is daytime, with fine weather;
- The antenna being tested or AUT is a type of Half Wave Open Dipole antenna;

Measurement of the radiation pattern of Half Wave Open Dipole antennas is done with various conditions, namely (1) without a transmitter signal; (2) with a transmitter signal where the position of the transmitter antenna is located 90° in front of the Half Wave Open Dipole antenna stretch (the transmitter antenna is perpendicular to the axis of the Half Wave Open Dipole stretch); (3) with a transmitter signal where the position of the transmitting antenna is 0° in front of the Half Wave Open Dipole antenna stretch (the transmitter antenna is in line with the axis of the Half Wave Open Dipole antenna); (4) simulation of electromagnetic wave interference by turning on a motorized vehicle near the measurement area (the effect of spark plugs on motorized vehicles on the measurement results).

5.2. Radiation pattern measurement results

Based on the results of the measurement of the radiation pattern by running the Antenna Radiation Pattern application without the RF signal from the transmitter shown in Figure-9.

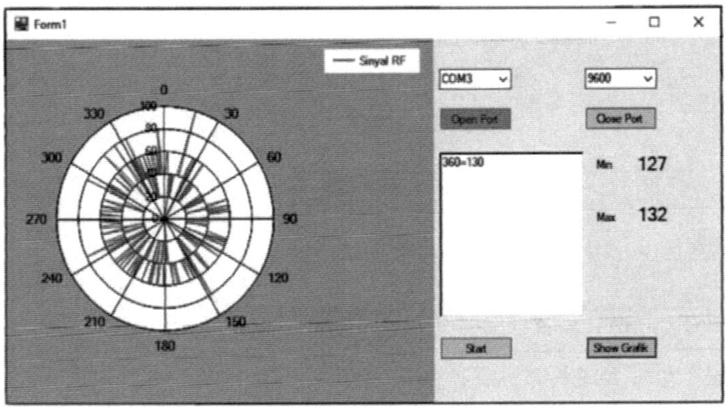

Figure 9. Display of radiation patterns when no RF signal is received by AUT

Measurement of radiation patterns without RF signals from the transmitter is used to describe the noise received by AUT and is used to describe the noise in the radiation pattern graph. The measurement results in Figure-9 show that the noise signal is spread evenly to all angles (from 0° to 360°), with a minimum value of 127 and a maximum value of 132. These values are used as a reference in the process of eliminating noise when signal processing is carried out by the application Radiaton Pattern Antennas.

The next measurement is to place the transmitting antenna located 90o from the axis of the Half Wave Open Dipole antenna. The results of the measurement of the radiation pattern are shown in Figure-10. The minimum and maximum values for the RF signal quantization are 104 and 188, respectively. The maximum value is much higher than the maximum value when there is no RF signal as shown in Figure 9 previously.

Based on Figure-10 it appears that the maximum radiation pattern is at an angle of 0° and 180°. This is due to the position of 0° and 180° perpendicular to the Half Wave Open Dipole antenna. Thus, the Half Wave Open Dipole antenna will receive the largest signal according to the characteristics of the Half Wave Open Dipole antenna having a maximum radiation pattern in front of and behind the stretch of the antenna element. The results of this measurement show that the function of RF to DC components, the microcontroller ADC quantization process, the data sent by Module-2 to the computer, and the application in displaying a radiation pattern graph are in accordance with the AUT characteristics of the Half Wave Open Dipole antenna.

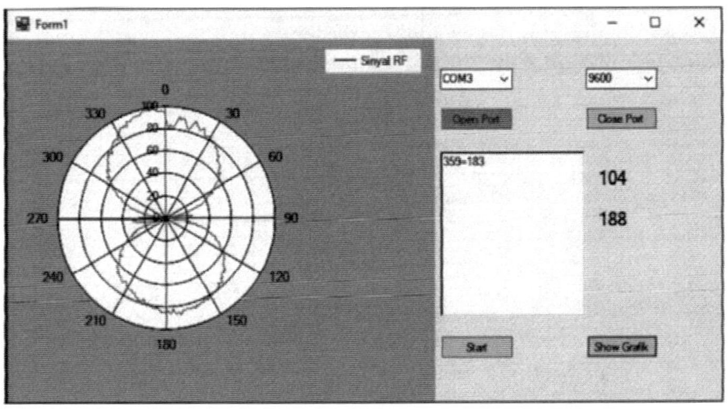

Figure 10. AUT radiation pattern when RF signal source is located 90°.

The next measurement is to place the transmitter antenna (RF signal source) at an angle of 0° (in line with the stretch of the Half Wave Open Dipole antenna element). The results of measurements of radiation patterns with these positions are shown in Figure-11 below.

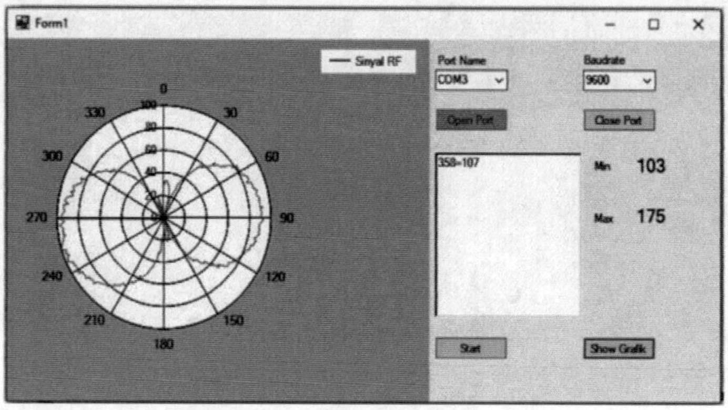

Figure 11. AUT radiation pattern when RF signal source is located 0°

Based on Figure-11 it appears that the maximum radiation pattern occurs when the AUT rotates as far as 90° and 270°. The results of this measurement indicate that the maximum radiation pattern occurs when the transmitting antenna is perpendicular (90° or 270°) to the stretch of the Half Wave Open Dipole antenna element. The minimum radiation pattern occurs when the transmitting antenna is located next to the stretch of the Half Wave Open Dipole antenna (at an angle of 0o or 180o). The results of this radiation pattern measurement are consistent with the characteristics of the Half Wave Open Dipole antenna, which has maximum radiation in front of or behind the stretch of the antenna element and has minimum radiation on the left or right side of the stretch of the Half Wave Open Dipole antenna. Thus, the hardware functions of Module-1, Module-2, and the Antenna Radiation Pattern application are functioning properly.

The last improvement of the Antenna Radiation Pattern application is a facility to eliminate very high amplitude surges due to the arrival of noise whose value exceeds the maximum value of the RF signal. This phenomenon can be conditioned by giving a disturbance signal originating from the spark plugs of motor vehicles located near the AUT during the process of measuring the radiation pattern. Correction facility is needed to eliminate values that far exceed the average RF signal received, namely by setting the Maximum Limit value when displaying images of radiation pattern measurements. Thus, if interference occurs when taking radiation measurements, then the display of radiation pattern images can be corrected by eliminating noise that interferes with the measurement process.

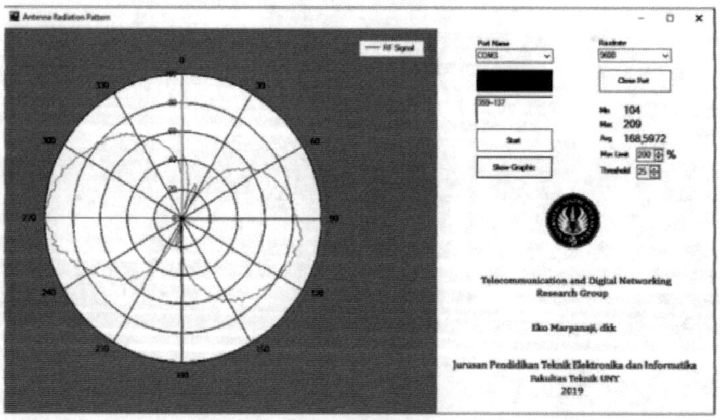

Figure 12. Improving the application by adding the Maximum Limit adjustment facility as a data correction factor in displaying the radiation pattern graph.

In addition to the correction facility, the application is enhanced with other accessories to complement the appearance of the Antenna Radiation Pattern application so that users can more easily understand the values of radiation pattern measurements. Improvements to the Antenna Radiation Pattern application appearance are shown in Figure 12.

Based on the results of experiments measuring antenna radiation patterns as a testing process for developing Module-1 hardware, Module-2, and Antenna Radiation Pattern application software with AUT using the Half Wave Open Dipole antenna, it can generally be stated that the hardware and software created have suitable for use in measuring radiation patterns. Testing should be continued for other types of antennas for further exploration of the performance of Antenna Radiation Pattern hardware and software, as well as testing the reliability and validity of Antenna Radiation Pattern hardware and software.

6. Conclusion and Future Works

6.1. Conclusion

Based on the experimental results of radiation pattern measurements using the Antenna Radiation Pattern device that has been made, it can be concluded as follows:

(1) Antenna Radiation Pattern device consisting of hardware Module-1 and Module-2 as well as application software for displaying radiation pattern graphs have functioned well for the measurement of radiation patterns.
(2) Measurement of the radiation pattern is strongly influenced by environmental conditions where the measurement is carried out, especially the interference of electromagnetic waves generated by devices other than the RF signal transmitter (RF signal source) used, for example noise signals due to spark plugs in motor vehicles.
(3) Display of radiation patterns in accordance with the ideal radiation pattern graph (according to theory) can be obtained by processing digital signals resulting from the quantization of RF to DC signals before being displayed in the form of radiation patterns.

6.2. Future Works

Further work based on the results of this study are:

(1) Improvement of digital signal processing resulting from RF to DC signal quantization so that the graphical display of the antenna radiation patterns measured is more accurate.
(2) It is necessary to test various types of antennas to test the reliability and validity of measurements of antenna radiation patterns.
(3) Further development of the Antenna Radiation Pattern device for measuring antenna parameters other than radiation patterns, such as Gain, Beam Width, Directivity, and Aperture.

7. References

[1] Hendrik, D. P., dkk. 2013. *Perancangan Alat Bantu Pengukuran Otomatis Pola Radiasi,Polarisasi*, Gain , dan Direktivitas Pada Antena. Online di: http://repository.telkomuniversity.ac.id/pustaka/93449/perancangan-alat-bantu-pengukuran-otomatis-pola-radiasi-polarisasi-gain-dan-direktivitas-pada-antena.html. Accessed February, 19th 2018.

[2] Kirbi, T. N., dkk. 2015. *Perancangan Sistem Untuk Pengukuran Parameter Antena Otomatis, Realtime Berbasis Mikrokontroller*. Online di http://snti.untar.ac.id/images/ prosiding/2015/C09-Perancangan-sistem-untuk-pengukuran-parameter-antena-otomatisrealtime-berbasis-mikrokontroller.pdf. Accessed February, 19th 2018.

[3] Kraus, John Daniel., 2003. *Antennas: For All Applications. 3rd Edition.* New Delhi: McGraw Hill. pp 39-40.

[4] Balanis, Constantine A., 1997. *Antenna Theory: Analysis and Design. Second Edition*. New York: John Wiley & Sons, Ltd. Chapter 2 pp 28-100.

[5] Saunders, Simon R., Zavala, Alejandro Arago'N., 2007. *Antennas And Propagation for*

Wireless Communication Systems. West Sussex: John Wiley & Sons, Ltd. pp 28-34.

[6] Visser, Hubregt J., 2012. *Antenna Theory And Applications*. West Sussex: John Wiley & Sons, Ltd. Chapter 2 pp 15 – 46.

[7] Carr, Joseph J., Hippisley, George W., 2012. *Practical Antenna Handbook. Fifth Edition*. New York: McGraw-Hill, Inc. pp 96 – 98.

[8] Pressman, G. S., and Maxim, B. R., 2015. Software Engineering. A Practitioner's Approach. 8th Edition. New York: Mc-Graw Hill Education. Pp 41-42.

Acknowledgments

We would like to thank the Faculty of Engineering, Yogyakarta State University for giving the opportunity to carry out this research. We would also like to thank the Telecommunications Communication and Digital Networking team for helping to carry out the research so that we can present this paper. We expect criticism and suggestions for the improvement of this paper.

Development of Intelligent Floor Cleaning Robot

S Yatmono, M Khairudin, H S Pramono and A Asmara
Electrical Engineering Education Department, Universitas Negeri Yogyakarta, Yogyakarta, Indonesia

E-mail: s_yatmono@uny.ac.id

Abstract. Cleaning the floor from dust is one of the routine activities carried out every day. This activity is not only at home but also in offices or shopping centers. Sometimes due to dust cleaning activities take a long time then there are other activities that are overlooked. For this reason, we are trying to develop a smart floor cleaning robot that can navigate, clean dust, and polish floors automatically. The study was conducted using Pressman's research and development methods which included the following phases: analysis, design, implementation and testing. The robot, developed using an omniwheel wheel and is equipped with a vacuum cleaner and floor polishing motor. The control system used is based on the Arduino microcontroller, the robot is also equipped with a Bluetooth communication system so that it can be controlled via an Android smartphone. The developed robot is able to move according to navigation controls via an android smartphone. Besides this robot is also able to move to avoid obstacles if the distance between the robot and the barrier is less than 15 cm. Floor cleaning performance on various types of dirt is quite good with only leaving dirt on the floor less than 20%.

1. Introduction

Cleaning the floor from dust is one of the routine activities carried out every day. This activity is not only at home but also in offices and shopping centers. Sometimes because the activity of cleaning the floor takes a long time then there are other activities that are overlooked. Floor cleaning activities in public spaces even have to be done all the time is not enough to be scheduled in the morning or evening, this requires the janitor to always be ready at all times to clean the floor.

Various types of technology have been created to facilitate human activities in their work every day. One of them is the technology developed in robotics technology. Robots are able to replace human activities that regularly or occasionally have to be done at unscheduled times. In terms of cleaning the floor, it is necessary to develop robots as a substitute for floor cleaning officers.

In 2002 in the United States a robot named Roomba was developed. Roomba's automatic robot vacuum cleaner is produced by iRobot. Introduced in September 2002, Roomba has several sensors that enable it to do the job of cleaning the floor. For example, Roomba is able to change direction when facing obstacles and detect dirty spots on the floor. But there are some things that need to be improved from Roomba's robot, which is because the robot's body is circular, there are some points on the corner of the floor that cannot be reached. Moreover, this Roomba robot can only suck up dust and does not have components that polish the floor so that there are still dirty stains on the floor left behind.

Ajay P John (2017) makes Arduino-based floor cleaning robots and is controlled using a GSM telephone[4]. In that study, the robot that was developed was equipped with a vacuum cleaner motor, sensors and control systems based on SMS messages from mobile phones. After receiving a message to turn on the robot, the robot will move forward and if there is a barrier in front of the robot will turn right. The robot will rotate 180 degrees and move forward if it does not receive navigation control messages so robot's navigation controls are inflexible. The robot will stop if there is a Stop message from a GSM mobile phone.

Departing from this, in this study developed an intelligent robot floor cleaners shaped box with an obtuse angle at the edges so that it can reach a wider angle of space. In addition to the motor is attached to a vacuum cleaner, a robot developed also include floor polishes motors so that not only vacuuming but also clean the stain. Developed operating mode there are two moving robot automatically follow the pattern already set and will navigate with the aid of ultrasonic sensors to avoid obstacles and semi-automatic mode with the help of Android smartphone based control. The communication system developed using Bluetooth and come with Android GUI display that can be easily operated via Android smartphones.

2. Research Method

The development of floor cleaning robot using research and development (R & D) method, which refers to the method developed by Pressman[6]. The stages that must be passed are: analysis, design, implementation and testing. Analysis phase is to identify and obtain data about the needs of what is required in the design and implementation of systems.

a. Analysis of user needs, analysis of what needs are needed by users to be applied to the robot system that was developed such as a vacuum cleaner, motor polish, and robot navigation control components.
b. Job analysis, i.e. analysis of data performance to be carried out by the system to be designed.
c. Data analysis, i.e. analysis of what data will be processed either as input or as an output.
d. Technology analysis, i.e. analysis of what technology will be used in the system to be designed.

Design is the stage carried out to achieve the most effective and efficient system implementation with the help of data obtained at the analysis stage. In the design we will get a framework for implementing the system. There are several stages in design, namely:

a. The design of a mechanical system consisting of a robot body chassis, a controller module, a Bluetooth module and a smartphone interface with Arduino.
b. The design of electronic circuits that will be used in the design of robot motion navigation systems.
c. Program flowchart design is a chart with special symbols that describe the sequence of processes and the relationships between processes in detail in an Arduino-based robot control program. In this case also made an Android-based GUI programming application design.

Implementation is the stage of translating the design into an actual circuit. The design of the placement of the wheel motor, vacuum cleaner and polishing motor are arranged in the most optimal placement configuration. Supporting circuits such as Arduino microcontroller circuits, motor drivers, and Bluetooth receivers are placed in the robot's body without disturbing the robot's movements. The implementation of the Android-based robot navigation control application is also made as needed.

Hardware testing is done by activating the robot's navigation sensor (ultrasonic sensor) and then measuring the distance of the robot's navigation barriers which are then compared to a predetermined distance, whether the distance obtained is still within the specified tolerance limits. The next test is the Omni Wheel Control Performance Test, Vacuum Cleaner Performance Test, and Floor Polish Performance Test.

Software testing is the process of implementing a program to find errors. Before a program is implemented, it must be error free. Therefore, the program must be tested to find errors that might occur. Testing is done for each module and continued with testing for all modules that have been assembled.

Journal of Physics: Conference Series **1413** (2019) 012014 doi:10.1088/1742-6596/1413/1/012014

3. Result and Discussion

The results of the analysis of the needs of the floor cleaning robot system developed include:

a. Robot can work with maneuvering in all directions with the conditions of various shapes of the room
b. Robots that are built have a cleaning system in the form of polishing machines and vacuum cleaners
c. The robot motor operates with a 12V DC voltage
d. Minimum battery voltage of 12 V with a current of 20Ah
e. Wireless control is needed to reach in all directions

The hardware and software design developed in the floor cleaning robot includes the design of locating the drive wheel, vacuum cleaner motor and floor polish. It also developed the design of Bluetooth communication devices between robots and smartphones along with the design of the Android application interface.

Robot drive wheel consists of three motors that form a triangular configuration. The motor is placed at the end of the triangle to move the angle of the omniwheel wheel so it is easy to move. The design of the placement of the robot motor is shown in Figure 1.

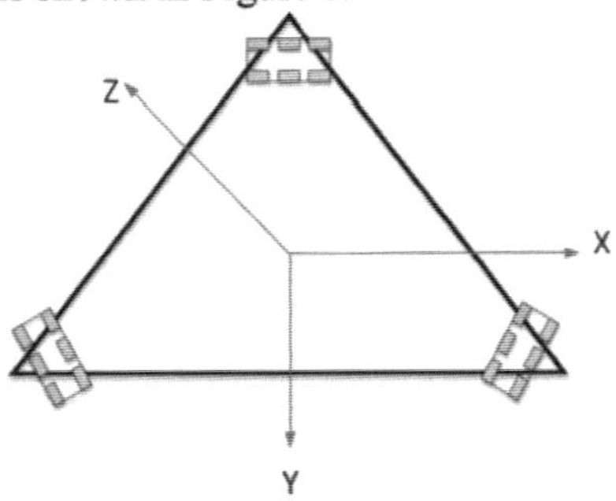

Figure 1. The design of robot's wheels placement

The results of the placement of the wheel design are also used as a reference to place a vacuum cleaner and motor polish. The placement of the vacuum motor and the polishing motor are placed on the back or base of the triangle from where the wheels are placed. A vacuum cleaner is placed in front of two polishing motors so that the floor is free of dust before being polished by a polishing motor. The results of the placement of the motor wheel, motor vacuum cleaner and second floor polisher motor are shown in Figure 2.

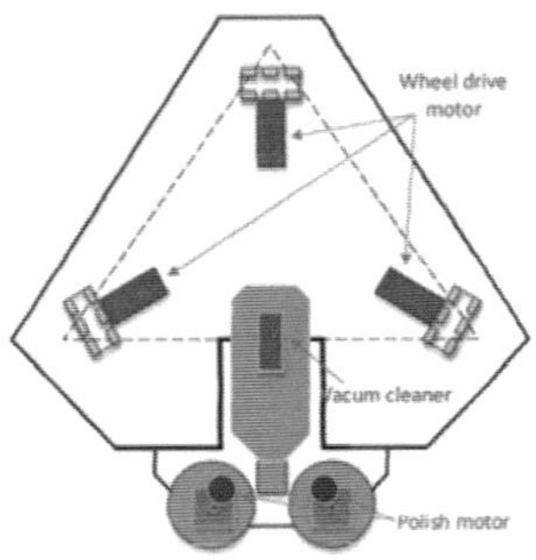

Figure 2. The design of the vacuum motors and polishes

In accordance with the results of the design, implementation placement of wheel drive motors, vacuum cleaner and the polisher motor can be seen in Figure 3.

Figure 3. Display layout robot motors

The hardware placement such as the Arduino microcontroller circuit, motor driver, LCD, and power supply are designed so that together the Bluetooth receiver is divided into two sub-systems. The first system is the master block which consists of Arduino control, rotary encoder input and wheel motor. While the second sub-system is a slave block consisting of Arduino control, sensors, Bluetooth communication, motor polish and vacuum cleaner. The component placement scheme design can be seen in Figure 4.

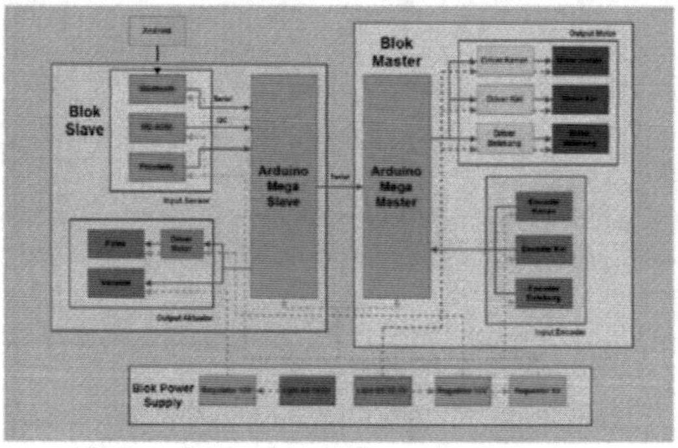

Figure 4. Design series robotic system

The implementation of robot motion control application design uses an Android-based application that is based on predetermined needs. There are two system condition settings that must be met for the application to function properly. The first arrangement is determining the length and area of the room to be cleaned. The second is the determination of buttons to control the direction of movement and activate vacuum and polish. Implementation of the appearance of the developed Android application can be seen in Figure 5.

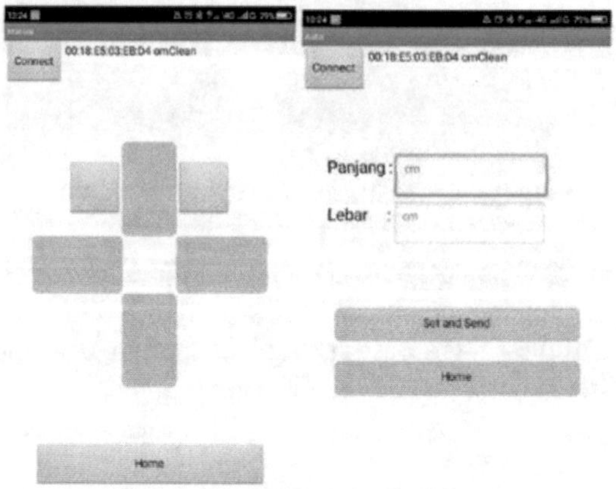

Figure 5. Display Android app

The results of hardware and software implementation of floor cleaning robots are tested using the blackbox testing method. System function tests and floor cleaning robot movements are performed via an Android-based control application input. The test is carried out on the basis of the robot's navigation movement that is forward, backward, right and left shear and motor pole activation. Test results are shown in Table 1.

Table 1. Functional test application button

Button	Action	Result				
		1st	2nd	3th	4th	5th
Forward	The robot moves forward by always maintaining heading 0 degrees	Ok	Ok	Ok	Ok	Ok
Backward	The robot moves backward by always maintaining heading 0 degrees	Ok	Ok	Ok	Ok	Ok
Swipe right	The robot moves swipe right by always maintaining heading 0 degrees	Ok	Ok	Ok	Ok	Ok
Swipe left	The robot moves swipe left by always maintaining heading 0 degrees	Ok	Ok	Ok	Ok	Ok
turn right	rotate right and when stopped will set heading 0 degrees	Ok	Ok	Ok	Ok	Ok
turn left	rotate left and when stopped will set heading 0 degrees	Ok	Ok	Ok	Ok	Ok
power on	turn on the arduino and bluetooth systems	Ok	Ok	Ok	Ok	Ok
On /Off motor	turn on and off vacum cleaner and polish motor	Ok	Ok	Ok	Ok	Ok

The experimental results show that the robotic navigation control can run well according to the keypress input developed on the Android application. During the five trials showing the same thing, the robot movement was observed in accordance with the specified key control function. Full activation of the motor and polishing system also gets the same result that the main system and the vacuum motor can be turned on and off remotely via the Android application control button.

The next test results are the results of the robot motion algorithm test if it finds a barrier in front of the robot. In accordance with the algorithm that has been implanted in the robot's slave motor control block, then if during moving the robot finds objects that block its movement, then the robot will move to avoid these obstacles. Table 2 shows the results of testing the robot maneuvering movement towards the presence of solid objects.

Table 2. Test distance of barrier

Barrier Distance	Test Result	Maneuver
5 cm	able to avoid	move to the left of the barrier
10 cm	able to avoid	move to the left of the barrier
15 cm	able to avoid	move to the left of the barrier
20 cm	can't escape	
25 cm	can't escape	
30 cm	can't escape	
35 cm	can't escape	
40 cm	can't escape	

The test results show that the robot will detect the distance of the object between the front of the robot and the barrier. If the distance of the barrier object is still more than 15 cm, then the robot will move straight ahead. If the detected distance is less than 15 cm the robot will move to the left of the barrier. The results of this test are carried out if the robot uses an automatic control mode based on ultrasonic sensor data placed on the robot's body.

Subsequent tests were carried out to get the performance of the floor cleaning robot. Testing is done by putting a kind of dirt in front of the robot. Performance test results obtained are shown in Table 3 below.

Table 3. Test performance floor cleaning robot

Type of Dirt	Before	Result	After
Dust	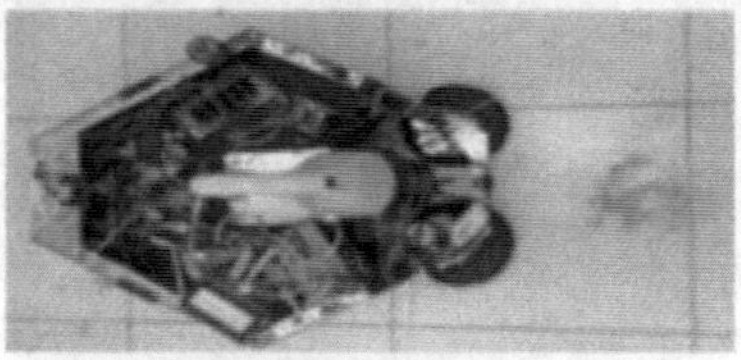	suck less power so there is still dust left	
Sand	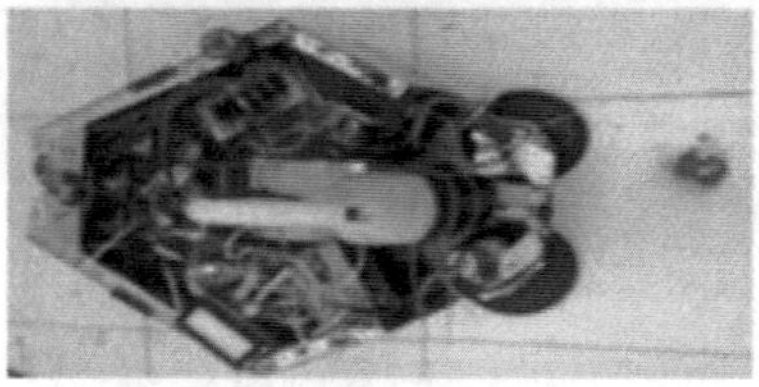	suck less power so there is still sand left	
Paper	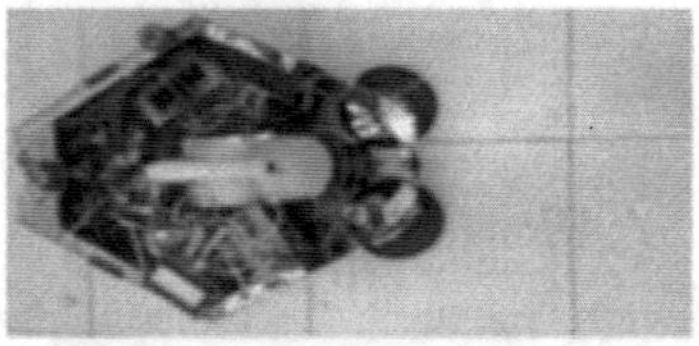	suck less power so there is still paper left	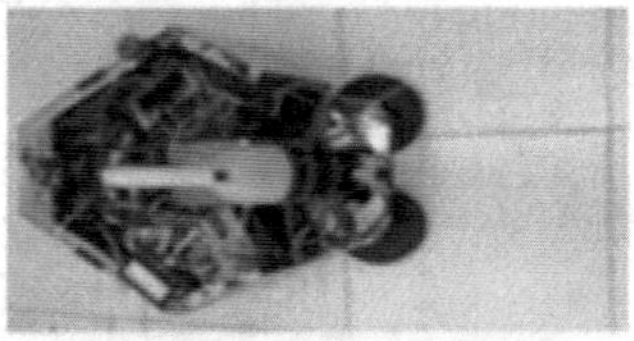
Carpet fur		suck less power so there is still fur left	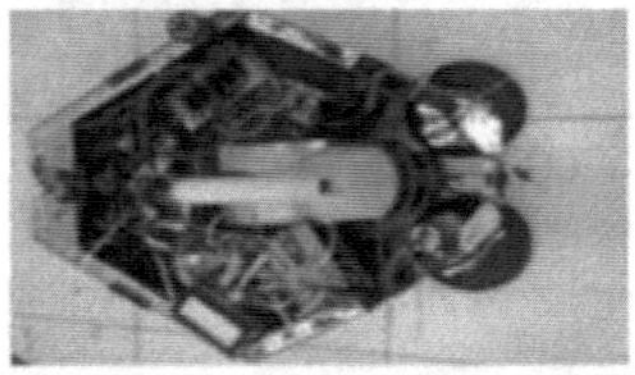

The results of the floor cleaning robot performance tests on the types of impurities such as dust, fine sand, pieces of paper and fur rugs show good results. There is still some dirt left after the cleaning robot has passed the floor. This is possible because the motor power is not strong enough. In addition, there is no algorithm to detect the presence of dirt that is still left on the floor. However, based on observations of the pictures shown in Table 3, it shows that the amount of dirts left is only around less than 20% .

The test results, can be used as a basis for stating the performance and specifications of the floor cleaning robot that has been developed. Based on the results of tests that have been done, the floor cleaning robot specifications developed in this study have the following product specification values:

- Weight robot: 10.2 kg
- Size Length, width, height robot: 55cm x 48cm x 36cm
- Robot battery capacity: 5200mAh LiPo 4S and 6S LiPo 5200mAh
- The ability of the robot working life time: ± 45 minutes
- Total current consumption of the robot: Motor 2A
- The current consumption control system and sensor 2A
- Polishing motor current consumption 4A (2 x 2A) + Vacuum 8A = 16A
- Robot Working Voltage: Motor 24Volt,
- Working Voltage Vacuum and Polishing 12volt, control systems and sensors 5volt

4. Conclusion

Based on the results of the design and testing of the floor cleaning robot system developed, it can be concluded as follows:

a. Floor cleaning robot consists of hardware which is composed of a series of robot control and mechanical components and Android application software as a robot motion controller. The robot control hardware component consists of two master and slave sub-systems, each of which consists of an Arduino control component, sensors as inputs and motor drivers. The motor wheel is controlled by the master block and the cleaner and polishing motor are controlled by the slave block.
b. Android's performance as a manual robot controller can function according to the set of buttons. To use the automatic mode the ultrasonic sensor robot can avoid obstacles with an object distance of less than 15 cm.
c. Floor cleaning performance for various types of dusty dirt, fine sand, pieces of paper and fur carpet shows that the remaining less than 20% of the dirt on the floor.

5. References

[1] Anonymous, Built-in Examples, Arduino.cc, 2017
[2] Hari Santoso, Arduino Practical Guide for Beginners, www.elangsakti.com , 2015
[3] Hari Santoso, Monster Arduino, www.elangsakti.com, 2017
[4] John, Ajay P, Implementation of an Automated Smart Robotic Floor Cleaner, International Journal of Advanced Research in Electrical, Electronics and Instrumentation Engineering Vol. 6, Issue 4, April 2017
[5] Singh, Satyinder, Robotic Vacuum Cleaner, International Journal of Engineering Science and Computing, May 2017
[6] Pressman, Roger S., "Software Engineering: a practitioner approach (Book 1)", Andi, Yogyakarta, 2002
[7] Shemanuev, Evgeny, Manage the Arduino-robot using the G-sensor on your smartphone, remotexy.com, 2017

Journal of Physics: Conference Series **1413** (2019) 012015 doi:10.1088/1742-6596/1413/1/012015

Development of Arduino Assisted Microcontroller Instructional Devices in Vocational High Schools

H A Wijaya and M Khairudin
Electrical Engineering Education Department, Universitas Negeri Yogyakarta, Yogyakarta, Indonesia

Email: husain.asyari2015@student.uny.ac.id

Abstract. This development research aims to find out: (1) feasibility, (2) functioning, and (3) the effectiveness of the arduino-assisted microcontroller assisted devices with the arduino simulator Xevro version 1.5 as a learning media for the material programming of microcontroller programming skills in Mechatronics Engineering at the Vocational High School. The subjects of this study were the XI grade students of the Mechatronics Engineering expertise program at vocational high schools N 1 Bawang Banjarnegara and Leonardo Klaten Vocational High School. The results of this study are an Arduino-assisted microcontroller simulation for microcontroller programming material. The results of the feasibility assessment of the Arduino-assisted microcontroller simulation according to the validator are 83.33 so that it is categorized as "Very Worthy" and according to students is 70.98 of a scale of 100 so that it is categorized as "Eligible". The results of the assessment of the function of Arduino-assisted microcontroller simulations according to the validator are 82.92 so that it is categorized as "Very Worthy" and according to students is 71.55 of a scale of 100 so it is categorized as "Eligible". This Arduino-assisted microcontroller simulation has an effective impact, because the significance value when tested using the t test is t count = 11.075> t table = 1.983 and Asymp. Sig (2-tailed) <0.05. The level of effectiveness can be seen from the magnitude of the gain value.

1. Introduction

Improving the quality of vocational education can be done by sharpening the cognitive abilities and skills of students in line with the growing work competency needs. Work competency is a basic characteristic possessed by someone who indicates how to think and act for various situations and for a long period of time [1].

Vocational education [2] is a form of education that is equivalent to high school but historically, vocational education has focused on work preparation and can be defined as courses and educational programs offered at less than baccalaureate level.

Vocational education has a variety of objectives. [3] states that the purpose of vocational education is to focus on (a) informing someone about job selection; (b) individual initial preparation for work life;

(c) developing one's competencies to face the world of work; and (d) educational experience as an effort to adapt someone to the world of work. The selection of majors is based on students' choices, students' skills in utilizing the choices they already have, and students' skills whether their chosen competencies are in line with opportunities in the future [4].

The success of vocational education can be assessed from four aspects, namely the number of students, infrastructure and teaching staff, curriculum, and graduate results [5]. Vocational education begins to provide training for skilled workers and teachers try to find ways to engage students in classrooms and laboratories so students can be skilled when working in the industry [6].
Great vocational teachers are dedicated, great facilitators, learning leaders, great communicators, motivators, positive thinkers, creative problem solvers capable of utilizing ICT, respecting students, caring for students, being able to identify the needs of students, listeners and pedagogically very competent in developing various teaching and learning methods [7].

Learning media is everything that is used to channel messages and can stimulate thoughts, feelings, attention, and willingness to learn so that it can encourage the occurrence of a deliberate, purposeful and controlled learning process [8]. The criteria for media selection according to [9, 10] are as follows: the accuracy of the teaching objectives; support for the content of study material; ease of obtaining media; teacher skills in using it; available time to use it; and in accordance with the level of thinking of students. Simulation is an imitation of the operation of real-world processes or systems over time [11]. Anderson [12], real objects or model objects that are very similar to real objects, will provide a very important stimulus for students in learning tasks that involve psychomotor skills. Simulator is included in the context of M-learning with a new concept and contains content that is easy to use, and does not involve the location where the learner does it [13-14]. Simulation of industrial multi-process control as a medium or means of practice has good performance in the use of learning in class [15]. Digital simulations include teaching how to use information, communication and technology that is very much needed in the world of work. Skills include digital information management through processing software in the form of data, numbers and presentations [16].

Based on the description previously explained, the formulation of the problem in this study is how the feasibility, functioning and effectiveness of the arduino-assisted microcontroller assisted with Xevro simulator version 1.5 which is equipped with material modules on the competence of Mechatronics expertise in vocational high schools.

2. Research Method

Learning material module Arduino assisted microcontroller was developed using the ADDIE model developed by Lee and Owens (2004). The ADDIE development model is a learning media product development model which consists of four stages, namely (1) needs analysis and assessment (needs assessment and analysis), (2) design (design), (3) development and implementation, and (4) evaluation (evaluation). Programming media simulations (simulators) in this study was developed using the waterfall method developed by Pressmann (2012). Waterfall method consists of: (1) system / information engineering and modeling; (2) software requirements analysis; (3) design; (4) coding; (5) testing; and (6) maintenance.

Needs analysis is the stage of gathering information to find out the gap between reality and ideality. Needs analysis consists of two stages, namely needs assessment and front-end analysis Arduino-assisted microcontroller products consist of two types of media, namely media objects (simulation) and print media (Aruino assisted Microcontroller material modules).

Testing is done in 2 vocational schools with mechatronics expertise competency with 3 stages, namely (a) black box testing to test portability and functionality, (b) material and media validation testing by experts, (c) testing is limited to 2 classes of competency student classes mechatronics expertise at vocational high schools N 1 Bawang and vocational high schools Leonardo.

3. Results and Discussion

This study analyzes the feasibility of an Arduino-assisted microcontractor system instructional device as a result of the development of instructional devices. The function of Arduino simulation is done by black box testing. The trial consists of two dimensions, namely portability and functionality. The results of the black box test can be seen in Table 1 below.

Table 1. Black Box testing

Respondent	Dimension		Category
	Portability	Functionality	
1	100	100	Eligible
2	100	100	Eligible
3	100	100	Eligible
Average	100	100	Eligible

Table 1 presents the feasibility of an Arduino-assisted microcontractor system instructional device to be used in Vocational Schools. The feasibility of this Arduino-assisted microcontractor system instructional device was developed from the point of view of the substance aspects of the material and the design of learning..

The aspects developed based on functionality are assessed from aspects of functionality, portability, usability and layout. Rating scales from 1 to 4 are applied in the questionnaire statement, the scores obtained from the next questionnaire are changed to a scale of 1 to 100 for determining the assessment category. Table 2 shows the results of the assessment of the function of the Arduino-assisted microcontractor system.

Table 2. Results of assessment of the functioning of the microcontroller system

No.	Respondent	F		P		U		LA		Average Total	
		Val	Catg.	Val	Catg.	Val	Catg.	Val	Catg.	Val	Category
1	Expert 1	75	L	75	L	75	L	78.57	L	76	SL
2	Expert 2	79.17	SL	83.3	SL	83.33	SL	82.14	SL	82	SL
3	Expert 3	87.5	SL	91.7	SL	79.17	SL	89.28	L	87	SL
4	Expert 4	83.33	L	91.7	SL	79.17	SL	89.28	SL	86	SL
5	Expert 5	91.67	L	83.3	SL	87.5	SL	89.28	L	88	SL
6	Expert 6	83.33	L	75	L	75	L	85.71	SL	80	SL
7	Expert 7	87.5	L	91.7	SL	83.33	SL	85.71	L	87	SL
8	Expert 8	79.17	SL	79.2	SL	75	L	78.57	L	78	SL
Average		83.33	SL	83.9	SL	79.69	SL	84.82	SL	83	SL

L: Eligible, SL: Very Eligible

Student learning outcomes from cognitive aspects between before and after operating the Arduino simulation are used as a reference in terms of the effectiveness of Arduino-assisted microcontroller instructional devices with simulations developed.

Furthermore, the assessment of the microcontroller instructional system media to the user in this case is the VOCATIONAL HIGH SCHOOLS students. The results of the student response assessment consisted of two aspects of assessment, namely the feasibility and functioning aspects of the instructional system of the microcontractor assisted by Arduino simulation.

4. Conclusion

First, the results of testing instructional devices Arduino-assisted microcontrollers found that Arduino simulations have feasibility and functionality in the "Eligible" category. Second, instructional devices Arduino-assisted microcontrollers can be used in learning because the display is acceptable (interesting), the operation is considered easy, and has a stable performance in practicum, complete material, easy to understand, coherent from simple to complex, this shows Arduino simulation is very useful in facilitating the learning process (for teachers and students), because it can motivate and improve student learning outcomes. Third, the effectiveness of Arduino simulations of microcontroller subjects between the posttest mean values of the pretest mean was declared effective, because the value of $t_{hitung} = 11.075 > t_{tabel} = 1.983$ and Asymp. Sig (2-tailed) <0.05. This shows that the use of Arduino-assisted Microcontroller simulation has significant effectiveness on increasing the value of student learning outcomes.

5. References

[1] Spencer, Lyle & Signe M. Spencer. 1993. Competence at Work, Models For Superior Performance. Canada : John Wiley & Sons, Inc.

[2] Zirkle, Chris. (1998). Perception of Vocational Educators And Human Resource/ Training and Development Profesisonals Regarding Skills Dimension of School to Work transsitin Programs. Journal of Vocational and Technical Education, Vol 15, 1, 4.

[3] Billet. S. (2011). Vocational Education: Purpose, Traditions and Prospect. New York: Spinger.

[4] Gough, S. (2009). TVET as Sustainable Investment. Technical and Vocational Education and Training: Issues, Concerns and Prospects. Journal Vocational, Vol.8, 107-116. doi: 10.1007/978-1-4020-8194-1

[5] Guo, Z. & Lamb, S. (2010). International Comparisons of China's Technical and Vocational Education and Training System. New York: Springer

[6] Zirkle, C. & Martin, L. (2012). The Future of Vocational Education and Training in a Changing World. New York: Springer

[7] Lucas, B. (2014). Vocational Pedagogy: What It Is, Why It Matters and How to Put It into Practice. Report of UNESCO-UNEVOC Virtual Conference, 12-26 Mei 2014. Retrieved from www.unevoc.unesco.org.

[8] Miarso, Yusufhadi. (2009). *Menyemai Benih Teknologi Pendidikan.* Jakarta: Kencana Prenada media Group.

[9] Sudjana, N. (2016). Penilaian Hasil Proses Belajar Mengajar. Bandung: Remaja Rosdakarya.

[10] Khairudin M., Triatmaja A.K., Istanto W.J., Azman M.N.A., (2018). Mobile Virtual Reality to Develop a Virtual Laboratorium for the Subject of Digital Enggineering. *International Journal of Interactive Mobile Technologies. Vol 13, No 04 (2019)*

[11] Banks, J., Nelson, B. L. (2004). Discrete-Event System Simulation. New York: Prentice Hall

[12] Anderson, L.W dan Krathwohl, D.R. 2010. Kerangka Landasan untuk Pembelajaran, Pengajaran dan Asesmen (Revisi Taksonomi Pendidikan Bloom). Yogyakarta: Pustaka Pelajar.

[13] *Hulme, A.K dan Traxler, John. (2005). Mobile Learning, A Handbook for Educators and Trainers. USA: Routledge.*

[14] Triatmaja, A. K. & Khairudin, M. (2018).Study on Skill Improvement of -Digital Electronics Using Virtual Laboratorium With Mobile Virtual Reality. Journal of Physics: Conference Series. Vol. 1140, DOI: 10.1088/1742-6596/1140/1/012021

[15] Sukir. (2010). Simulasi pengendalian multiproses industri dengan Programmable logic controller sebagai sarana Dan bahan ajar praktik instalasi listrik. Jurnal Pendidikan Teknologi dan Kejuruan. Vol 19, No.1, Mei 2010.pp. 81-104.

[16] R.W Komang & M.M Abubakar, (2018). The Usage Effectivity Of Project Based Interactive E-Module In Improving Students' Achievement. Jurnal Pendidikan Teknologi dan Kejuruan, Vol. 24, No. 2, October 2018, 198-202

Variations in the Number of Layers and the Number of Neurons in Artificial Neural Networks: Case Study of Pattern Recognition

F Arifin[1], H Robbani[1], T Annisa[1] and N N M I Ma'arof[2]
[1]Department of Electronic Engineering, Engineering Faculty, Universitas Negeri Yogyakarta, Yogyakarta, Indonesia
[2]Department of Engineering Technology, Faculty of Technical and Vocational, Universiti Pendidikan Sultan Idris, Tanjong Malim, Perak, Malaysia

Email: fatchul@uny.ac.id

Abstract. This paper presents the analyst the number of layers and the number of neurons in the hidden layer of the Artificial Neural Network. In this study, case studies were taken in the recognition of alphabet patterns and shape patterns. First, the number of layers is varied to get the best number of layers. Furthermore, the number of neurons is varied to get the best number of neurons. The results showed that the best number of layers was 1-5 layers in the hidden layer, with validation values from the recognition system 96-100%. While the best number of neurons is obtained with 19 neurons, with an average accuracy percentage of 81%.

1. Introduction

Artificial Neural Network (ANN) is a programming algorithm that is applied to the process of computer computing by adapting the structure of human biological neural networks. Artificial neural networks (ANN) are information processing systems that have characteristics similar to biological neural networks. Artificial neural networks resemble the human brain in gaining knowledge that is by learning and storing knowledge gained in the strength of connections between neurons[1], [2]. The accuracy of ANN is influenced by three things, namely: the pattern of relationships between neurons (network architecture), methods for determining weights (learning methods) and activation functions [3].

According to Peterson & Rognvaldsson artificial neural networks have three main layers, namely the input layer, hidden layer and output layer **[4]**. The input layer is connected to the hidden layer, the hidden layer is connected to the output layer. The final calculation will produce a value at the output layer. In simple terms neural networks have the structure as shown in Figure 1. "W" represents the weights on each path, "x" represents the input layer, "h" represents the hidden layer, and "o" represents the output layer.

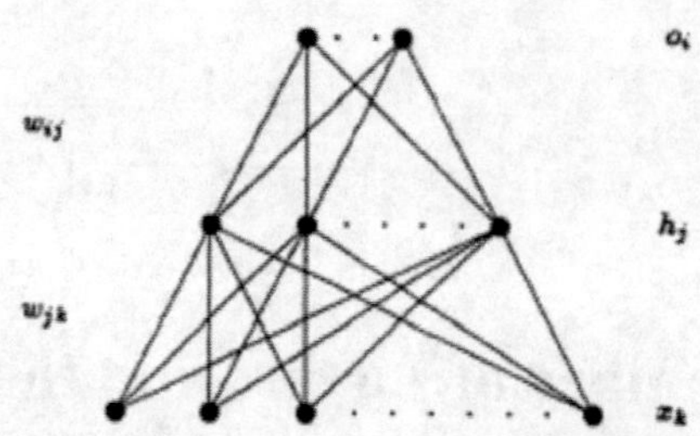

Figure 1. Basic Structure of Artificial Neural Networks [4]

The input layer will receive input data from outside of ANN system. All neurons in this input layer are connected to the neurons in the next layer (hidden layer). Hidden layer consists of a number of neurons that receive data from neurons in the input layer. Furthermore, the neurons in this hidden layer will provide the data to the neurons in the output layer [5].

Hidden layer has a function to increase network capabilities to be more full power, so that it will be easier to do their jobs. ANN that does not have hidden layers, has very limited capabilities. Hidden layers can consist of one layer or more. Each layer has a number of neurons. Until now, there were no definitive rules about the number of hidden layers, and the best number of neurons in the hidden layer in ANN. Users still fined it difficult to determine the number of hidden layers and the ideal number of neurons in the hidden layer of the ANN system.

In this paper, the author will present the results of the study related to the analysis of the number of hidden layers, and the number of neurons that should be used in designing ANN. In this study, case studies of letter patterns A, B, C, D, and E were used as well as the pattern recognition of shape patterns (triangles, squares, and circles).

2. Research Methods

2.1. Variation the Number of Layers in ANN

This paper presents a study of the analysis of variations in the number of layers and variations in the number of hidden layer neurons in artificial neural networks. The type of ANN that was used in this study is multi-layer perceptron (MLP) with sigmoid activation function. The number of neurons in the input is 35, while the number of neurons in the output layer is 5 for alphabetical pattern recognition, and 4 for shape pattern recognition.

The analysis of the variation in the number of layers is done by taking a case study of alphabetical pattern recognition consisting of letters A to E, with a pattern as shown in Figure 2. The number of hidden layers will vary from amount 1 to 7. First the ANN system is used with 1 layer on the hidden layer. Before being used, ANN needs to be trained first. Some parameters in ANN training include: setting the goal error value is 0.1, the initial value is 0.01, the maximum epoch is 5000, the constant momentum is 0.95, and the minimum gradient is 1e-10. The number of neurons for each layer hidden is 10. After the training phase, and find the best weight results, the system will then be tested. Testing is done using 5 alphabetical A-E patterns, where each alphabet is provided with 6 kinds of variations (by giv`ing noise).

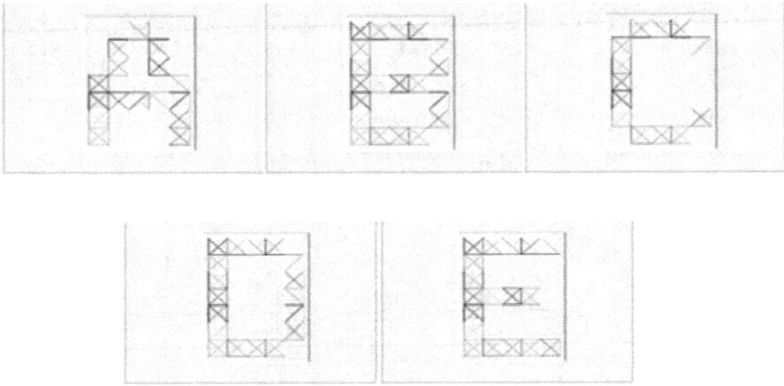

Figure 2. A-E Pattern

Then this process is repeated (in the same way as above) for the number of hidden layers 2, 3, 4, etc. From testing, this will get 600 results of test data. Next will be analyzed how the system validity is obtained.

2.2. Variation in the Number of Neurons in the Hidden Layer

In the above section, the method for analyzing the best number of layers in the MLP hidden layer is presented. In this session the method will be explained in the analysis of the best number of neurons in the MLP hidden layer. Case studies of the square, triangular, round shape and parallelogram patterns recognition are are used in this study. The ANN used is MLP with one hidden layer with the sigmoid activation function. The number of neurons in the input layer is 35, while the number of neurons in the output layer is 4. Then the ANN training system is done with the setting of the goal error value is 0.1, the initial value is 0.01, the maximum epoch is 5000, the momentum constant is 0.95, and the gradient minimum is 1e-10. The number of neurons in the hidden layer is varied by 25 kinds (1, 2, 3, 5, 6, 7, 8, 9, 10, 11, 12, 13, 14, 15, 16, 17, 18, 19, 20, 50, 75, 100, 250, 500 and 1000).

After training, testing is carried out. Each layer variation, tested 4 patterns (Squares, triangles, rounds, and parallelograms), where each pattern is given 6 variations (by voting), so that the total is 600 times.

3. Result and Discussion

3.1. Variation the Number of Layers in ANN

As explained in the sub discussion of research methods, variations in layers were tested through case studies of Abjad pattern recognition, which in this case represented 5 alphabets A, B, C, D, E which were repeated with the addition of different noise 6 times. Then the layer starts to be varied, starting from 1 layer, 2 layers etc. The test data can be seen in **Table 1** and **Table 2**.

Table 1. Result of Layer Variation (1-5)

Abjad	Hidden layer 1		Hidden layer 2		Hidden layer 3		Hidden layer 4		Hidden layer 5	
	identified	recognition value.	identified	recognition value.	Identified	recognition value.	Identified	recognition value.	identified	recognition value.
A	A	0.9261	A	0.8764	A	0.8831	A	0.9182	A	0.9106
A'	A	0.9376	A	0.8764	A	0.8831	A	0.9182	A	0.9016
A''	A	0.9569	A	0.9250	A	0.9693	A	0.8188	A	0.7654
A'''	A	0.8581	A	0.8524	A	0.9164	A	0.9039	A	0.9135
A''''	A	0.9308	A	0.9636	A	0.9101	A	0.9138	A	0.9695
A'''''	A	0.9300	A	0.8399	A	0.9351	A	0.8925	A	0.8416
B	B	0.8363	B	0.8973	B	0.9002	B	0.8910	B	0.9610
B''	B	0.8255	B	0.7446	B	0.9838	B	0.8730	B	0.8877
B'''	B	0.9046	B	0.9313	B	0.9280	B	0.8971	B	0.6979
B''''	B	0.8000	B	0.8974	B	0.3870	B	0.8707	B	0.9080
B'''''	B	0.7851	B	0.6676	B	0.9677	B	0.7750	B	0.8898

C	C	0.8796	C	0.8974	C	0.9185	C	0.8866	C	0.9585
C’	C	0,8131	C	0.9464	C	0.8237	C	0.9045	C	0.9569
C’’	C	0.9341	C	0.9453	C	0.9312	C	0.8816	C	0.8734
C’’’	C	0.8876	C	0.8777	C	0.9148	C	0.9321	C	0.9277
C’’’’	C	0.7919	C	0.8457	C	0.8067	C	0.7294	C	0.8459
C’’’’’	C	0.9162	C	0.8793	C	0.8259	C	0.9732	C	0.8906
D	D	0.7323	D	0.9341	D	0.9263	D	0.9060	D	0.9230
D’	D	0.8713	D	0.8494	D	0.9447	D	0.9581	D	0.9019
D’’	D	0.8285	D	0.7851	D	0.7704	D	0.8650	B	0.8891
D’’’	D	0.8478	B		C		D	0.8474	D	0.8568
D’’’’	D	0.8242	D	0.8785	D	0.8957	D	0.9651	D	0.9248
D’’’’’	D	0.7635	D	0.8701	D	0.9079	D	0.9072	D	0.9070
E	E	0.8976	E	0.9373	E	0.9098	E	0.9141	E	0.8218
E’	E	0.9132	E	0.9386	E	0.9092	E	0.9294	E	0.9428
E’’	E	0.9315	E	0.8810	E	0.9569	E	0.9909	E	0.9193
E’’’	E	0.9005	E	0.9228	E	0.9344	E	0.9472	E	0.9491
E’’’’	E	0.8980	E	0.9295	E	0.8800	E	0.9089	E	0.8781
E’’’’’	E	0.8988	E	0.9021	E	0.8533	E	0.9659	E	0.7673
Rata-Rata		0.8732		0.8209		0.8642		0.8460		0.8482
Persentase		100%		96%		96%		100%		96%

Table 2. Result of Layer Variation (6-10)

Abjad	Hidden layer 6		Hidden layer 7		Hidden layer 8		Hidden layer 9		Hidden layer 10	
	identified	recognition value.	identified	recognition value.	identified	recognition value.	Identified	recognition value.	identified	recognition value.
A	A	0.912	A	0.8593	A	0.8979	A	0.9001	A	0.8951
A’	D		A	0.8593	A	0.8796	E		A	0.8753
A’’	E		A	0.9312	B		A	0.8573	C	
A’’’	A	0.945	A	0.7822	A	0.9441	E		D	
A’’’’	A	0.911	A	0.8915	A	0.8376	A	0.9062	A	0.8999
A’’’’’	A	0.819	C		E		A	0.9918	C	0.6722
B	B	0.939	B	0.9379	B	0.9012	B	0.8899	B	0.9330
B’	D		D	0.1103	B	0.9094	B	0.8674	B	0.8926
B’’	B	0.445	B	0.9389	D		B	0.9230	D	
B’’’	B	0.851	B	0.7727	B	0.8793	B	0.8497	B	0.9223
B’’’’	B	0.939	C		C		D		B	0.9503
B’’’’’	A		C		B	0.8864	B	0.8749	A	
C	C	0.866	C	0.8927	C	0.8979	C	0.8885	C	0.9519
C’	D		A		C	0.8931	C	0.8031	C	0.9421
C’’	C	0.810	C	0.8927	C	0.9007	C	0.9381	A	
C’’’	C	0.890	C	0.8404	C	0.8993	C	0.9028	C	0.9820
C’’’’	C	0.866	C	0.5194	C	0.9028	E		C	0.9538
C’’’’’	E		C	0.2485	C	0.9189	C	0.8903	C	0.8638
D	D	0.984	D	0.8671	D	0.9637	D	0.9271	D	0.9117
D’	D	0.984	A		A		D	0.9283	C	
D’’	D	0.807	D	0.8329	D	0.8709	D	0.9129	D	0.9283
D’’’	D	0.906	D	0.8043	A		D	0.9467	D	0.9048
D’’’’	D	0.931	D	0.8556	D	0.7938	E		D	0.8963
D’’’’’	D	0.782	C		D	0.9687	C		E	
E	E	0.984	E	0.9351	E	0.9583	E	0.9682	E	0.9395
E’	D		E	0.8028	E	0.9439	E	0.9199	E	0.9119
E’’	E	0.928	E	0.8807	C		A		E	0.9318
E’’’	E	0.889	B		E	0.9371	E	0.9212	E	0.8376
E’’’’	E	0.887	E	0.5431	B		E	0.9539	D	
E’’’’’	C		C		E	0.8978	C		C	
Rata-Rata		0.859		0.7772		0.8078		0.8165		0.7832
Persentase		68%		64%		68%		68%		60%

Table 1 and Table 2 show that when the number of layers in hidden layer 1, the system has a pattern recognition accuracy of 100%, when the number of layers in the hidden layer 2, the accuracy is 96%, when the number of layers in hidden layer 3, the accuracy is 96%, when the number of layers in the hidden layer 4, the accuracy is 100%, when the number of layers in the hidden layer 5, the accuracy is 96%. Accuracy will decrease when the number of layers in the Hidden layer rises above 5. When the number of layers 6, the accuracy is 68%, when the number of layers 7, the accuracy is 64%, When the number of layers is 8, the accuracy is 68%, when the number is 9, the accuracy is 68% , when the number of layers 10, the accuracy is 60%.

It appears that the best hidden layer is achieved when the number of layers ranges from 1-5. Fausset [4] in his book entitled "Fundamental of Neural Network: Architecture Algorithm and Application" said that the number of ideal hidden layers used in backpropagation is 1 hidden layer. However, from the experimental results it is known that the hidden layer still has a high percentage of truth and ideal range between 1- 5.

3.2. Variation the Number of Neurons in the Hidden Layer

The ideal number of neurons will be analysed in the hidden layer. From the previous explanation it has been found that the ideal number of layers is between 1- 5. Whereas Fausset [4] reinforces that it is best for one layer. Therefore to analyse the number of these neurons, the hidden layer with the number 1 is selected. The number of neurons varied 25 times (1, 2, 3, 5, 6, 7, 8, 9, 10, 11, 12, 13, 14, 15, 16, 17, 18, 19, 20, 50, 75, 100, 250, 500 and 1000). Each layer is tested with 4 types of shapes (square, triangle, round, parallelogram). Each shape is given a variation of noise 6 times. The complete data on the results of testing neuron variations can be seen inTable 3.

The results of testing the effect of the number of neurons in the hidden layer on identifying pattern shapes found that the best results of using the number of neurons is 19 with a validation value of 81%. From the test results it also appears that the worst value is obtained when the use of the number of neurons is 1-3 with the average percentage validation rate of 62%. This indicates that the number of neurons below 4 is not recommended. In other words, it's best to use a minimum number of neurons 4 and above. The best value is obtained when the number of neurons is 19. More than 19, the validation of pattern recognition again decreases. This indicates that the number of neuron in ANN affects the level of accuracy of shape pattern recognition.

Table 3. The results of testing the variation in the number of ANN neurons in shape pattern recognition

Variation In The Number of ANN Neurons In Shape Pattern Recognition

	1	2	3	4	5	6	7	8	9	10	11	12	13	14	15	16	17	18	19	20	21	22	23	24	25
Number Of Neuron	1	2	3	5	6	7	8	9	10	11	12	13	14	15	16	17	18	19	20	50	75	100	250	500	1000
Square																									
Test 1	0.887	0.924	0.894	0.883	0.921	0.864	0.878	0.856	0.875	0.883	0.871	0.888	0.897	0.882	0.885	0.877	0.849	0.86	0.904	0.864	0.862	0.885	0.835	0.832	0.814
Test 2	0.806	[illegible]	[illegible]	0.781	0.913	0.771	0.666	0.794	0.738	0.545	0.434	0.786	0.794	0.769	0.791	0.703	0.734	0.833	0.787	0.739	0.684	0.656	0.743	0.629	0.597
Test 3	0.887	0.914	0.895	0.855	0.936	0.793	0.831	0.802	0.808	0.9	0.848	0.909	0.876	0.784	0.761	0.853	0.849	0.768	0.868	0.852	0.841	0.82	0.769	0.783	0.769
Test 4	0.881	0.918	0.239	0.662	0.759	0.763	0.695	0.832	0.568	0.506	0.72	0.574	0.474	0.803	0.659	0.867	0.376	0.597	0.759	0.503	0.635	0.629	0.613	0.509	0.496
Test 5	0.721	[illegible]	[illegible]	[illegible]	0.919	0.701	[illegible]	0.729	0.413	0.234	0.464	0.524	[illegible]	0.54	0.693	0.817	0.378	0.742	0.621	0.435	0.429	0.487	0.553	0.366	0.364
Test 6	0.721	[illegible]	[illegible]	[illegible]	0.919	0.701	[illegible]	0.729	0.413	0.234	0.464	0.524	[illegible]	0.54	0.693	0.817	0.378	0.742	0.621	0.435	0.429	0.487	0.553	0.366	0.364
Average	0.817	0.633	0.459	0.565	0.894	0.765	0.656	0.79	0.636	0.55	0.683	0.701	0.568	0.719	0.747	0.822	0.594	0.757	0.76	0.638	0.647	0.66	0.678	0.581	0.567
Triangle																									
Test 1	0.894	0.923	0.848	0.896	0.871	0.885	0.921	0.902	0.903	0.901	0.907	0.908	0.896	0.909	0.929	0.926	0.914	0.939	0.921	0.9	0.899	0.913	0.971	0.935	0.935
Test 2	0.894	0.919	0.823	0.901	0.876	0.847	0.947	0.865	0.885	0.911	0.897	0.893	0.89	0.9	0.855	0.886	0.895	0.916	0.91	0.887	0.86	0.921	0.948	0.914	0.915
Test 3	0.894	0.388	0.805	0.892	0.787	0.83	0.871	0.839	0.909	0.914	0.812	0.8	0.653	0.725	0.797	0.906	0.812	0.848	0.894	0.824	0.796	0.851	0.944	0.87	0.883
Test 4	0.893	[illegible]	0.567	0.9	0.745	0.812	0.875	0.583	0.921	0.923	0.789	0.758	0.637	0.749	0.646	0.826	0.79	0.724	0.866	0.85	0.708	0.824	0.899	0.842	0.861
Test 5	0.876	[illegible]	0.334	0.882	0.609	0.758	0.923	0.724	0.908	0.918	0.75	0.5	0.771	0.653	0.536	0.758	0.8	0.604	0.84	0.812	0.729	0.82	0.866	0.818	0.834
Test 6	0.793	[illegible]	[illegible]	0.871	0.329	0.768	0.94	0.524	0.869	0.921	0.716	0.346	0.687	0.551	0.448	0.712	0.786	0.501	0.866	0.694	0.692	0.764	0.829	0.764	0.812
Average	0.874	0.483	0.615	0.89	0.703	0.817	0.911	0.739	0.899	0.915	0.812	0.701	0.756	0.748	0.702	0.837	0.833	0.754	0.883	0.828	[illegible]	[illegible]	[illegible]	[illegible]	[illegible]
Round																									
Test 1	0.421	0.878	0.946	0.87	0.9	0.886	0.894	0.89	0.889	0.897	0.911	0.871	0.871	0.882	0.833	0.861	0.915	0.869	0.833	0.863	0.88	0.884	0.893	0.831	0.821
Test 2	[illegible]	0.878	0.766	0.855	0.905	0.83	0.904	0.726	0.62	0.533	0.902	0.842	0.817	0.838	0.843	0.756	0.481	0.858	0.57	0.821	0.706	0.766	0.723	0.696	0.74
Test 3	0.405	0.845	0.424	0.852	0.891	0.569	0.83	0.602	0.658	0.275	0.804	0.779	0.869	0.763	0.789	0.705	[illegible]	0.833	0.565	0.713	0.598	0.666	0.601	0.596	0.612
Test 4	0.366	0.662	[illegible]	0.851	0.785	0.574	0.888	0.259	0.471	0.408	0.715	0.715	0.82	0.677	0.695	0.557	0.443	0.847	0.662	0.503	0.546	0.625	0.513	0.582	0.579
Test 5	0.375	0.598	[illegible]	0.862	[illegible]	0.411	0.845	[illegible]	0.26	[illegible]	0.46	0.51	0.398	[illegible]	[illegible]	[illegible]	[illegible]	0.841	0.372	0.213	0.27	0.436	0.304	0.303	0.289
Test 6	0.375	0.598	[illegible]	0.862	[illegible]	0.411	0.845	[illegible]	0.26	[illegible]	0.46	0.51	0.398	[illegible]	[illegible]	[illegible]	[illegible]	0.841	0.372	0.213	0.27	0.436	0.304	0.303	0.289
Average	0.402	0.743	0.435	0.859	0.619	0.614	0.868	0.487	0.526	0.405	0.708	0.705	0.665	0.598	0.585	0.573	0.393	0.846	0.566	0.594	0.545	0.605	0.552	0.552	0.555
Paralellogram																									
Test 1	0.404	0.85	0.883	0.938	0.907	0.945	0.897	0.932	0.955	0.893	0.876	0.907	0.914	0.919	0.942	0.91	0.934	0.949	0.92	0.899	0.886	0.913	0.97	0.884	0.924
Test 2	0.275	0.852	0.88	0.931	0.901	0.934	0.839	0.882	0.934	0.888	0.855	0.851	0.918	0.825	0.882	0.913	0.901	0.923	0.897	0.884	0.833	0.92	0.955	0.861	0.906
Test 3	0.297	0.851	0.864	0.878	0.864	0.914	0.715	0.908	0.913	0.895	0.787	0.856	0.922	0.866	0.842	0.903	0.875	0.912	0.888	0.852	0.84	0.905	0.935	0.852	0.874
Test 4	0.287	0.832	0.725	0.743	0.751	0.914	0.585	0.931	0.876	0.884	0.762	0.845	0.919	0.865	0.861	0.764	0.854	0.884	0.863	0.79	0.817	0.873	0.894	0.807	0.836
Test 5	0.408	0.847	0.639	0.686	[illegible]	0.901	[illegible]	0.871	0.846	0.868	0.721	0.841	0.848	0.864	0.902	[illegible]	0.761	0.842	0.551	0.802	0.755	0.804	0.834	0.685	0.724
Test 6	0.373	0.844	0.607	0.861	[illegible]	0.844	[illegible]	0.803	0.895	0.861	0.696	0.793	0.86	0.723	0.765	[illegible]	0.587	0.794	0.329	0.656	0.704	0.731	0.787	0.542	0.638
Average	0.341	0.846	0.766	0.839	0.6	0.909	0.602	0.888	0.903	0.882	0.766	0.849	0.897	0.844	0.865	0.689	0.819	0.884	0.741	0.814	0.806	0.858	0.896	0.771	0.817
Average Total	0.61	0.68	0.57	0.79	0.70	0.78	0.76	0.73	0.74	0.69	0.73	0.74	0.73	0.73	0.72	0.73	0.66	0.81	0.74	0.71	0.69	0.75	0.76	0.69	0.70
Procentage	61%	68%	57%	79%	70%	78%	76%	73%	74%	69%	73%	74%	73%	73%	72%	73%	66%	81%	74%	71%	69%	75%	76%	69%	70%

4. Conclusion

From the results of the study, it was found that:

(1) the more hidden layers that were used, the more inaccurate results would be obtained. The ideal number of layers that can be used in MLP is less than 5 (<5).

(2) The best number of neorons in the MLP hidden layer is 19 with an accuracy percentage of 81%. The test results also found that the use of the minimum number of neurons was 4. This was proven by the results of the study that for the number of neurons 1, 2 and 3 had the lowest accuracy percentage of 61%, 68% and 57%.

5. References

[1] F. S. *Rahmahwati* and F. Arifin, "Developing Artificial Neural Network Based on Visual Studio for Dance Assessment," *J. Pendidik. Teknol. Dan Kejuru.*, vol. 23, no. 4, pp. 402–411, Oct. 2017.

[2] Hasanah, D. Irmawati, and F. Arifin, "Speech Recognizing for Presentation Tool Navigation Using Back Propagation Artificial Neural Network," in *Artificial Intelligence*, Hongkong, 2016, vol. Volume 75, 2016, p. 05.

[3] S. N. Luthfie, "Implementasi Jaringan Saraf Tiruan Backpropagation Pada Aplikasi Pengenalan Wajah Dengan Jarak Yang Berbeda Menggunakan MATLAB 7.0," p. 7.

[4] C. *Peterson* and T. Rognvaldsson, *An Introduction to Artifical Neuron Network.* Departement of Theoretical Physic, Cern School of Computing, Ystad, Sweden, 1991.

[5] G. Dhaneswara and V. S. Moertini, "Implementasi Jaringan Syaraf Tiruan Tipe Multilayer Feed-Forward Menggunakan Algoritma Backpropagation dengan Momentum untuk Klasifikasi Data," *J. Ilmu Komput. Univ. Katolik Parahyangan.*

Develop of Control Water Supply Pump Based on *Supervisory Control and Data Acquisition* on Instrumentation and Power Generator Control in Grade X Students Majoring In Electric Power in the PGRI 3 Vocation School Malang

Y Rahmawati, M Zainuri, A N Afandi
S1 Electrical Engineering Education, Electrical Engineering Department, Technical Faculty, Universitas Negeri Malang, Indonesia

Email : muchzainuri29@gmail.com, yuni.rahmawati.ft@um.ac.id, an.afandi@um.ac.id

Abstract. The quality of education can affect the quality of the learners were produced. The importance to improve the quality of education so learners produced increased anyway. According to Asyhar, one way to improve the quality of education is the use of media of instruction. Instrumentation and generator control the Generator operation of SCADA on power plant aims to train students designing and mengoperasiakan the existing SCADA system at the power plant. One operation of SCADA in electricity generation at the WTP (Water Treatment Plant), namely on Water Supply Pump to control water pump. SCADA learning media packed in prototype trainer Water Supply Pump -based Supervisory Control And Data Acquisition SCADA learning to support. Aiming to understand the way design of SCADA, create, communicate, SCADA program and operate it. The model used is the development model of development Sugiyono. Stages of development model Sugiyono are: (a) the potential and problems, (b) data collection, (c) product design, (d)design validation, (e) the revised design, (f) a trial product, (g)product revision, (h) a trial use, (i) the revision of the product, and (j) production mass. The results of the research and development shows that the eligibility level trainer control water supply pump has a percentage of 91.9% of the 3 (three) validator experts which means decent products used with little revision, 83.72% of test small groups by 10 (ten) grade X PBB meaning decent products used without revision, 90.53% of tests by a large group of thirty (30) students of class X PBA product which means it is very worthy of being used without any revisions. The conclusions of this research and development are decent products used in the learning process. The product can help learning activities can help so that the students of class X students majoring in electric power can be better understand the operation of scada in electricity generation units.

1. Introduction

From UU RI No. 20 on year 2003 about education national system article 1 paragraph 1 is conscious and planned effort to realize the process learning for students develop their potential and skill that benefit themselves and the community. With this education have skill and to be competent human resource. This can be achieved if the quality of education is good. According of Rayandra Asyhar (2012) to upgrade the quality of education can be use instructional media. The facility and infrastructure educational is instructional media that important to be considered. Especially in the vocation school instructional media support to train students skill. Vocation school used the power point, video, and others of instructional mediafirst.

Vocation school PGRI 3 of Malang have several technical fields of expertise, from Electrical engineering and technical informatics, Mechaniacal engineering, Otomotif Engineering. In Electrical engineering and technical informatics have a Power Plant class that prepare someone expert in the field of Power Plan Engineering. On Power Plant class the subject Instrumentation and Control Power Generator is taught in accordance with the existing curriculum. Subject Instrumentation and Control Power Generator related to Supervisory Control and Data Acquisition. According to Amar (2012) Supervisory Control and Data Acquisition is a system that can control, supervise, retrieve and record dataparagraphs are.

The observations I made on Nevember 22, 2018 through interviews with teacher of these subject. Student's knowledge of the Instrumentation and Control subjects is already good and outcomes obtained with average value meeting the graduation stnadart. However, it was stated by teacher that skill of student in the field of operating SCADA, which is one of the skill learned in the Instrumentation and Control subjects is still low. Because the media used so far only use instructional media in the form of power points. So anything instructional media that supports. Therefore, it is necessary to develop a trainer to train SCADA operating skills. The trainer needs to be desaigned according to the main material listed in the instrumentations and control syllabus, which is installation of control units based on SCADA. The subject matter explain the use of the SCADA for control of power plant equipment. So that trainer developed is a water supply pump control using SCADA.

Trainer of control water supply pump using SCADA will using students for practice. The trainer is important as a media to train student's skill in designing SCADA from the start and communicating it with a water supply pump trainer. After successfully communicating the SCADA with the RTU device in the from of Arduino. Student can practice SCADA skills for control, supervision, retrieval and recording of data on the water supply pump trainer.

2. Development Method

This research and development is useful for developing products that can be used to support the practical learning process in the subjects of Generator Instrumentation and Control. This water supply pump control trainer can be used to support practical learning operations on the generator unit. The use of instructional media makes it easier for students to understand the operation in the generating unit because students will make it from the beginning of the display to connecting to the trainer.first.

The research and development model used to develop the product is Sugiyono's research and development model. The model was chosen as the basis for developing the trainer's products because the steps were systematic in accordance with the needs to be used, as well as good and flexible rules to be used to develop a water supply pump control trainer.

Sugiyono's research and development model consists of 10 steps, namely: (1) Potential and problems; (2) Data collection; (3) Product design; (4) Design validation; (5) Design revisions, (6) Product trials, (7) Product revisions, (8) Usage trials, (9) Product revisions, (10) Productiont

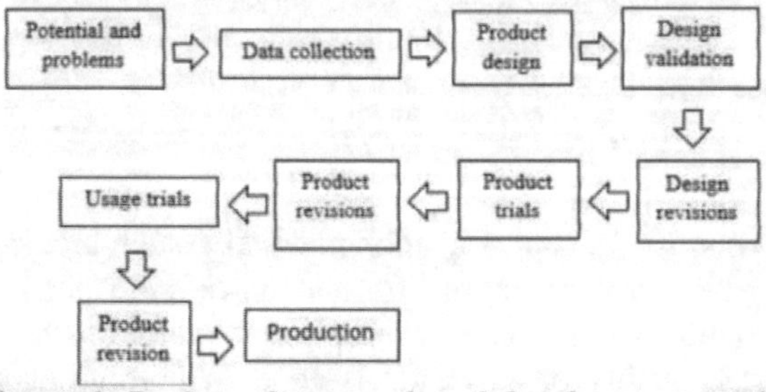

Figure 1. Sugiyono's research and development model

2.1. Product Design
The product design developed was divided into 2, which is design of Control Water Supply Pump in PLTGU using SCADA trainer and jobsheet design. The product design is as follows:

2.1.1. Trainer Design

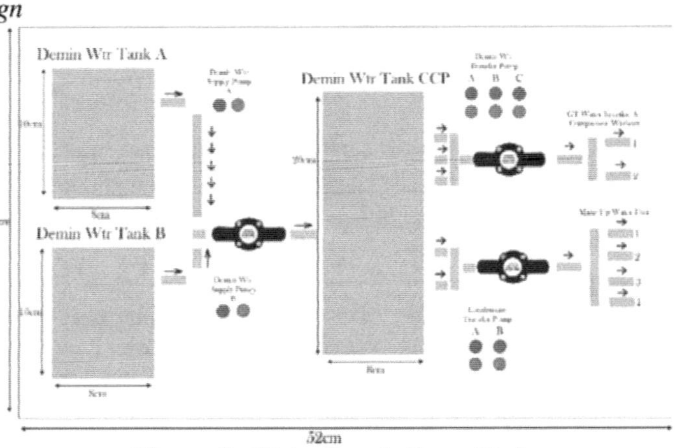

Figure 2. Water Supply Pump Trainer

Figure 2. front view water supply pump trainer is a trainer design, which refers to the design of the HMI water supply plant. The trainer measures 103cm in length, 54.5cm in width, and 45cm in height with plastic tanks, Arduino, flow water sensors, water level sensors, terminal blocks, relay modules, indicator lights, and mini water pumps. There are 3 plastic tubes with a volume size of 400cm3 for the tank, with tanks A and B, while 750cm3 for demanding the WTR tank CCPP. The Arduino used is Arduino Uno as the device. The flow water sensor consists of 3 pieces placed on 3 channels, the first channel that leads to the demin tube, the ccpp water tank, the second channel from the condensate transfer pump, and the third channel from the demin water transfer pump. Water level sensors that will be placed in each tank which number 3. The relay module used contains 4 channels and consists of 2 relay modules. The indicator light is red as many as 7 and the indicator light is green as many as 7. Mini pumps are 7 pieces with a voltage of 6-12 VDC.

2.1.2. Jobsheet Design
In development research, jobsheets will also be created to support practical learning activities. The design of the worksheet consists of several points in it such as: (1) title, (2) objectives, (3) Basic theory, (4) tools and materials, (5) work safety that must be considered, (6) experimental steps, (7) experimental results, (8) data analysis, and (9) conclusions, (10) problem training.

2.2. Validation and Trial Techniques
The validation and trial or product evaluation techniques in this research and development are carried out in 3 (three) stages, that is product validation by experts, small group trials and large group trials. The results of the validation and trial are intended to data collect related to and lack of teaching materials. The three stages are explained as follows.

2.2.1. Expert Validation
In the product validation stage, the experts involved 3 (three) validators, they were TE UM lecturers and teachers in the Power Generation Vocational School PGRI 3 Malang as reviewers. This evaluation aims to assess the appropriateness of the contents, whether or not the contents are relevant to the learning objectives, and the quality of teaching materials before being tested on students. After validating the product by the expert, then the product is revised to the errors that have been found. Data collection instruments in this evaluation used a questionnaire.

2.2.2. Small Group Trials

The small groups evaluate to find out the feasibility of the water supply pump control trainer product by selecting 10 (ten) class X PBB students who have taken the subject matter contained in the basic competencies of the Plant Instrumentation and Control subjects. Through this trial can be known quality, products interesting and identify problems that may arise. Data collection instruments in this evaluation used a questionnaire.

2.2.3. Large Group Trials

A large group trial was conducted to determine the feasibility of the learning media that had been made involving 30 (thirty) PBA class X students. The purpose of this evaluation is to determine the feasibility of the product. Data collection instruments in this evaluation used a questionnaire.

2.3. Data Collection Instruments

The evaluation instrument for testing validity or feasibity of teaching materials uses a questionnaire. Product feasibility test questionnaire contains material assessment and media assessment. Before being used, product validity questionnaire was validated by the research supervisory.

Material and media assessment instrument lattices are used as guidelines for the preparation of the research questionnaire. Through a research questionnaire that will be filled out by validator I, validator II, validator III, students of small group trial subjects and large group trial students can find errors from the products that have been developed. Thus it can correct errors so that the product that the researcher develops can be used for learning activities in Instrumentation and Control Generators.

Products that are assessed on the media eligibility assessment consist of: (1) Trainer; (2) Software; (3) Manual books; and (4) jobsheet. Questionnaire grids for assessing the feasibility of instructional media are shown in **Table 1.** Assessing the feasibility of instructional media.

Table 1. Assessing the feasibility of instructional media

No	Product	Indicator	Source
1	Trainer	The writing on the Trainer can be read and understood	Wahono, 2006
		Simple and attractive trainer	
		Reliable Trainer	
2	Software	Efficient and effective trainer in use	
		Easy to use Trainer	
		Simple software in operation	Arsyad, 2013
3	Jobsheet	Easy to use software	Wahono, 2006
		Reliable software	
		Clarity of steps in the jobsheet	Wahono, 2006
		Use of standard language in the jobsheet	
		Image is clear	
		The text on the jobsheet is clear	
4	Manual Book	The language used is standard	Wahono, 2006
		Image is clear	
		The text on the manual book is clear	

Products that are assessed on the media eligibility assessment consist of: (1) Trainer; (2) Manual bools; and (3) jobsheet. Questionnaire grids for assessing the feasibility of material are shown in **Table 2.** Questionnaire grids feasibility of material.

Table 2. Questionnaire grids feasibility of material

No	Product	Indicator	Source
1	Trainer	Trainer In accordance with the objectives to be achieved	Arsyad, 2013
		Trainer in accordance with KD / Curriculum	
		The writing on the Trainer can be read and understood	Wahono, 2006
2	Jobsheet	The suitability of the material with the jobsheet	Arsyad, 2013
		The suitability of the material with the objectives on the jobsheet	
		The depth of the jobsheet is in accordance with the level of student development	Akbar, 2013
		The presented of material in the jobsheet is according to scientific truth (actuality)	Wahono, 2006
		The presented of material in the jobsheet is in accordance with the latest developments (contextual)	
		Jobsheet presents learning objectives	
		Jobsheet presents material description	Akbar, 2013
		Jobsheet presents work steps	
		Jobsheet presents assignments	
3	Manual Book	Use standard language	Wahono, 2006
		Accuracy in structuring sentences in the jobsheet	
		The size of the writing on the jobsheet is suitable so it is easy to read	
		Use standard language	
		The writing is clear and legible	
		The relevance of information conveyed with purpose	

Making a research questionnaire using 4 (four) levels of assessment according to Rating scale. Table 3 shows the assessment level criteria in the questionnaire for media experts, material experts and students using Rating scale (Sugiyono, 2016: 135).

Table 3. Skala Likert

No	Number	Qualitative Meaning
1	4	The subject matter is very good / very decent / very interesting / very appropriate / very clear.
2	3	The subject matter is quite good / decent enough / interesting enough / quite appropriate / clear enough.
3	2	Sub subject is not good / not suitable / less attractive / not suitable / less clear.
4	1	The subject matter is not very good / very improper / very unattractive / very unclear.

2.4. Data Analysis

The type of data obtained from this development research is in the form of qualitative and quantitative data. Qualitative data in the form of responses, comments, criticisms, suggestions and recommendations about the product by validator I, validator II, validator III and students of the test subject are written in the suggestion box on the final sheet of the trainer validation questionnaire. While quantitative data were obtained from questionnaire scores given by test subjects. To determine the appropriateness of the trainer developed by conducting an evaluation or trial. The initial step is by finding the percentage score obtained through the results of the product validation test questionnaire, which is by formula 1.

$$V = \frac{Tse}{Tsh} x\ 100\% \qquad (1)$$

Information:
V = Validity (feasibility)
Tse = Total empirical score
Tsh = Expected maximum total score

After knowing the results of the percentage of the questionnaire, then grouping into the trainer eligibility criteria shown in **Table 4.** Feasibility Criteria.

Table 4. Feasibility Criteria.

No.	Value Scale (100%)	Information
1.	85,01% - 100,00%	Very valid, or can be used without revision.
2.	70,01% - 85,00%	Quite valid, or can be used but needs minor revisions.
3.	50,01% - 70,00%	Invalid, it is recommended not to be used, it needs a big revision.
4.	01,00% - 50,00%	Invalid, or may not be used.

3. Discussion Result

3.1. Product Manufacture Results

The results of this product development in the form of a control water supply pump trainer, manual books, and jobsheets for class X students majoring in Electric Power Manufacturing. This trainer is intended for practicum scada operation. After the product is finished being made, then the product is validated by 2 (two) TE UM lecturers and 1 teacher majoring in Electric Power Generation Vocational School PGRI 3 Malang who are competent to know the product validity.

After passing the product validation stage by the expert, the product trial phase is continued in a small group involving 10 (ten) grade X PBB students who are taking the Generating Instrumentation and Control subjects. The 10 (ten) grade X PBB students will use the trainer developed by the researcher, then the students fill out a research questionnaire. Through the research questionnaire the researcher can find out the feasibility of the product being developed.

The next trial was a large group trial involving 30 (thirty) students of class X PBA. At this stage a demonstration of the use of a water supply pump control trainer was developed. After that students fill out a questionnaire to determine the level of product feasibility and student response after using the trainer that has been developed.

3.2. Product Feasibility Test Results

Presentation of data in this research consists of data: (1) the results of individual trials; (2) results of small group trials; and (3) results of large group trials. Individual trials are conducted by 3 (three) expert validators, that is TE UM lecturers and Power Plant majors teachers to test the feasibility of teaching materials. While the small group trial was conducted by 10 (ten) grade X PBB students who were taking the Instrumentation and Control subjects, this trial was conducted to see how feasible the product had been validated by the expert and improved according to expert advice. The final test is a large group trial that aims to find out the level of product feasibility when used in learning activities. Research subjects in large group trials were students of class X PBA.

Based on the data obtained, the final results of the percentage of eligibility for validation by experts of 91.9% obtained an average percentage of validator I, validator II, and validator III. The final result of the feasibility percentage in small group trials is 83.72% and the last average eligibility percentage from the results of large group trials is 90.53%. The results of the validation are shown in **Figure 3**.

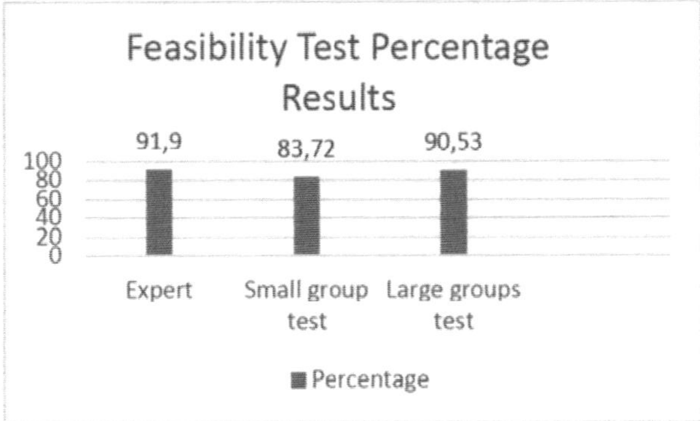

Figure 3. Feasibility Test Results Graph

4. Conclusion

Research and development of Trainer use research and development Sugiyono have 10 (ten) stage is: (1) Potensial problem; (2) Data collection; (3) Product design; (4) Validation design; (5) Design revision; (6) Trial Product; (7) Product revision; (8) Usage product; (9)Product revision; (10) Production. Beside on data result, on result persentase of expert validation product is 91,9% obtained average persentase from validator I, validator II, dan validator III. The result from small group of trial persentase appropriateness is 83,72% end then average on big group of usage persentase is 90,53%. The conclusion trainer to lecture is appropriates.

5. References

[1] Sumiharsono, Rudy & Hisbiyatul Hasanah. 2017. *Media Pembelajaran*. Jember: Pustaka Abadi.reference.

[2] Andrianto, Heri & Aan Darmawan. 2016. *Arduino Belajar Cepat dan Pemrograman*. Bandung: Informatika Bandung.

[3] Asyhar, H.Rayandra. 2012. *Kreatif Mengembangkan Media Pembelajaran*. Jakarta: Referensi Jakarta.

[4] Pribadi, Benny. A. 2009. *Model Desain Sistem Pembelajaran*. Jakarta: Dian Rakyat.

[5] Sugiyono. 2016. *Metode Penelitian Kuantitatif, Kualitatif, dan R&D*. Bandung: Alfabeta.

[6] Akbar, S. 2016. *Instrumen Perangkat Pembelajaran*. Bandung: PT Remaja Rosdakarya. 57.

[7] Arsyad, Azhar. 2013. *Media Pembelajaran*. Jakarta: PT Raja Grafindo Persada. 74-76.

[8] Arikunto, S. 2013. *Prosedur Penelitian Suatu Pendekatan Praktik*. Jakarta: Rineka Cipta. 194.

[9] Sudjana, Nana. 2014. *Dasar-Dasar Proses Belajar Mengajar*. Bandung: Sinar Baru Algensindo. 99-100.

[10] Jauzie, Arif. 2015. Laporan Akhir Praktikum Kendali Elektrk Modul 3 SCADA. *Laporan Akhir*. Bandung. Institut Teknologi Nasional Bandung

[11] Nugroho, S.A. 2013. Desain SCADA Untuk Sisten PArkir Berbasis PLC. *Tugas Akhir*. Diponegoro. Universitas Diponegoro.

[12] Effendy, Asnal dan Robby Wirza. 2013. Perancangan Sistem SCADA Cooling Tower Menggunakan Siemens Simatic Step 7 dan WINCC. **1.** *Jurnal TEKNOIF*, 8.

[13] Ghozali, H.R.B.Moch. 2011. Desain Perancangan kwh Meter Digital Berbasis Sistem SCADA Pada Pelanggan Tegangan Menengah 20 kV di PT. PLN (Persero) APJ Jember. **3.** *Jurnal Rekayasa Elektrika*, 117

[14] Syufrijal. 2015. Aplikasi SCADA Pada Sistem Pengendalian Trainer Lift Menggunakan OPC Berbasis PLC. **2.** *Jurnal Autocracy*, 65‾73

[15] Anshori, M.L. 2015. Pengembangan *Prototype Water Level Control and Monitoring System* Sebagai Media Pembelajaran Pada Mata Diktlat Pengoperasian SCADA. **2.** *E-Journal Universitas Negeri Yogyakarta*, 137‾150

An Asynchronous Serial Communication Learning Media: Usability Evaluation

D Hariyanto[1], A C Nugraha[1], A Asmara[2] and H Liu[3]
[1,2]Department of Electrical Engineering Education, Faculty of Engineering, Yogyakarta State University, Indonesia
[2]Department of Technological and Vocational Education, College of Humanities and Social Sciences, National Yunlin University of Science and Technology, Taiwan
[3]Institute of Vocational Education and Vocational Didactics, Faculty of Education, TU Dresden, Germany

Email: didik_hr@uny.ac.id, ariadie@uny.ac.id, andikasmara@uny.ac.id, huan.liu1@mailbox.tu-dresden.de

Abstract. The availability of a good acceptable learning media is a vital part of the learning process. This paper presents a usability evaluation of the asynchronous serial communication module, which can exhibit the level of acceptance of Data Communication's students. The survey was performed based on ten simple statements of the System Usability Scale (SUS) questionnaire. The research's findings have been analyzed in three different approaches, i.e., visual "five-pointed star" shape assessment, Bangor et al. model verification, and curved grading scale justification. The analysis showed that the serial communication module is in the acceptable category.

1. Introduction

The learning media is one essential component in the learning process to enhance the student's outcome and performance. In the Data Communication Practice course, one important topic is discussing the serial communication. Different from parallel communication, which is relatively easy to understand, the serial communication is rather challenging to figure out. This situation is getting worse with the absence of the appropriate serial module. The old-version serial module has some limitations [1]; there is no information concerning serial data bits in detail. It cannot show the typical data in serial communication such as the start bit, data bits, and stop bit. The previous study proposed the new serial communication module equipped with graphical serial data information [1].

Although, it was reported that the serial module could functionally perform as designed [1]. However, it is important to evaluate the module in terms of the usability collected from the user's perspective. Usability evaluation is one of the methods allowing the user to evaluate the usability of the products through any approaches, e.g., field observations [2], interviews [3], focus groups [4], and questionnaire [5]. The questionnaire method is widely used to assess usability because of its simplicity, economic, and reliable. Therefore, the primary goal of this article is to explore the usability evaluation of the asynchronous serial communication module through the most widely used SUS (System Usability Scale) instrument.

2. Asynchronous Serial Communication Learning Media

The serial module used in this study is the asynchronous serial communication learning media designed and developed in the previous study [1]. This serial module proposes a visualization of a serial bit data in order to facilitate the students to comprehend the serial communication easily. This module should be connected to the computer, either a desktop-based or notebook through a USB port that complies with the serial data characteristics.

Basically, this serial module consists of hardware and software. The hardware comprises of one graphic LCD and two sub-circuits, i.e., the main controller circuit and the serial interface circuit (see Figure 1). The LCD functions for displaying the serial data transferred from/to serial module. Meanwhile, the serial interface circuit provides the interface port to establish communication with the computer. The main process to interpret the non-visual serial data into the animated graphic is managed by the main controller circuit.

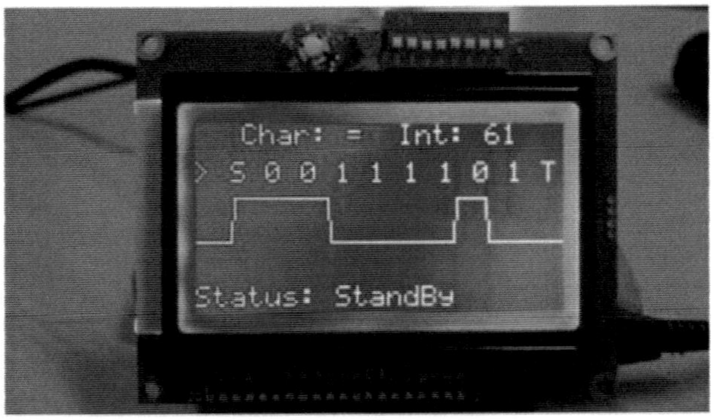

Figure 1. The serial communication hardware module

Concerning the software, two different programs should be installed into one in computer and the other in the serial module. The program attached in the serial module creates based on the Code Vision AVR programming. Meanwhile, the program installed on the computer develops through the Delphi IDE (Integrated Development Environment). Figure 2 shows the GUI (Graphical User Interface) of the computer-side software.

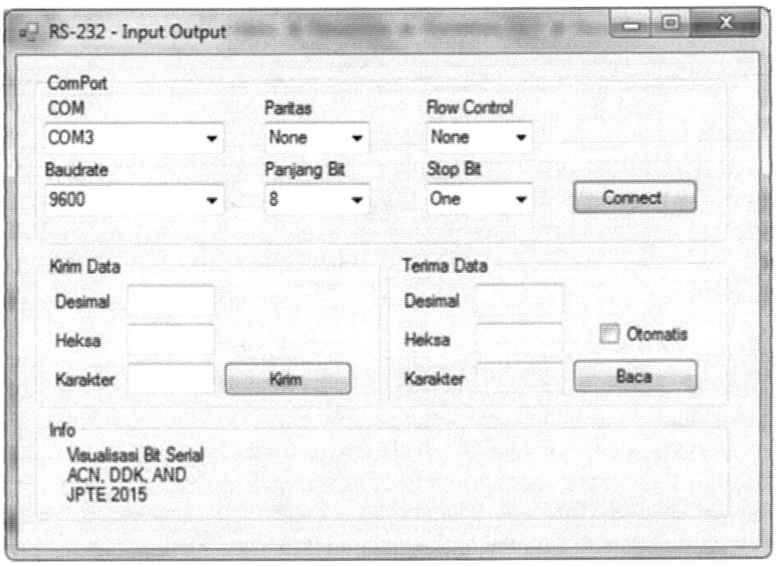

Figure 2. The serial communication software GUI

In terms of the hardware and software test, the thorough examination was conducted in the context of the functional-based test in order to assess the performance of the serial module. Four aspects of basic functions, receiving data, sending data, and visualizing data were tested carefully. The test results were reported that the serial module could perform as good as expected [1].

3. Usability Evaluation

The International Standards Organization (ISO) 9241 defined usability as the degree to which a specified product can be used by specified users to accomplish specified goals with considering effectiveness, efficiency, and satisfaction in a specified context of use [6]. Although there are no specific guidelines on how to overcome the measurement of the three aspects of effectiveness, efficiency, and satisfaction, a large number of usability tests have been studied and developed by practitioners and researchers. There are numerous well-known questionnaires available for evaluating the usability of the technological-based products based upon the subjective user assessment, e.g., the Questionnaire for User Interaction and Satisfaction (QUIS) [7], the Software Usability Measurement Inventory (SUMI) [8], the After-Scenario Questionnaire (ASQ) [9], the Computer System Usability Questionnaire (CSUQ) [9], the Post-Study System Usability Questionnaire (PSSUQ) [9], the questionnaire System Usability Scale (SUS) [10], the Website Analysis and Measurement Inventory (WAMMI) [11], and the USE questionnaire [12].

From those mentioned questionnaires, the one which offers many advantages in terms of simple, reliable, and low-cost assessment is the SUS. First, the SUS is constructed in a "quick and dirty" method [10]. This method made in order to meet the demands of evaluating usability in the industrial context that needed a quick and simple, but also reliable enough to assess the products. The SUS comprises only ten statements with a five-point possible response spanning from strongly disagree to strongly agree. With the very short and simple statements, it doesn't need much time and effort for respondents to complete the surveys.

Secondly, the SUS is technological agnostic [13] that is possible to be used to evaluate a wide variety of products. Some of the studies using SUS are websites [14], Learning Management Systems (LMSs) [15,16], adaptive e-learning [17,18], computer hardware [19], multimodal systems [20], interactive voice response systems [21], voting systems [22,23], mobile applications [24,25], mobile/PDA devices [26,27], safety signs [28], and medical systems [29].

Thirdly, the SUS is available free and has a public domain license. The public domain license means that the one who wants to use this instrument should maintain the attribution to the original developer. Fourthly, the SUS is proved a robust and valid tool for assessing the usability of products [16,30–33], even for a small number of respondents [34]. The detail about the SUS validity and reliability will be explained in the next section.

4. Research Design

In this section, the research method used in the study is explained. There are some points that importantly considered for discussing in order to provide a good understanding of this research. First of all, describing the psychometric instrument used to collect the user's perspective towards the serial module. Then, defining the participants involved in this work. Finally, the procedure conducted in this study is also described in detail.

4.1. Instrumentation

The instrument used in this study is derived from the SUS questionnaire [10]. Originally, the SUS questionnaire was written in the English version. Since the respondents in this study are the students from Indonesia, thus the questionnaire needs to be translated into the Indonesian language. The back-and-forth translation method was chosen to address this process. This translation method started by converting the original version of the questionnaire into the Indonesian version. Then, the Indonesian version was re-translated or back-translated to the original language. The back-and-forth translation was performed as literally as possible in order to obtain the exact meaning of the translation in the target language. Certain aspects of clarity and readability were also considered in the translation process. Few modifications had been made by replacing the word 'system' with the word 'serial module' throughout the questionnaire. This replacement was performed in order to suit the context of this study as implemented by Bangor et al. [31]. With consideration of the precise meaning, the final version of the instrument was compiled in bi-language; the Indonesian and original text (see Appendix A).

The SUS comprises of 10 questions with a Likert scale form. A 5-point Likert scale is made to indicate the degree of agreement or disagreement of each item. Scale 1 states the strong disagreement (strongly disagree) to the intended statement, and scale 5 represents the strong agreement (strongly agree). It should be taken into consideration when providing the SUS to the respondents that there is a difference wording format for odd and even statements. The odd statements (1, 3, 5, 7, and 9) are associated with positive wording questions while the even statements (2, 4, 6, 8, and 10) are rather constructed in negative wording format [16].

The calculation process of the SUS score starts by summing the score contributions from each item. For the odd-numbered items (1, 3, 5, 7, and 9), the score contribution is getting by subtracting 1 from the scale position. For the even-numbered items (2, 4, 6, 8, and 10), the contribution is 5 minus the scale position. The score contribution for each item ranges from 0 to 4 (with four being the most positive response). After that, sum the score contributions for all items. Then multiply the summed score by 2.5 to get the total score. The SUS scores were ranged in between 0 to 100 [10].

4.2. Validity and Reliability of the SUS

Validity refers to how well an instrument can measure what it is intended to measure. Meanwhile, reliability refers to how consistent the instrument when it used to measure repeatedly. Concerning the validity and reliability of the SUS, many studies indicate that SUS is a robust and valid tool for usability evaluation.

Kirakowsky in 1994 reported that the SUS has a reliability of 0.85 [30]. It is categorized in an acceptable range of reliability. In 2004, Tullis & Stetson used five different tools, including QUIS, SUS, CSUQ, and two others to measure the usability of the website and found that the SUS was the most reliable tool even with a small sample, as low as 12 participants [34]. Additionally, extensive research managed by Bangor et al. (2008) involving more than 2300 surveys from more than 200 studies has demonstrated that the Cronbach's alpha was 0.91 [31]. This finding showed highly acceptable criteria for reliability. In 2009, Bangor et al. continued their research by adding adjective rating scale to nearly 1000 SUS surveys and found that a significant correlation (r=0.822) [32]. This finding strengthened the validity of the SUS questionnaire. Orfanou et al., in 2015, conducted the usability evaluation of two LMSs (eClass and Moodle) involving 769 students [16]. Analysis of the results indicated that the SUS was a considerably valid and reliable tool for assessment. Recent research conducted by Sauro & Lewis in 2016 has also confirmed that the SUS has excellent reliability and validity [33].

4.3. Participants

There were 43 second-year students of the Electrical Engineering Education Department, Yogyakarta State University, participating in this study. The research was conducted in one of the particular class meetings in the odd semester of the academic year 2018/2019. Among the respondents, 33 were male (76.74%) while 10 were female (23.26%).

4.4. Procedure

The typical characteristic of the Data Communication Practice course is the existence of computer as a tool to establish the data communication. This situation may lead to organize the course in the computer laboratory conveniently. Therefore, the usability evaluation study was conducted in the computer laboratory of the Electrical Engineering Education Department, Yogyakarta State University. Since the course has two credits, thus the total hour for this study is roughly four meeting hours.

In the first session, the explanation of the main objectives of the study, the serial module including its features, and the operating instructions were briefly given to the students. Then, a simple example was demonstrated to the participants. In this period, the question and answer session was open for discussion. The next session was distributing the serial module to the students. It was followed by asking the students to use the module for the serial communication practice. The experiment program was following the instructional operating procedure written in the manual sheet. The students were allowed to try different functions offered by the serial module. The problems and shortcomings emerged along with the practice may be asked to the instructor. Around 15 minutes prior to the end of the meeting, the paper-based SUS questionnaires were distributed to the students. A short instruction to fill out the questionnaire was explained as clearly as possible. Then, the session was closed by collecting the fulfilled questionnaire.

5. Research Result and Discussion

To get the research results, the first process is preparing the data sets. From the collected questionnaires, the raw data were pre-processed by coding and tabulating to a data set. The final data sets as can be seen in Table 1 comprises of the single score for each individual statements, including the total average score. The "RAW Score" column displays the original mean scores. These scores are associated with positive and negative tone statements to distinguish the odd and even items. Meanwhile, the "SUS Score" column exhibits the mean scores from the standard calculation of SUS ranges in 0-4 and 0-100 format. These scores in principle try to adjust the scores of the even statements (2, 4, 6, 8, and 10) into the positive responses.

Although, the principal value of SUS is providing the single total score, however, it is still necessary to look into detail the individual score for each statement [31]. To address this, Caglarca suggested taking into account the evaluation by verifying the shape of a "five-pointed star" [35]. For doing that, the RAW scores from Table 1 have been transformed into the radar chart, as shown in Figure 3, Caglarca also concluded that the more the "five-pointed star" looks like, the more positive result of the usability will get [35]. As can be seen in the chart, although it tends to be subjective assessment, however, it indicates that the "five-pointed star" shape is almost in perfect form.

Table 1. Responses to individual statements of the SUS

Statements	RAW Score	SUS Score	
		(0-4)	(0-100)
1. I think that I would like to use this serial module frequently.	3.44	2.44	61.05
2. I found the serial module unnecessarily complex.	2.60	2.40	59.88
3. I thought the serial module was easy to use.	4.47	3.47	86.63
4. I think that I would need the support of a technical person to be able to use this serial module.	2.72	2.28	56.98
5. I found the various functions in this serial module were well integrated.	3.93	2.93	73.26
6. I thought there was too much inconsistency in this serial module.	2.16	2.84	70.93
7. I would imagine that most people would learn to use this serial module very quickly.	4.14	3.14	78.49
8. I found the serial module very cumbersome to use.	1.51	3.49	87.21
9. I felt very confident using the serial module.	3.84	2.84	70.93
10. I needed to learn a lot of things before I could get going with this serial module.	2.72	2.28	56.98

	Average	2.81	70.23

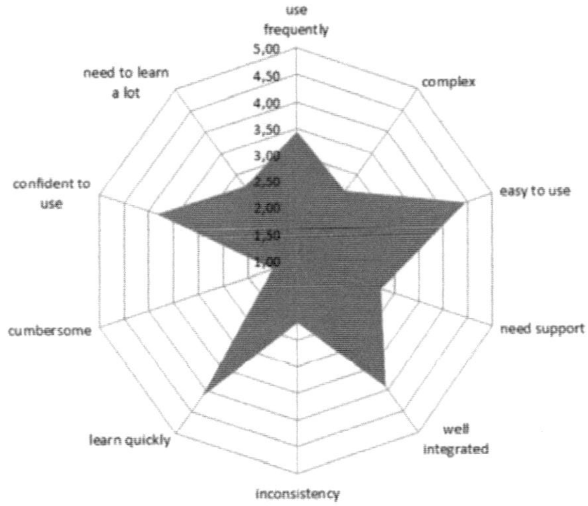

Figure 3. SUS score chart

As there is no standard to interpret whether the particular score represents the acceptance of users for specific products or not, Bangor et al. attempted to provide a grading scheme for the SUS score interpretation [31]. They determined the SUS score into the traditional school grading scale (from 0 to 100 with 100 is the most positive score) [32]. Furthermore, based upon the average mean score and considering the traditional passing score, they proposed score 70 as an acceptable threshold. The detailed SUS score interpretation, including the acceptability ranges, grade scale, and adjective rating, can be seen in Figure 4.

From Table 1, it is interesting to see that the minimum score is 56.98 (statement 4) and the maximum score achieved by statement 8 (reaches 87.21). Concerning the acceptability ranges, score 56.98 is categorized as marginal low, and score 87.21 is in the acceptable range. Looking at the average rating (70.23) achieved, it is slightly higher than the acceptable limit set by Bangor et al. It is also indicated that the score 70.23 is in grade "C" and somewhat below the "Good" of adjective rating.

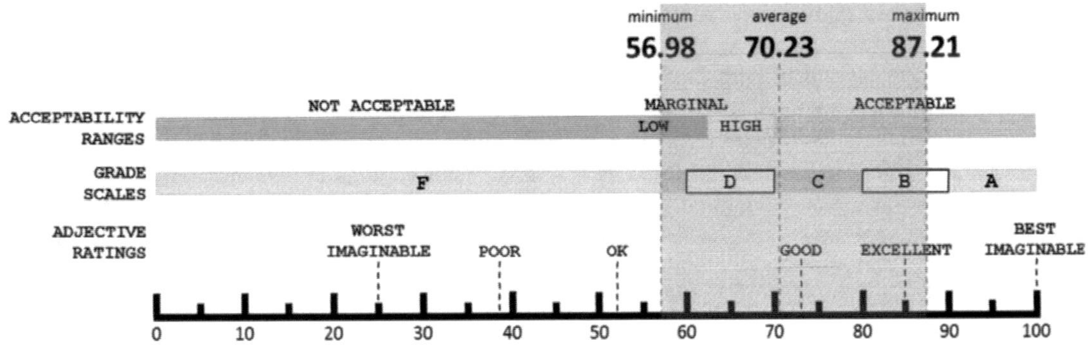

Figure 4. SUS score categorization

Another approach for determining the SUS score was conducted by Sauro & Lewis [33]. They develop the Curved Grading Scale (CGS) to interpret the SUS score into the grade and percentile rank (%). Based on the data from over 400 studies and more than 5000 individual SUS responses, they found that the overall mean score of the SUS is 68 and decided to use it as an acceptable limit [33]. This average score is a little bit below the score limit (70) suggested by Bangor et al.

As can be seen in Figure 5, the SUS score achieved in this study (70.23) is above the acceptable criteria suggested. The score of 70.23 also represents a percentile rank of around 56%. This means that score 70.23 has higher perceived usability than 56% of all products tested in the database of [36]. It also confirms that score 70.23 is classified as a grade C.

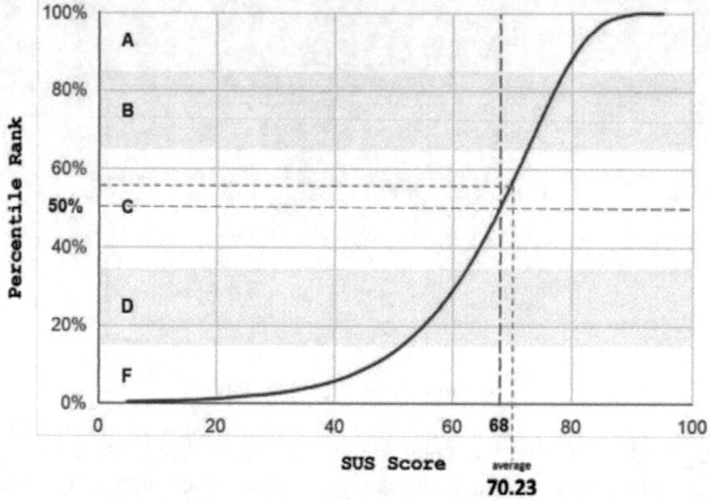

Figure 5. SUS percentile rank

6. Conclusion

In the Data Communication Practice course, one essential part is the availability of the serial communication module to support the students for easily grasping the intended knowledge. It is also important to assure that the learning media may be usable for the students. In this paper, empirical usability evaluation of the serial module has performed through the popular, proved valid, and reliable SUS questionnaire.

The average SUS score achieved in this survey reaches as high as 70.23 (out of 100). Though, there is no standard guidance to decide whether the SUS scores reported in this study classified as acceptable. Hence, the findings are discussed with respect to three different acceptance criteria. First, taking into account the visual judgment of the "five-pointed star" shape suggested by Caglarca [35], it was found that the star shape associated with the SUS score of this study indicated a relatively good contour. It means that there is a positive acceptance from the students for the serial module.

Second, based upon the acceptance limit suggested by Bangor et al. [31] the SUS score in this study was classified as acceptable and indicated as grade "C" and slightly below the "Good" of adjective rating. Third, according to the curved grading scale of Sauro & Lewis [33], the finding in this study could exceed the acceptance threshold. It also confirms that the SUS score of this study has the same grade as Bangor et al. have.

All in all, it can be concluded that the existence of the serial communication module might be accepted by the students and may lead to the improvement of the learning outcome particularly in the topic of serial communication. However, this study only digs the usability evaluation from the subjective student's perception. Therefore, a further investigation from other sorts of perspective for instance interview or expert-based evaluation may bring to the comprehensive results.

7. Appendix A

Angket Penilaian Modul Serial oleh Pengguna

Questionnaire of Serial Module Evaluation by User (Students)

	strongly disagree 1	2	3	4	strongly agree 5
1. Saya pikir saya akan sering menggunakan modul serial ini. *I think that I would like to use this serial module frequently.*					
2. Saya merasa modul serial ini tidak perlu dibuat kompleks. *I found the serial module unnecessarily complex.*					
3. Saya pikir modul serial ini mudah untuk digunakan. *I thought the serial module was easy to use.*					
4. Saya pikir saya akan membutuhkan bantuan dari teknisi untuk dapat menggunakan modul serial ini. *I think that I would need the support of a technical person to be able to use this serial module.*					
5. Saya merasa berbagai fungsi dalam modul serial ini terintegrasi dengan baik. *I found the various functions in this serial module were well integrated.*					
6. Saya pikir ada terlalu banyak ketidakkonsistenan dalam modul serial ini. *I thought there was too much inconsistency in this serial module.*					
7. Saya bayangkan kalau banyak pengguna yang akan belajar menggunakan modul serial ini dengan sangat cepat. *I would imagine that most people would learn to use this serial module very quickly.*					
8. Saya merasa modul serial ini sangat sulit untuk digunakan. *I found the serial module very cumbersome to use.*					
9. Saya merasa sangat percaya diri menggunakan modul serial ini. *I felt very confident using the serial module.*					

10. Saya butuh belajar banyak hal sebelum saya dapat menggunakan modul serial ini. *I needed to learn a lot of things before I could get going with this serial module.*	1	2	3	4	5

8. References

[1] Hariyanto D, Nugraha AC, Asmara A. Design and Development of an Asynchronous Serial Communication Learning Media to Visualize the Bit Data. In: Journal of Physics: Conference Series. 2018. p. 12010.

[2] Holzinger A. Usability engineering methods for software developers. Commun ACM [Internet]. 2005;48(1): 71–4. Available from: http://portal.acm.org/citation.cfm?doid=1039539.1039541

[3] Olsen H. An evaluation of Danish qualitative interview investigations. Nord Psykol.2002; 54(2):145–72.

[4] Nielsen J. The use and misuse of focus groups. IEEE Softw. 1997;14(1):94–5.

[5] Assila A, Oliveira KM, Ezzedine H. Standardized usability questionnaires: Features and quality focus. Electron J Comput Sci Inf Technol eJCIST. 2016;6(1).

[6] ISO. ISO 9241-11: Guidance on Usability. Int Organ Stand [Internet]. 1998;1998(2):28.Available.from:http://www.iso.org/iso/iso_catalogue/catalogue_tc/catalogue_detail.htm?csnumber=16883

[7] Chin JP, Diehl VA, Norman KL. Development of an instrument measuring user satisfaction of the human-computer interface. In: Proceedings of the SIGCHI conference on Human factors in computing systems. 1988. p. 213–8.

[8] Kirakowski J, Corbett M. SUMI: The software usability measurement inventory. Br J EducTechnol. 1993;24(3):210–2.

[9] Lewis JR. IBM computer usability satisfaction questionnaires: psychometric evaluation and instructions for use. Int J Hum Comput Interact. 1995;7(1):57–78.

[10] Brooke J. SUS - A quick and dirty usability scale. Usability Eval Ind [Internet] 1996;189(194):4–7. Available from: http://hell.meiert.org/core/pdf/sus.pdf

[11] Kirakowski J, Cierlik B. Measuring the usability of web sites. In: Proceedings of the Human Factors and Ergonomics Society annual meeting. 1998. p. 424–8.

[12] Lund AM. Measuring usability with the USE questionnaire. Usability interface. 2001;8(2):3–6.

[13] Kortum PT, Bangor A. Usability Ratings for Everyday Products Measured With the System Usability Scale. Int J Hum Comput Interact [Internet]. 2013 Jan [cited 2018 Jul 7];29(2):67–76. Available from: http://www.tandfonline.com/doi/abs/10.1080/10447318.2012.681221

[14] Flavián C, Guinaliu M, Gurrea R. The role played by perceived usability, satisfaction and consumer trust on website loyalty. Inf Manag. 2006;43(1):1–14.

[15] Katsanos C, Tselios N, Xenos M. Perceived usability evaluation of learning management systems: a first step towards standardization of the System Usability Scale in Greek. In: 2012 16th Panhellenic Conference on Informatics. 2012. p. 302–7.

[16] Orfanou K, Tselios N, Katsanos C. Perceived usability evaluation of learning management systems: Empirical evaluation of the System Usability Scale. Int Rev Res Open Distrib Learn [Internet]. 2015;16(2). Available from: http://www.irrodl.org/index.php/irrodl/article/view/1955

[17] Alshammari M, Anane R, Hendley RJ. Design and usability evaluation of adaptive e-learning systems based on learner knowledge and learning style. In: IFIP Conference on Human-Computer Interaction. 2015. p. 584–91.

[18] Alshammari M, Anane R, Hendley RJ. Usability and effectiveness evaluation of adaptivity

in e-learning systems. In: Proceedings of the 2016 CHI conference extended abstracts on human factors in computing systems. 2016. p. 2984–91.
[19] Copas GM, Elder L. Comparing data input methods on handheld computers. Usability News. 2004;6(2).
[20] Wechsung I, Naumann AB. Evaluation methods for multimodal systems: A comparison of standardized usability questionnaires. In: International Tutorial and Research Workshop on Perception and Interactive Technologies for Speech-Based Systems. 2008. p. 276–84.
[21] Evans RE, Kortum P. The impact of voice characteristics on user response in an interactive voice response system. Interact Comput. 2010;22(6):606–14.
[22] Everett SP, Byrne MD, Greene KK. Measuring the usability of paper ballots: Efficiency, effectiveness, and satisfaction. In: Proceedings of the Human Factors and Ergonomics Society Annual Meeting. 2006. p. 2547–51.
[23] Greene KK, Byrne MD, Everett SP. A Comparison of Usability Between Voting Methods. In: EVT. 2006.
[24] Campbell BA, Tossell CC, Byrne MD, Kortum P. Voting on a smartphone: Evaluating the usability of an optimized voting system for handheld mobile devices. In: Proceedings of the Human Factors and Ergonomics Society Annual Meeting. 2011. p. 1100–4.
[25] Kortum P, Sorber M. Measuring the usability of mobile applications for phones and tablets. Int J Hum Comput Interact. 2015;31(8):518–29.
[26] Költringer T, Grechenig T. Comparing the immediate usability of graffiti 2 and virtual keyboard. In: CHI'04 Extended Abstracts on Human Factors in Computing Systems. 2004. p. 1175–8.
[27] Nithyanand R, Saxena N, Tsudik G, Uzun E. Groupthink: usability of secure group association for wireless devices. In: Proceedings of the 12th ACM international conference on Ubiquitous computing. 2010. p. 331–40.
[28] Ng AWY, Lo H, Chan A. Measuring the Usability of Safety Signs: A use of system usability scale (SUS). In: proceedings of the International MultiConference of Engineers and Computer Scientists. 2011. p. 1296–301.
[29] Trafton J, Martins S, Michel M, Lewis E, Wang D, Combs A, et al. Evaluation of the acceptability and usability of a decision support system to encourage safe and effective use of opioid therapy for chronic, noncancer pain by primary care providers. Pain Med. 2010;11(4):575–85.
[30] Kirakowski J. The use of questionnaire methods for usability assessment. Unpubl manuscript Recuper el. 1994;12.
[31] Bangor A, Kortum PT, Miller JT. An empirical evaluation of the system usability scale. Int J Hum Comput Interact [Internet]. 2008 Jul 29 [cited 2018 Jul 7];24(6):574–94. Available from: http://www.tandfonline.com/doi/abs/10.1080/10447310802205776
[32] Bangor A, Kortum P, Miller J. Determining what individual SUS scores mean: Adding an adjective rating scale. J usability Stud [Internet]. 2009 [cited 2018 Jul 7];4(3):114–23. Available from: http://66.39.39.113/upa_publications/jus/2009may/JUS_Bangor_May2009.pdf
[33] Sauro J, Lewis JR. Quantifying the user experience: Practical statistics for user research. Morgan Kaufmann; 2016.
[34] Tullis TS, Stetson JN. A comparison of questionnaires for assessing website usability. In: Usability professional association conference. 2004.
[35] Caglarca S. Usability Testing [Internet]. [cited 2019 Jul 30]. Available from: http://www.sercancaglarca.com/portfolio/destinationuppsala.html
[36] Sauro J. A Practical Guide to the System Usability Scale: Background, Benchmarks & Best Practices. Measuring Usability LLC, Denver, CO.; 2011.

Implementation of Web-Based Weighted Product Use Decision Support System to Determine the Post-Disaster Damage and Loss

A Bachriwindi[1], E K Putra[1], U M Munawaroh[1] and A T W Almais[2]
[1]Student, Department Informatics Technology, UIN Maulana Malik Ibrahim Malang, Indonesia
[2]Lecture, Department Informatics Technology, UIN Maulana Malik Ibrahim Malang, Indonesia

Email: aniqohbachriwindi@gmail.com, erwindchristian@gmail.com, umi.madinatul@gmail.com, agung.twa@gmail.com

Abstract. Determination of damage and losses after natural disasters is carried out to determine the type of damage and the amount of loss after natural disasters that must be borne by the government. In order for the type of damage and the magnitude of the loss after natural disasters according to the data in the field, a study is carried out that implements the Decision Support System (DSS) with the Weighted Product (WP) method. The results obtained are tests of pattern data which are calculated using Weighted Product (WP) compared to 3 different test data, namely using damage and loss data after the natural disaster in East Java in 2010, 2011. The results of the 3 test data used the first amount of 373 data, which produces precision 56.50%, recall 50.50%, accuracy 53.30%, and f-measure 39.10%. For the second data using the data 77, which produces precision 52.80%, recall 36.30%, accuracy 43.02%, and f-measure 36.30%. And for the third data using data from data 24, which produces 50% precision, 50% recall, 50% accuracy, and 50% f-measure. From the 3 types of test data used can be concluded that the Weighted Product (WP) method has a high level of precision, f-measure, recall, and accuracy if the amount of data used is increasing.

1. Introduction

Natural disasters are phenomenon caused by events or a series of events caused by nature, including earthquakes, tsunamis, volcanic eruptions, floods, droughts, hurricanes, and landslides [1]. There are 3 natural disaster managements in Indonesia, namely mitigation (prevention before natural disasters), during natural disasters, and post natural disasters. In Law 27 of 24 regarding to Disaster Management and Presidential Regulation number 8 of 2008 regarding to the National Disaster Management Agency, post-disaster handling includes emergency response, reconstruction rehabilitation, and sustainable development. This research is focused on post-disaster on the rehabilitation and reconstruction of the settlement sector.

The success of the "*Perencanaan dan Pengendalian Penanganan Bencana* (P3B)" Team in the post-disaster recovery program is largely determined by the planning of good recovery from accurate data and information. The number of problems in the field is caused by data inaccuracies. This is happened because when the initial data collection criteria used by the surveyors in the field are perceived to be different so that the categorization of existing data is different. Thus, a problem arises and the disaster data that goes to the Provincial BPBD is different from the situation in the field since the data criteria for the preparation of post-disaster rehabilitation and reconstruction actions used by each surveyor are perceived to be different.

From the identification of the above problems, a system is needed to assist the surveyor team in the process of preparing post-natural disaster rehabilitation and reconstruction actions. The use of a system equipped with a decision support system is expected to accelerate the surveyor team in processing the post-natural disaster rehabilitation and reconstruction actions, and can improve accuracy in obtaining the right results. [2]

One of the Decision Support System (DSS) models used to determine the level of damage and loss of post-natural infrastructure requires research and development of a decision support system that uses the Weighted Product (WP) method. This system is expected to be a decision support tool for related parties, specifically for surveyors from the P3B team in planning post-disaster rehabilitation and protection actions as a basis for decisions to assist surveyors in determining the types of damage and losses after natural disasters into the category of "Severe Damage" ," Moderately Broken" , and " Lightly Broken".

2. State of the Art

2.1. Decision Support System (DSS)

To create a dynamic and real-time irrigation management system, an irrigation management system is created using a web-based Decision Support System (DSS). According to [3] Web-based, Decision Support System (DSS) can be a solution to manage irrigation channels in large areas. The results obtained from managing the slice channel system using a web-based Decision Support System are the acquisition of data and data detection functions, real-time irrigation estimates, dynamic water allocation decisions, and irrigation information management.

From the journal [4] it was explained that a Decision Support System (DSS) could be used for scheduling emergency ambulances. There are two criterias used for the emergency ambulance scheduling system (DSS), namely the average response time and the percentage of ambulance requests that are responded to within 15 minutes. It is explained in this journal that two criterias are ignored in traditional ambulance delivery since the traditional scheduling, those two criterias are very dynamic depends on time and other parameters.

2.2. Metode Weighted Product (WP)

[5] WP is a finite set of alternative decisions which is explained in terms of several criteria decision. Weighted Product Method or commonly called WP is one of the Decision Support System (DSS). This method is used to make a decision by considering criteria and weight. The Weighted Product (WP) method is also called dimensionless analysis since its solution uses a mathematical structure that removes the measurement unit of objects in a data. In its completion, the Weighted Product (WP) method uses multiplication to connect the attribute rating. Each attribute rating is raised first with the weight of the attribute in question (such as the normalization process). Formula:

$$S_I = \prod_{j=1}^{n} X_{i_j}^{w_j} \qquad (1)$$

Preferences for alternative S are given as follows:

S : Alternative preference is analogous to vector S
X : Criteria value
W : Criteria weight / sub-criteria
I : Alternative with i = 1, 2,..., n;
J : Criteria
N : Number of criteria

2.3. Natural Disaster

Disaster management activities are usually focused only on one activity such as evacuation, logistics, or coordination and only limited to one specific activity such as natural disaster response or disaster alertness. According to [6] the research is to find a solution, in order that disaster management activities do not only focus on one specific activity, a Decision Support System is needed to handle all disaster management activities. This Decision Support System (DSS) requires an adequate representation of knowledge because the results presented differ according to stakeholder needs in the decision-making process.

In the journal [7] it is explained that the Decision Support System (DSS) can be used to evacuate fire victims. The journal describes the development of the Supplies Decision fire evacuation system which is integrated with social media, census surveys, Geographic Information System (GIS) data, voluntary advice, and remote sensing data. The integration is based on a variety of scientific and multi-disciplinary experts in the fields of expertise including civil engineering, computer science, geographic information systems, transportation, social media, and communication. To develop an integrated Decision Support System (DSS) fire evacuation, four core modules are needed: dynamic population estimation, stage-based strong evacuation planning, social perception analysis, and web-based geomatics analytic platforms.

3. Research Methods

3.1. System Architecture

Figure 1 shows the relationship between the components of the Decision Support System (DSS) and the problems that occur in the system. In the component parts of the Decision Support System (DSS) it is divided into four environments, namely database management, knowledge base, engine, and interface. The database contains data used for pattern data and test data in the system. The knowledge base contains all knowledge from libraries or from related parties. The engine is described as two parts; the first one is the Weighted Product (WP), which is included to a Decision Support System (DSS) method. This engine interacts with interface components that contain input and output data.

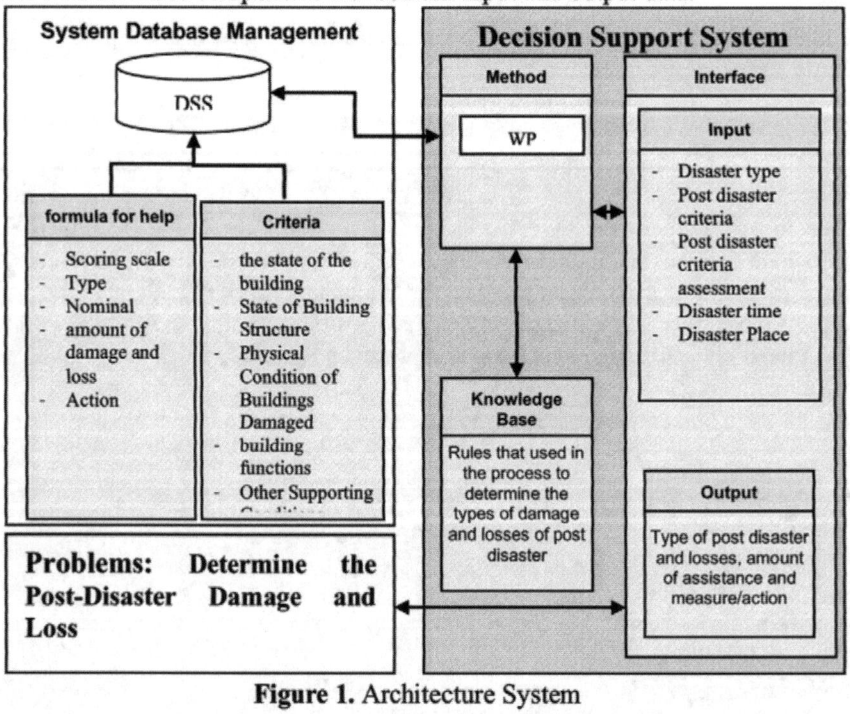

Figure 1. Architecture System

3.2. Weighted Product as Pattern Forming Data

According to Yoon **[8]** the weighted product method uses multiplication techniques to connect the attribute rating, where the rating of each attribute must be raised first with the weight of the attribute in inquiry. The first step to create pattern data is by determining the criteria. After getting the criteria, the next step is to determine the compatibility rating . Then normalize the weight. After getting the normalized value from the weight, determine the value of vector S and the value of vector V. The last is the vector V value ranking.

$$\sum_j^n = 1\ W_{j=1} \quad (2)$$

$$S_i = \prod_j^n = 1\ x_{ij}{}^{wj} \quad i=1,2,\ldots,m \quad (3)$$

$$V_i = \frac{\prod_{j=1}^n X_{ij}{}^{wj}}{\prod_{j=1}^n (x_j)^{wj}}; i = 1,2, \ldots, m \quad (4)$$

3.3. Data flow on the System

In general, the forming components of the Decision Support System (DSS) include alternatives, criteria, weights and values. Each component has its own purpose. Alternatives are used for output or results from the Decision Support System (DSS) as in table 3. Whereas criteria is requirements or assessment components to achieve an alternative as in table 2, while weights are the importance of each criterion as in table 3 , and value is an assessment of each criterion.

Table 1. Alternative

No.	Alternative
1.	Light Damage
2.	Moderate Damage
3.	Heavy Damage

Table 2. Criteria Data

No	Criteria
1	Building's state
2	Building structure's state
3	Physical Condition of Damaged Building
4	Building's Functions
5	Other Building Conditions

(Source: BPBD Jawa Timur, tahun 2010, 2011, 2013 and Data Direktorat Jenderal Cipta Karya, DPU, 2006, Pedoman Teknis Rumah dan Bangunan Gedung Tahan Gempa Data)

Table 3. Damage and Value Level Scale

No.	Criteria	Scale of Interest	Scale Rating
1.	Building's state	Light	Standing
		Moderate	Sloping
		Heavy	Collapsed
2.	Building structure's state	Light	A small part of the building was lightly damaged
		Moderate	A small part of the building is damaged
		Heavy	Most buildings are damaged

No.	Criteria	Scale of Interest	Scale Rating
3.	Physical Condition of Damaged Building	Light	< 30%
		Moderate	30-50 %
		Heavy	> 50%
4.	Building's Functions	Light	No danger
		Moderate	relatively dangerous
		Heavy	Danger
5.	Other Building Conditions	Light	A small part of the building is damaged
		Moderate	Most buildings are damaged
		Heavy	Totally damaged

This study uses the Weighted Product (WP) method that is used as an engine in the system. The test will use two types of data, namely pattern data and test data. Pattern data is data that is processed using the Weighted Product (WP) method. If the pattern data has been formed, the user is easier to use this system. While the data test is data that is input by the user (user) to get the results in the pattern data. Users only enter the type of disaster, criteria and criteria values to get the expected results. So users don't use weights to look for expected results, weights are obtained from pattern data based on data entered by the user. As in Figure 2 which illustrates the flow of the system.

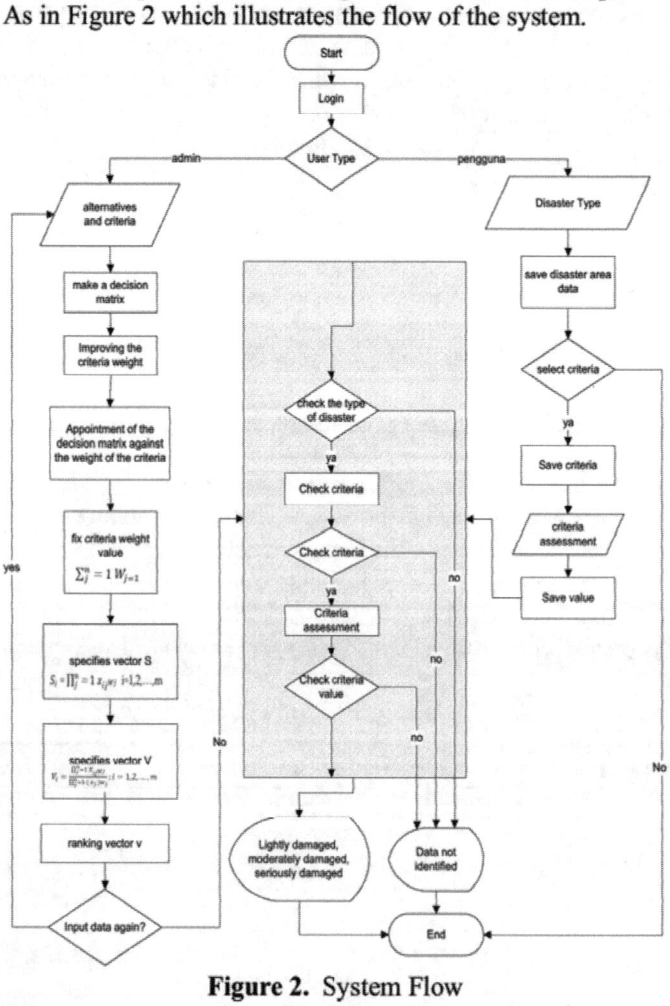

Figure 2. System Flow

3.4. System Design

To make a system, first we have to analyze the needs and existing systems. From the results of the analysis, it will be normalized into a 1NF database design by removing duplicate columns from the same table and creating separate tables for each group of related data and identifying each row with a unique column (primary key).

After the normalization results of 1NF is gained, further it will be normalized to 2NF by deleting some subset of data in the table and placing them in separate tables, creating relationships between new tables and old tables by creating foreign keys, and no attributes in tables that functionally depend on the candidate key table. After that, the third step is 3NF normalization which is fulfilling what is in normalization 2NF and deleting columns that are not dependent on the primary key. The results of 3NF can be seen in Figure 3 which illustrates the relationship between tables in the Entity Relational Database (ERD).

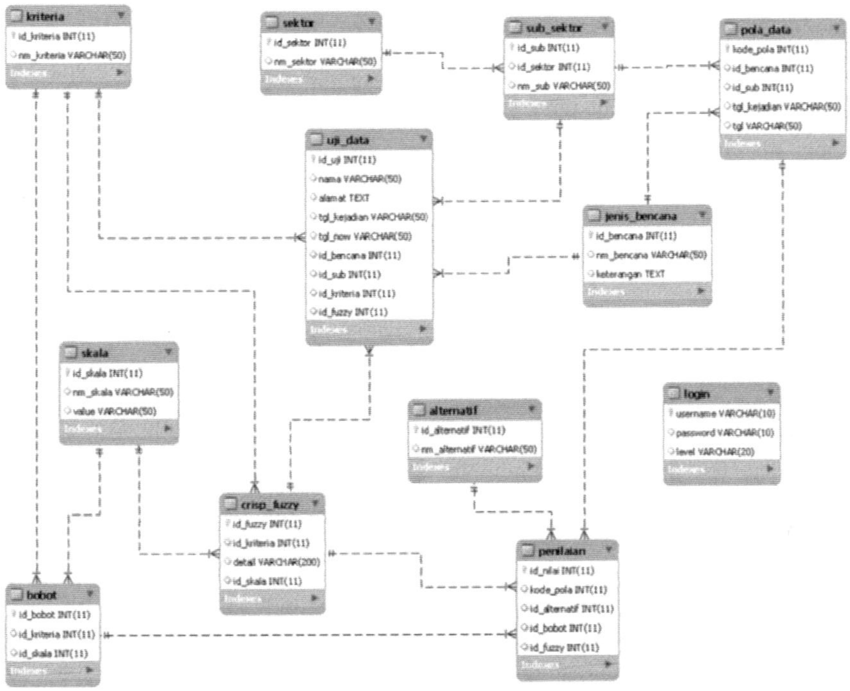

Figure 3. Entity Relational Database (ERD)

The results of implementing the above Entity Relational Database (ERD) can be seen in Figure 4 which describes the process of the Weighted Product (WP) method that has been implemented in a web-based Decision Support System (DSS). In Figure 4, there are steps to calculate the Weighted Product (WP) equations.

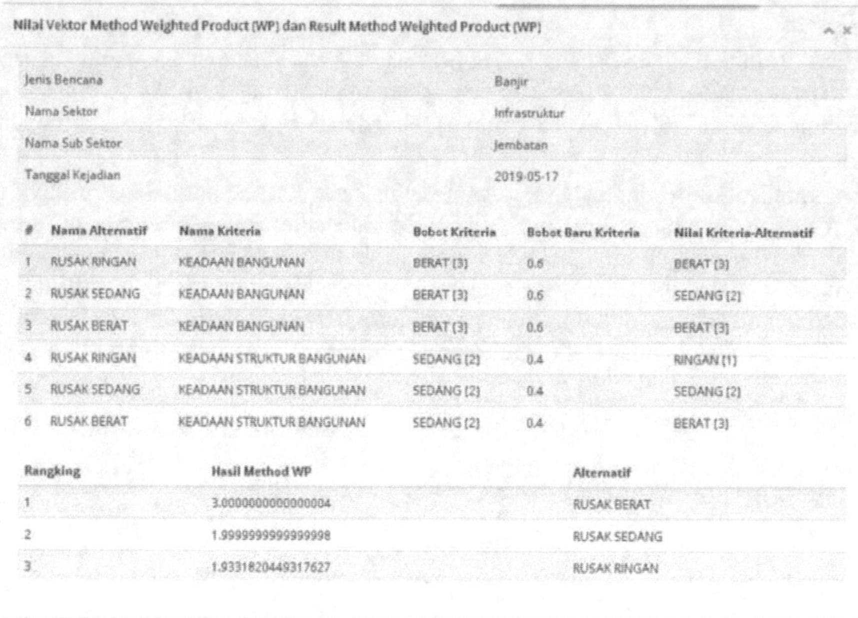
Nilai Vektor Method Weighted Product (WP) dan Result Method Weighted Product (WP)

Jenis Bencana	Banjir
Nama Sektor	Infrastruktur
Nama Sub Sektor	Jembatan
Tanggal Kejadian	2019-05-17

#	Nama Alternatif	Nama Kriteria	Bobot Kriteria	Bobot Baru Kriteria	Nilai Kriteria-Alternatif
1	RUSAK RINGAN	KEADAAN BANGUNAN	BERAT (3)	0.6	BERAT (3)
2	RUSAK SEDANG	KEADAAN BANGUNAN	BERAT (3)	0.6	SEDANG (2)
3	RUSAK BERAT	KEADAAN BANGUNAN	BERAT (3)	0.6	BERAT (3)
4	RUSAK RINGAN	KEADAAN STRUKTUR BANGUNAN	SEDANG (2)	0.4	RINGAN (1)
5	RUSAK SEDANG	KEADAAN STRUKTUR BANGUNAN	SEDANG (2)	0.4	SEDANG (2)
6	RUSAK BERAT	KEADAAN STRUKTUR BANGUNAN	SEDANG (2)	0.4	BERAT (3)

Rangking	Hasil Method WP	Alternatif
1	3.0000000000000004	RUSAK BERAT
2	1.9999999999999998	RUSAK SEDANG
3	1.9331820449317627	RUSAK RINGAN

Figure 4. Implementation Result of Weighted Product Method

4. Results and Discussion

After the stages of designing entity relational diagrams and implementation in programming to create pattern data and test data, the next step is to test the system using confusion matrix calculations.

4.1. Testing Method Weighted Product (WP)

Data pattern formation in the Decision Support System (DSS) uses the Weighted Product (WP) method. The pattern data will be tested using test data to test the role of the pattern data. Pattern data testing will be carried out three times with the same type of test data used but different contents of the data. Table 4 is a description of the composition of the data used for pattern data and test data. The test data used is damage and loss data after the disaster in East Java province in 2010. While the pattern data uses post-disaster data in East Java province years 2010, 2011 and 2013.

Table 4. Data Composition

Method	Data	Testing Data		
	Amount of data	Case I 373 data	Case II 77 data	Case III 24 data
WP	36	146	28	12

Each experiment will produce a confusion matrix rule and calculation. From several experiments, there is the same rule, so that if the test data used is always the same then the value of precision, recall, f-measure, accuracy and response time are also the same. This can be seen in the results of the 1st, 2nd, and 3rd experiments. The similarity of the value of precision, recall, f-measure, accuracy and response time of some experiments is caused by the pattern data used to produce the same pattern of rules and the same test data used. In summary, the test results can be seen in Table 5.

Table 5. Result Testing method *Weighted Product* (WP)

Testing Method	Experiment			Average
	1	2	3	
Precision	56.50%	52.80%	50%	53.1%
Recall	50.50%	36.30%	50%	45.6%
Accuracy	53.30%	43.02%	50%	48.7%
F-Measure	39.10%	36.30%	50%	41.8%

4.2. Discussion

Based on the table 5, it can be seen that the more pattern data used, the greater the value of precision, recall, f-measure, and accuracy. The greater the value of precision, recall, f-measure, and accuracy produced, the method used is getting better. Therefore by using the test in table 5 above, it shows the performance of the Weighted Product (WP) method to be used to determine the damage and loss of settlements after natural disasters. To explain the test results from the Weighted Product (WP) method can be seen in Figure 5 illustrates the percentage of the results of the comparison between the value of precision, recall, f-measure, and accuracy.

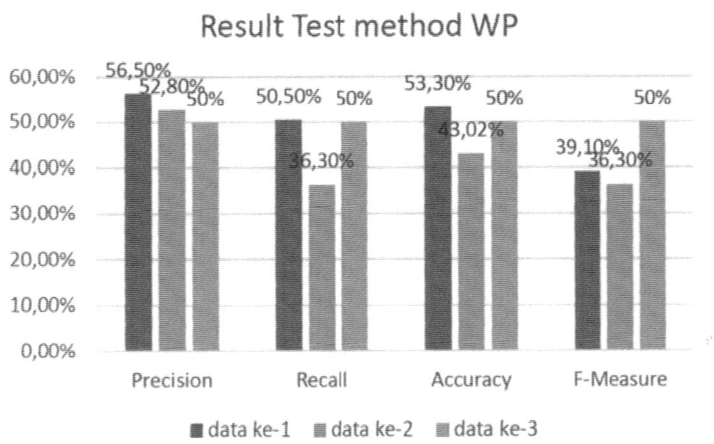

Figure 5. Test Result of Weighted Product Method

5. Conclusion

Based on the testing result that has been done, it can be concluded that the Weighted Product (WP) method can be implemented in the Decision Support System (DSS) to determine the damage and loss of housing after natural disasters properly since the testing pattern data carried out using three different test data can produce better data based on the value of precision, recall, f-measure, and accuracy. Thus, for further research more test data can be used in order to get better values of precision, recall, f-measure, and accuracy. It can also be developed using other types of Multi-Criteria Decision Making (MCDM) methods or Artificial Intelligent (AI) methods to get better precision, recall, f-measure, and accuracy results.

6. References

[1] P3B, Penilaian Kerusakan dan Kerugian, Jakarta, 2008.

[2] A. T. W. Almais, M. Sarosa and M. A. Muslim, "Implementation Of Multi Experts Multi Criteria Decision Making For Rehabilitation And Reconstruction Action After A Disaster," *Jurnal MATICS,* vol. 8, pp. 27-31, 2016.

[3] W. Wang, Y. Cui, Y. Luo, Z. Li and J. Tan, "Web-based decision support system for canal

irrigation management," *Computers and Electronics in Agriculture,* pp. 1-10, 2017.
[4] L. Zhen, S. Sheng, Z. Xie and K. Wang, "Decision rules for ambulance scheduling decision support systems," *Applied Soft Computing,* pp. 1-7, 2015.
[5] Basri, "METODE WEIGHTD PRODUCT (WP) DALAM SISTEM PENDUKUNG KEPUTUSAN PENERIMAAN BEASISWA PRESTASI," 2018.
[6] D. I. Inan, G. Beydoun and B. Pradhan, "Developing a Decision Support System for Disaster Management: Case study of an Indonesia volcano eruption," *International Journal of Disaster Risk Reduction,* pp. 1-17, 2018.
[7] A. Nara, X. Yang, S. G. Machiani and M.-H. Tsou, "An Integrated Evacuation Decision Support System Framework with Social Perception Analysis andDynamic Population Estimation," *International Journal of Disaster Risk Reduction,* pp. 1-33, 2017.
[8] S. Kusumadewi, S. Hartati, A. Harjoko and R. Wardoyo, Fuzzy Multi-Attribute Decision Making (MADM), Yogyakarta: Graha Ilmu, 2006.
[9] R. Yager, "Non-numeric multi-criteria multi-person decision mak-ing," *Kluwer Academic Publishers,* p. 81 – 93, 1993.
[10] D. E. Kartika, W. A. P. Dania and R. L. R. Silalahi, "Analisis Elemen Kunci dan Alternatif Perbaikan Mutu Susu Pasteurisasi dengan Metode ISM dan Fuzzy ME-MCDM (Studi Kasus KUD DAU, Malang)," pp. 1-11, 2014.
[11] M. Noor-E-Alam, T. F. Lipi, M. A. A. Hasin and A. Ullah, "Algorithms for fuzzy multi expert multi criteria decision making (ME-MCDM)," *Knowledge-Based Systems,* pp. 367-377, 2010.
[12] R. A. Hadiguna, I. Kamil, A. Delati and R. Reed, "Implementing a web-based decision support system for disaster logistics: A case study of an evacuation location assessment for Indonesia," *International Journal of Disaster Risk Reduction,* pp. 38-47, 2014.
[13] N. Nurhasanah, "PENENTUAN PRIORITAS ALTERNATIF KEBIJAKAN SISTEM PRODUKSI BERDASARKAN PENDEKATAN NON NUMERIC MULTIEXPERTS MULTICRITERIA DECISION MAKING: STUDI KASUS PT X," *INASEA,* 2006.

Development of Learning Media Based on Androidin Basic Algorithm and Programming Course for Students Grade X of Nasional Berbah Vocational High School

R Irfan and A Wulandari
Electronic and Informatics Engineering Education Department,
Universitas Negeri Yogyakarta, Yogyakarta, Indonesia

Email: irfan@uny.ac.id, agustina.wulandari@student.uny.ac.id

Abstract. The purpose of research are develop Basic Algorithm and Programming for student grade X of National Berbah Vocational High School Application on Android-based learning media and knowing the feasibility of the Android-based learning media. The research use ADDIE as development model. The result of the development is an Android Application that contains material, sample of program, video, and evaluation on learning and can use for self learning or in class. The product has been tested for feasibility based on the assessment of media experts is 95.34% with the category "Very Feasible", the assessment of material experts is 88.6% with categories "Very Feasible" and product testing by students with final result feasibility is 82.47% so that the learning media is very feasible to be used in the learning of Basic Algorithm and Programming in grade X of Nasional Berbah Vocational High School.

1. Introduction

Learning media which are packaged by utilizing information technology are considered to have provided convenience in the field of education. Various media that can help overcome difficulties in learning are now widely applied to certain learning in the form of applications, one of which is learning media based on Android applications.

Basics of Algorithms and Programming is one of the lessons learned in class X majoring in Computer Engineering and the Networking at Vocational School Berbah. Difficulties experienced by students are difficulties in learning the material because of the limitations of independent learning resources and the lack of learning media that supports the learning process. The teacher must strive to guide students in learning because students' interest and motivation to learn are still lacking because of the use of learning media factors that have not varied.

This is in line with the results of previous research by Rosiani [8] which states that the lack of interest and awareness of students majoring in Computer Engineering and Networking of the Berbah National Vocational School of Sleman in studying at school or at home has not yet achieved the KKM (Minimum Completeness Criteria). In that research there is a conclusion that the variation of media used in learning is very important to foster student interest in learning.

Based on these problems, it is necessary to apply learning media that are expected to be able to attract students' interest and motivation to learn and provide convenience in learning the material in the presence of learning media. As a learning resource that can be used in class or independently, almost all students have an Android smartphone that has not been used optimally as a tool for learning. Therefore, this study aims to develop Android-based learning media products on the Basics of Algorithms and

Programming for Class X students of the National Vocational School Berbah and find out the feasibility of the product being developed.

General characteristics in learning media need to be considered so that the resulting media products are good. This includes: (1) Introduction to the program in the form of learning media titles, (2) Student control of navigation buttons and menus, (3) Consistency of information provision, (4) Text, graphic and video presentation modes, (5) Providing assistance, and (6) Closing the program. [1]

Learning media developed need to be tested for eligibility before being applied in learning. Asyhar [5] describes one of the principles of learning media evaluation criteria which is very important, namely the aspect of presentation clarity. Presentation of content needs to be considered in order to facilitate students in learning the material and information presented.

As quoted by Arsyad [4] that the evaluation criteria for the quality of instructional media are divided into three, namely (1) Quality of content and objectives, (2) Instructional quality, and (3) Technical quality. In addition, Thorn [10] categorizes several fundamental aspects in evaluating the effectiveness of a media, namely: (1) Ease of Use and Navigation, (2) Cognitive load, (3) Knowledge space and information presentation, (4) Media Integration, (5) Aesthetics, and (6) Overall Functionality.

2. Research Method

The type of research is development research using ADDIE development models according to Branch [6] namely Analysis, Design, Development, Implementation, and Evaluation to develop learning media products. The study was conducted in the odd semester from August to October 2018 at Berbah National Vocational School which is located in Tanjungtirto, Kalitirto, Berbah, Sleman, Yogyakarta Special Province, Indonesia. The subjects in this study were two expert lecturers in the media of informatics engineering education in the assessment of instructional media in terms of media, one teacher in Basic National Vocational Program Berbah in material assessment, and 20 students in class X Computer and Network Engineering as respondents.

The research procedure adjusts the five stages of the ADDIE development model, namely Analysis, Design, Development, Implementation, and Evaluation. The purpose of the analysis phase is to determine the initial needs in developing this learning media. In developing this media analysis is needed, namely: (1) Analysis of User Needs, (2) Analysis of Content or Content, and (3) Analysis of Hardware and Software Requirements.

In the Design stage, the steps taken are to plan the elements that will be included in the learning media, namely making flowcharts that describe the sequence and structure of the learning media. Besides designing a storyboard that includes a template design plan and also material. After making the storyboard the interface is made or what is called the interface design. The development phase is carried out to produce learning media products. The product was developed using Adobe Flash CS6 software with Adobe Air for Android extension. After the product is developed, it needs to be assessed by expert judgment, namely media experts and material experts so that a product that is suitable for use is produced. After the product is finished and is declared feasible by the testing of media experts and material experts, the product is implemented to grade X students majoring in Computer Engineering and Networking at the National Vocational School Berbah. Every stage that is carried out needs to be evaluated to ensure that the product complies with the specifications. In addition, an evaluation is conducted to measure the feasibility of the learning media developed. The instrument used to collect data on the development of this product was a questionnaire / questionnaire distributed to media experts, material experts, and respondents to measure the level of media viability. The type of measurement scale used in this study is a Likert scale. According to Sugiyono [9], Likert scale is a scale that is widely used in questionnaires and is used to measure a person's perception, attitude or opinion of a product that has been developed.

Data collection techniques in this study are: (1) Observation to observe various problems in schools, (2) Interviews to find out the problems found in depth, and (3) Questionnaire to evaluate learning media products that have been developed. Before questionnaires are distributed to respondents, instruments that have been developed need to be tested for validity and reliability in order to meet the requirements of valid and reliable instruments. A valid instrument means that the instrument is able to measure what should be measured, while a reliable instrument is that the instrument can be trusted to be used in data collection. The validity test was carried out by the instrument experts followed by the analysis of each question by comparing the results of r arithmetic with r tables with a significance value of 5%. If the r table is less than r, the item is valid. The technique used in the item validity test is Product Moment correlation according to Arikunto [2] which is described in the following formula:

$$r_{xy} = \frac{n(\sum XY)-(\sum X)(\sum Y)}{\sqrt{\{n(\sum X^2)-(\sum X)^2\}\{n(\sum Y^2)-(\sum Y)^2\}}} \quad (1)$$

Description:
R_{xy} = correlation coefficient between item scores and total scores
N = total respondents
$\sum x$ = total of item scores
$\sum Y$ = total scores
$\sum XY$ = total of item scores multiplications with total scores

The reliability test using the Alpha formula as follows:

$$r_{11} = \left(\frac{k}{(k-1)}\right)\left(1 - \frac{\sum \sigma_{b^2}}{\sigma_1^2}\right) \quad (2)$$

Description:
r_{11} =reliabilitas instrumen instrument reliability
k = total of question
$\sum \sigma_{b^2}$= total of item varian
σ_1^2 = total varian

After the reliability coefficient is known, the calculation results are converted in the interpretation table r value according to Arikunto [2], as follows:

Tabel 1. Interpretation of r value

r value	Interpretation
0,800 – 1,00	very high
0,600 – 0,800	high
0,400 – 0,600	fair
0,200 – 0,400	low
0,000 – 0,200	very low

Data analysis in this research used three steps. Based on Arikunto [2], three steps in conducting data analysis: (1) Preparation to check the completeness of data and the kinds of data fields, (2) Tabulating data, and (3) Application of data according to the research approach. At the tabulation stage, data analysis activities include obtaining quantitative data by scoring items that need to be scored. The score given for each answer is presented in Table 2.

Tabel 2. Scoring for each item

answering	Skor
very agree	5
agree / good	4
fair	3
not agree	2
very not agree	1

Based on the scores that have been given, the data is applied in calculations to assess the feasibility in the form of a percentage which can be briefly described by the following formula [7]:

$$Appropriate\ percentage\ (\%) = \frac{obtained\ score}{ideal\ maximum\ score}\ x\ 100\% \quad (3)$$

The percentage of eligibility obtained, then converted into descriptive sentences to assess the feasibility of the media. The following is a guideline table for the media eligibility criteria according to Arikunto [3]

Tabel 3. Learning Media appropriateness criteria Kriteria Kelayakan Media Pembelajaran

Appropriateness percentage	interpretation
81-100%	very appropriateness
61-80%	appropriateness
41-60%	fair
21-40%	not appropriateness
<21%	very not appropriateness

3. Result and Discussion

This research produces learning media products in the form of applications that can be used on Android-based smartphone devices. Figure 1. example of the display of the product developed

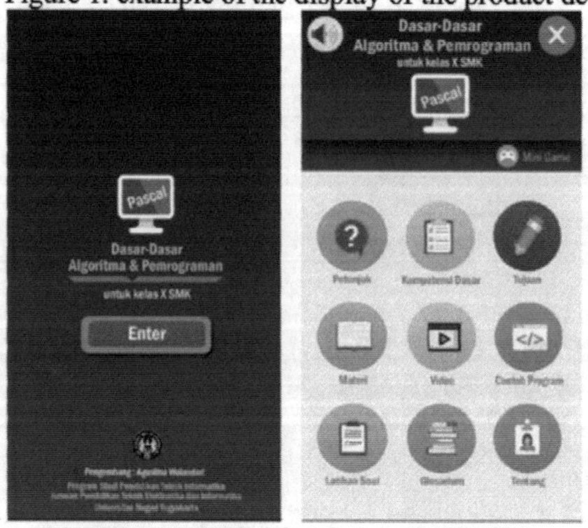

Figure 1. Product development results

Successfully developed products have been tested for suitability based on assessments by media experts and material experts. The following are the results of the assessment of media experts and material experts:

The results of the assessment by media experts are shown in Figure 2.

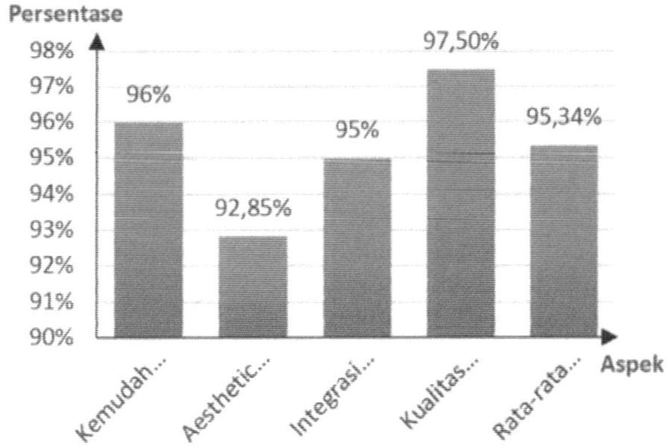

Figure 2. Bar graph of expert media evaluation

Based on the assessment of media experts the percentage value obtained from the ease of use and navigation aspect is 96%, the percentage value from the aesthetic aspect is 92.85%, the percentage value from the media integration aspect is 95%, and the percentage value from the technical quality aspect is 97.5%. Thus, obtained an average final percentage of 95.34% which stated that the media aspect of the Android-based learning media is included in the very feasible category.

The results of the assessment by content experts are shown in Figure 3.

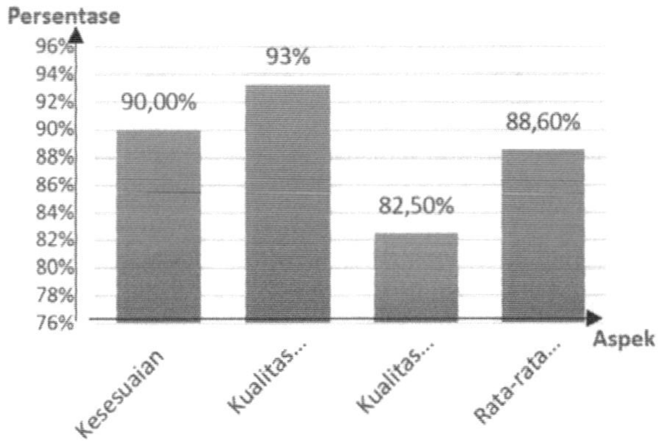

Figure 3. Bar graph of expert material evaluation

Based on Figure 3, the material eligibility from the suitability aspect is stated to be very feasible with a feasibility percentage of 90%, the aspect of content quality and objectives is declared to be very feasible with a eligibility percentage of 93.3%, and the instructional quality aspect is declared to be very feasible with a feasibility percentage of 82.5 %. The overall average percentage is 88.6% so that the material in instructional media is categorized as "Very Eligible".

Before the questionnaire is used for data retrieval to respondents, it needs to be tested for validity and reliability. The instrument validation was carried out by the instrument expert lecturer then trialed to the students. The results of testing the instrument on students showed that from a total of 23 question items, there was item question number 15 which was invalid because r count <r table. So that the items that can be used for data retrieval are 22 items.

The instrument reliability test was performed by calculating the Alpha value assisted with SPSS 16.0 software. The following results from the reliability test:

Table 4. Instrument Reliability Results

Cronbach's Alpha	N of Items
0,887	22

From the resulting reliability value compared to the interpretation table r value, the instrument reliability is included in the very high category, so that the instrument can be trusted to be used in testing the feasibility of Android-based learning media on respondents.

The results of the feasibility test by student are shown in Figure 4.

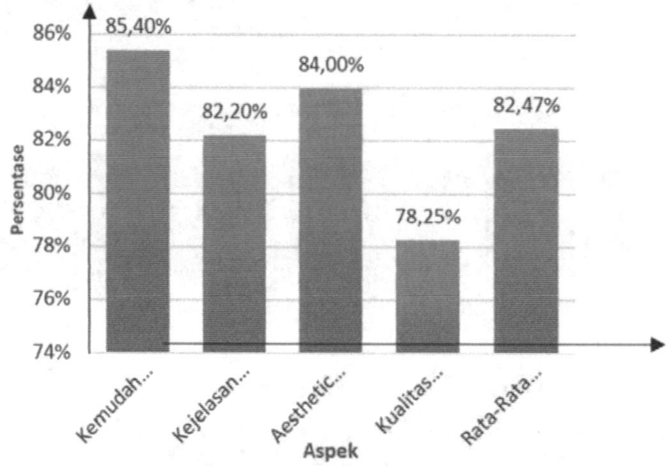

Figure 4. Results of the feasibility test by students

Based on testing to students in Figure 4. obtained the percentage of eligibility results from the aspects of ease of use and navigation of 85.4% which was declared very feasible, the aspect of presentation clarity of 82.2% which was declared very feasible, aesthetic aspects or beauty of 84% which was stated very feasible, and the instructional quality aspect of 78.25% which was declared feasible. Thus, the final score of 82.47 is obtained so that the Android-based learning media that are developed are very feasible to use.

4. Conclusion

The development of instructional media is carried out with the ADDIE development model, namely Analysis, Design, Development, Implementation, and Evaluation. The results of product development are in the form of applications that support classroom learning and independent learning by utilizing Android-based smartphones. The features provided in this application have been adjusted according to need, which includes instructions, material, videos, practice questions, and minigames.

The result of the feasibility test by the media expert was 95.34% which was stated in the very feasible category, the results of the test by the material expert was 88.6% which was declared very feasible, and the results of the feasibility test by the students amounted to 82.47% which was stated in the very feasible category. So as a whole the Android-based learning media is very feasible to use in learning the Basics of Algorithms and Programming of Class X National Vocational School Berbah.

Researcher's suggestion for further research related to the development of similar learning media is: (1) Android-based learning media on the Basics of Algorithms and Programming can be developed by loading broader basic competencies, (2) Learning media is more developed with database systems so that the material and more dynamic exercises, and (3) Development of instructional media that provides more means of interaction between users.

5. References

[1] Alessi, S. M., & Trollip, S. R. (2001). *Multimedia for Learning Methods and Development.* Massachusetts: A Pearson Education Company.

[2] Arikunto, S. (2013). *Prosedur Penelitian Suatu Pendekatan Praktik.* Jakarta: Rineka Cipta.

[3] Arikunto, S., & Abdul Jabar, C. S. (2004). *Evaluasi Program Pendidikan: Pedoman Teoretis Praktis Bagi Praktisi Pendidikan.* Jakarta: Bumi Aksara.

[4] Arsyad, A. (2017). *Media Pembelajaran.* Jakarta: Rajawali Pers.

[5] Asyhar, R. (2012). *Kreatif Mengembangkan Media Pembelajaran.* Jakarta: Referensi.

[6] Branch, R. M. (2009). *Instructional Design: The ADDIE Aproach.* London: Springer.

[7] Purwanto, N. (1992). *Prinsip-Prinsip dan Teknik Evaluasi Pengajaran.* Bandung: Remaja Rosdakarya.

[8] Rosiani. (2013). Hubungan Minat Belajar dan Kebiasaan Belajar Dengan Prestasi Belajar Siswa Kelas X dan XI Mata Diklat Teknik Komputer dan Jaringan di SMK Nasional Berbah Sleman. 2-4.

[9] Sugiyono. (2017). *Metode Penelitian & Pengembangan: Research and Development* (3rd ed.). Bandung: Alfabeta.

[10] Thorn, W. J. (1995). Points to Consider when Evaluating Interactive Multimedia. *The Internet TESL Journal.* Retrieved May 14, 2018, dari http://iteslj.org/Articles/Thorn-EvalueConsider.html

Journal of Physics: Conference Series **1413** (2019) 012021 doi:10.1088/1742-6596/1413/1/012021

Real-Time Emotion Recognition System to Monitor Student's Mood in a Classroom

W B Putra and F Arifin
Education Electronics and Information Technology,
Universitas Negeri Yogyakarta, Yogyakarta, Indonesia

Email: werda.buana2016@student.uny.ac.id, fatchul@uny.ac.id

Abstract. Depression is a commonly unattended health problem, unbounded by age border, and greatly affecting student's performance in their study. To prevent it, the writer built a real-time facial emotion recognition system, so the teacher can monitor students' mood through class activity. The system should be reliable enough when running on mid-end computer specification. The student will be given questionnaire to measure their stress. The questionnaire result will be used to analyze whether the use of the system able to reduce student's stress or not. The result from the questionnaire shows that the use of the system able to detect student's mood early so the teacher may minimize student's stress.

1. Introduction

Human emotion recognition has been the spotlight for many researches to test various theories, algorithms, and new technologies. Emotion recognition discipline of Human–Computer Interaction relies on the algorithmic robustness and hardware sensitivity.

Social anxiety disorder (SAD) is one of the most common mental health disorders, with approximately 13% of the population meeting diagnostic criteria for SAD during their life. Hence, the need to quickly recognize such anomaly surfaced. If left untreated, SAD typically runs a chronic course and total remission is rare. Though depression is classified as an adult disorder, and depression in very young children is rare, middle to late adolescence is the most common age when depression symptoms first appear or a first major depressive chapter happens. Lewinsohn, Hops, Roberts, Seeley, and Andrews sampled random adolescents within a U.S. community and figured out that the average age of onset of major depression was 14 years old [1].

Though facial expressions obviously does not necessarily convey emotions, in computer vision community, the term of facial expression recognition often refers to the classification of human facial features into one of the six basic emotions: happiness, sadness, fear, disgust, surprise and anger [2] [3]. According to National Institute of Mental Health data in 2014, about 2,8 million American teenagers ranged 12-17 years old would at least suffered by one episode of major depression and only 30% of them get a proper treatment [4].

Depression is categorized as mental-emotional disorder in Indonesia and it's prevalence as high as 6% of total population [5]. A number of researches investigate the role some factors to depressive symptom, such as emotion regulation [6]. Illegal drugs usage is proven to be the byproduct of depression and because the bigger prevalence on men, depression and comorbidity affected by gender. Data from particular newspaper in Indonesia shows that 90% of drug users is on large, unattended [7]. Psychological researchers have proved that there are negative effects of depression on student's academic performance [8]

In response to that condition, the writer proposed a way to monitor student's mood through class activity. The teacher should be able to constantly aware on the fluctuation of their student's mood by using facial emotion recognition system. At this point, the problem only lies on the willingness of the system developer and the educator to collaborate. But here is where the problems lies, Indonesia, where the research will take place upon, is a developing country. As sad as it sounds, 55% of under-15 Indonesian is functionally unable to read [9]. Implementing artificial intelligence to recognize human facial expression can be a complex and difficult task. High quality datasets are hard to found, and there are various pitfalls to avoid when designing such system. It is proved that facial expression could help lecturers to identify the involvement and comprehension of the student [10].

2. System Design

Real-time facial emotion recognition, in general, is divided by several phases. The first phases would be detecting general area of human face, this process include tracking system which require the hardware to monitor the general movement of facial layout. Second phase would be Facial landmarking, which pointing out more accurate facial point to be extracted. Of course, the writers have to choose the most efficient way to compute and doing facial analysis to prevent computer's memory overload. On the other hand the system still need a feature which is able to extract the subject's facial expression and then further processed it into the emotion as the result.

The writer needs to be sure the computer able to properly do image acquisition, facial extraction, facial landmarking, and logic computation to get the emotion from the subject. In short, a working webcam, and sufficient computation power on the subjected computer is crucial. The location of facial features can be represented as landmarks on the face. The image acquisition of human face can be represented as coordinated vector landmark and Action Units (AU). Hence, facial feature offered geometric information of each and overall shape of particular object [11]. Facial expression extraction and localization has become pivotal point among image analyst [12] [13].

3. Research Method

Prior to research conduct in school, a research proposal was requested to university board. After 3 days, the proposal granted and ready to be forwarded to the school board, which is *SDN Percobaan* 2 Yogyakarta. The research proposal was submitted to Institutional Board and approved by the headmaster. Participant's selection began by consulting with supervisor and the school's teacher. The writers looking for students, who capable of basic computer operation and can be left by themselves to do their work. Another criterion is the subject shouldn't be burdened by external variables, such as the approaching National Exam, which could be applied to all of the sixth grade students. This narrowed the choice to fourth grader and fifth grader. After further discussion, the writer choose to pick students from fifth grade A class as test subjects due to their relatively docile nature compared to other classes on their grade or fourth grader.

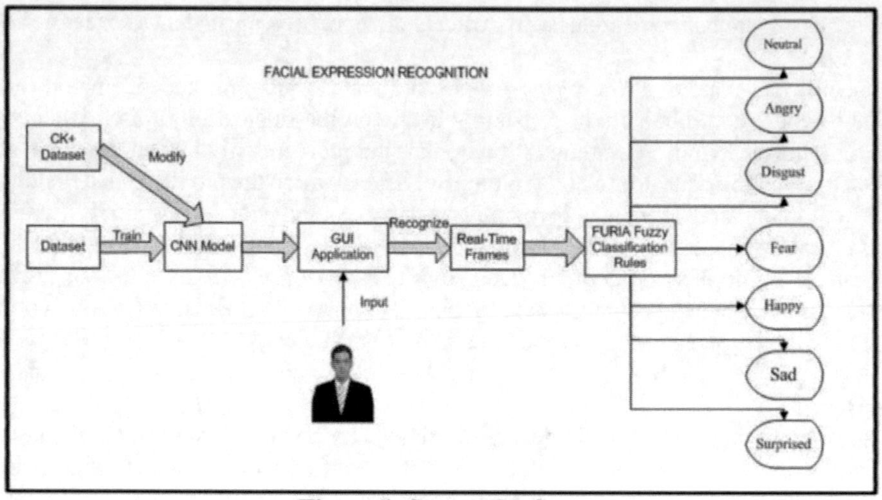

Figure 1. System Design.

To gather the research data, a particular device to capture images of each student is needed, in this case it would be USB webcam. The writer needs to be sure the computer able to properly do image acquisition, facial extraction, facial landmarking, and logic computation to get the emotion from the subject. In short, the research needs a working webcam, and sufficient computation power on the subjected computer. After consulting with the headmaster and IT teacher, it is confirmed that even though they have 20 computers in their lab, only 4 computers for student use that were installed with working webcam (installed by default on monitor display). Like that, there are four client computer and one server computer as a working system (see figure 11 for reference) prepared by the school.

The system produced systematical processes which allowed the teacher to monitor their student's mood through the class activity by using computer's webcam to acquire the image real-time. The system will use a pre-made library to acquire image(s) from webcam. The process to determine facial feature from the acquired image will utilize pre-trained dataset [14] [15] [16], which is trained using convolutional neural network model. While trying some method to recognize facial feature, the writers stumbled at some problems such as:

1. Difficulty to combine the metric with powerful image recognition technique such as SVM, LDA, PPA, etc is particularly high.
2. The measurement computation is very complicated.
3. The distance does not obey triangle inequality. So sometime, two similar images can be both similar to a completely unknown object.

With these points all covered, the system simplify the Euclidean Distance computation. The system uses Standardizing Transform to do domain smoothing [17]. This directly relates image to smoothing and indicates that smoothing noiseless images can still increase image recognition rate [18].

4. System Implementation

Table 1 through table 3 shows confusion matrix for pre-trained dataset that being used to build the system. The score that written in the table informs the reader how many images that represented an emotion that fit with the comparison. Table 1 is the comparison from the original database with the normalized shape feature benchmark. In table 2, the dataset will tested on Canonical Appearances benchmark. In table 3, the dataset will tested with the combination of both benchmarking system. This test is usually the most reliable of the three and the result often be made for the validity of the dataset.

Table 1. Confusion Matrix of Emotion Detection for the Similarity Normalized Shape [18]

	An	Di	Fe	Ha	Sa	Su	Co
An	**35**	40	0	5	5	15	0
Di	7.9	**68.4**	0	15.8	5.3	0	2.6
Fe	8.7	0	**21.7**	21.7	8.7	26.1	13
Ha	0	0	0	**98.4**	1.6	0	0
Sa	28	4	12	0	4	**28**	24
Su	0	0	0	0	0	**100**	0
Co	15.6	3.1	6.3	0	15.6	34.4	**25**

Table 2. Confusion Matrix of Emotion Detection for the Canonical Appearance

	An	Di	Fe	Ha	Sa	Su	Co
An	**70**	5	5	0	10	5	5
Di	5.3	**94.7**	0	0	0	0	0
Fe	8.7	0	**21.7**	21.7	8.7	26.1	13
Ha	0	0	0	**100**	0	0	0
Sa	16	4	8	0	**60**	4	8
Su	0	0	1.3	0	0	**98.7**	0
Co	12.5	12.5	3.1	0	28.1	21.9	**21.9**

Table 3. Confusion Matrix Of Emotion Detection For The Combination Of Features (SPTS+CAPP). The Fusion Of Both Systems Were Performed By Summing Up The Probabilities From Output Of The Multi-Class SVM

	An	Di	Fe	Ha	Sa	Su	Co
An	**75**	7.5	5	0	5	2.5	5
Di	5.3	**94.7**	0	0	0	0	0
Fe	4.4	0	**65.2**	8.7	0	13	8.7
Ha	0	0	0	**100**	0	0	0
Sa	12	4	4	0	**68**	4	8
Su	0	0	0	0	4	**96**	0
Co	3.1	3.1	0	6.3	3.1	0	**84.4**

JAFFE image set consisted of 213 images of Japanese female frontal facial pictures, classified by 7 basic emotions just like the classification used by the system. The writers will compare the result gotten from analyzing each of these 213 images using the system, with the default classification from JAFFE image set. In figure 2, the writers run the system to see how the emotion recognition system could detect facial features. The system proved to be able distinct emotion features of human expressions; this will be further tested using a specific expert-arranged image set which shown on table 4.

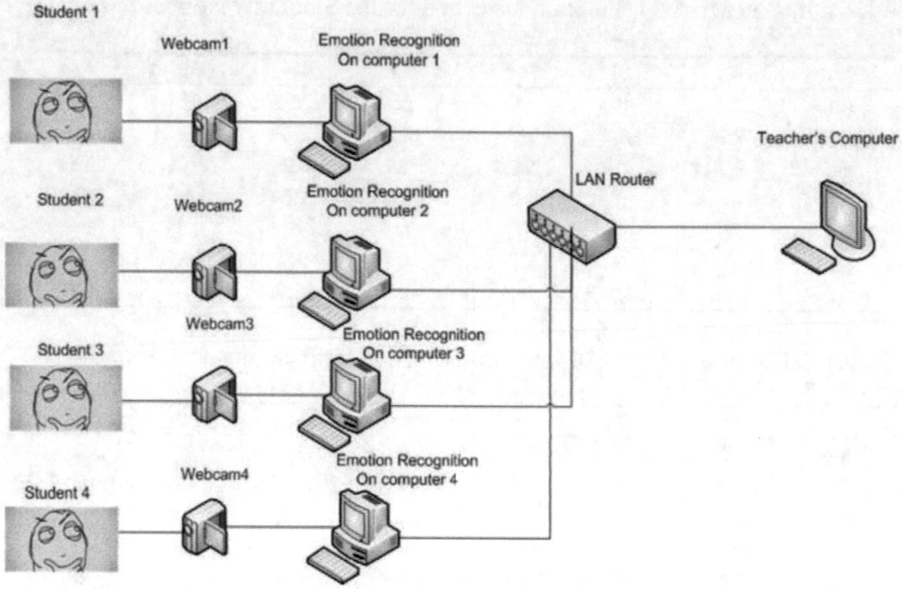

Figure 2. System Testing on Real Time Feed via Webcam

This suggests that an automated system can do just as a good job, if not better as a naive human observer and suffer from the same confusions due to the perceived ambiguity between subtle emotions. However, human observer ratings need to be performed on the CK+ database and automated results need to be conducted on the JAFFE database to test out the validity of these claims. Our tools also provide desirable result on the scanning and image analysis. Which could detect, locate, tracking, and analyze human expression that contained in captured image. This contributes greatly on this research mainly because this research is quite focused on the accuracy of the system. The writers have to measure the system's accuracy. The system uses JAFFE image set, which already classified based on specific emotions by expert. The writers will compare the measurement brought by our system, with the classification from the image set in a posed expression database; the database will display different basic emotional expressions simultaneously, while in spontaneous expression database, the expressions are natural. Spontaneous expressions are differed from posed ones remarkably in term of intensities, configurations, and durations. Apart from this, syntheses of some AUs are just barely achievable without the need of undergoing on associated emotional state. Therefore, at most cases, the posed expressions by human participants are exaggerated, while the spontaneous ones are quite subtle and differed in appearances.

The emotion annotation can be simply done in discrete emotion labels or on a continuous scale. Most of databases are usually based on the basic emotions theory proposed by Paul Ekman and Armindo Freitas-Magalhaes which assumes the existence of six discrete basic emotions (anger, fear, disgust, surprise, joy, sadness) [14] plus neutral emotion. JAFFEE database was planned and assembled by Michael Lyons, Miyuki Kamachi, and Jiro Gyoba [19].

The result of teacher's questionnaire will be processed with a simple formula, mainly because this questionnaire is only has two respondents, which is the IT teachers. The formula is simply required the total score collected, which ranged from 0 to 20, from the questionnaire with maximum value possible for the questionnaire on the question. This method of data analysis is called SUS (System Usability Scale).

Originally created by John Brooke, it allows an evaluation of vast variety of products, be it hardware, software, operating system, mobile devices, and websites. SUS has been used in more than 1300 science article and publications, or product quality assurance. There are 15 questions for each teacher, the score for each question will ranged from -2 to 2. The result is as follows:

Number of Questions : 15
Number of Respondent : 2
Score : 50
Maximum Score : 60

According to the acquired data which can be seen in Appendix I, the system achieved score of 50 out of 60. In percentage it means the system achieved 83,33% of total functionality. It proved the system can fulfilled its role as a tool to collect emotional information and monitor student's mood in a classroom activity.

To estimate the confidence, calculate t-value, and then lookup the inverse of CDF of Student's t-distribution with $(N_a - 1) + (N_b - 1)$ degrees of freedom. N_a is the size of sample A and N_b is the size of sample B. To find t-value you start from calculating the mean M_x and sum of squared deviations, or sum of squares $SS = \sum(X_i - M_x)^2$ for each sample. Then, estimate the variance of the source population as $\{s_p^2\} = \frac{SS_a + SS_b}{(N_a-1)+(N_b-1)}$. This estimation is called pooled variance, a method for estimating variance of several different populations when the mean of each population may be different, but one may assume that the variance of each population is the same.

The writer estimate the standard error of $M_{Xa} - M_{Xb}$as $est.\sigma_{M-M} = \sqrt{\frac{\{s_p^2\}}{N_a} + \frac{\{s_p^2\}}{N_a}}$. Finally, $t = \frac{M_{Xa} - M_{Xb}}{est.\sigma_{M-M}}$ calculate t as.

Index :
t_x : Size of the difference relative to the variation in sample data x
SS_x : Sum of Squares of x
N_x : Size of sample x
M_x : Sum of squared deviations from sample x

Here is the result from the two sample that run by testing the average of pre-test results and the average of post-test results as follows:

- Mean of samples a : 0.5
- Mean of samples b : 0.2
- t-value : 10
- Level of Confidence on one and two tall's test of significance : 100%

5. Conclusion

From the result collected in Data result and analysis, the research can be concluded to be able to answer the research purpose and research question while conducted in the area of problem limitation and research limitation. The conclusion can be summed as follows:

(1) The Facial Expression recognition tools provided desirable results on the scanning and image analysis. The system also successfully implemented Euclidian Distance formula and FURIA fuzzy rules to extract necessary information from the acquired image or image sequence into one of 7 basic human emotions.

(2) By comparing the result when the system detect the emotion from JAFFE image set, with the one that already classified by experts, the system able to reaches the accuracy of 90%.

It can be seen that when used in a classroom activity, the system could make a significant difference to student's mood. This result could greatly vary from school to school, not to mention when used on different grades. Because there will be many other external variable that could affected the state of the students. However, from this research, the writer concluded that it is possible to have a scientific measurement toward student's stress level by using a system which could detect human emotions.

6. References

[1] Peter Lewinsohn, Hyman Hops, Robert Roberts, and John Seeley, "Adolescent Psychopathology: I. Prevalence and Incidence of Depression and Other DSM-III-R Disorders in High School Students," *Journal of Abnormal Psychology*, vol. 102, no. 1, pp. 133-144, March 1993.

[2] Shivakumar Gundappa and P.A.Vijaya, "Facial Expression Based Human Emotion Recognition with Live Computer Response," *IJCSIT International Journal of Computer Science and Information Technology*, vol. 4, no. 2, pp. 81-84, December 2011.

[3] Paul Ekman and Wallace Friesen, *Pictures of Facial Affect*. Palo Alto: CA: Consulting Pshycologist, 1976.

[4] NIH. (2014, January) The National Institute of Mental Health (NIMH). [Online]. https://www.nimh.nih.gov/about/index.shtml

[5] Ministry of Health Republic of Indonesia. (2016, October) Ministry of Health Republic of Indonesia official website. [Online]. http://www.depkes.go.id/article/print/16100700005/peran-keluarga-dukung-kesehatan-jiwa-masyarakat.html

[6] Brett Marroquín, Chloe C. Boyle, and Susan Nolen-Hoeksema, "Using Emotion as Information in Future-Oriented Cognition: Individual Differences in the Context of State Negative Affect," *Pers Individ Dif*, pp. 121-126, 2016.

[7] Abu Hanifah and Nunung Unayah, "MENCEGAH DAN MENANGGULANGI PENYALAHGUNAAN NAPZA," *Informasi*, pp. 33-46, 2011.

[8] Shumaila Khurshid, Qaisara Parveen, M. Imran Yousuf, and Abid Ghafoor Chaudhry, "EFFECTS OF DEPRESSION ON STUDENTS' ACADEMIC PERFORMANCE," *Sci.int.(Lahore)*, p. 27(2), 2015.

[9] Yuli Yanna Fauzie, "Bank Dunia: Kualitas Pendidikan Indonesia Masih Rendah," 2018.

[10] Mohamed Sathik and G.Sofia Jonathan, "Effect of facial expressions on student's comprehension recognition in virtual," *SpringerPlus*, pp. 1-9, 2013.

[11] Staff Writer. (2017) The Mountain Ear. [Online]. https://themtnear.com/2017/12/tech-trends-facial-recognition-history-and-privacy/

[12] Hamit Soyel and Hasan Demirel, "Facial Expression Recognition Using 3D Facial Feature Distances," *Affective Computing, Focus on Emotion Expression, Synthesis and Recognition*, 2008.

[13] Seyedehsamaneh Shojaeilangari, Wei-Yun Yau, Karthik Nandakumar, Jun Li, and Eam Khwang Teoh, "Robust Representation and Recognition of Facial Emotions Using Extreme Sparse Learning," *IEEE Transactions on Image Processing*, vol. 24, no. 7, pp. 2140 - 2152, July 2015.

[14] Paul Ekman and Wallace Friesen. (1993) PaulEkmanGroup Website. [Online]. https://www.paulekman.com/product/pictures-of-facial-affect-pofa/

[15] Braathen, B. Bartlett, M. Littlewort, G. Smith, and E. Movellan, "An approach to automatic recognition of spontaneous facial actions.," in *International Conferance on FGR*, 2002, pp. 345-350.

[16] Patrick Lucey et al., "The Extended Cohn-Kanade Dataset (CK+): A complete dataset for action unit and emotion-specified expression," *IEEE*, 2010.

[17] Liwei Wangy, Yan Zhangy, and Jufu Fengy, "On the Euclidean Distance of Images," *IEEE Transactions on Pattern Analysis and Machine Intelligence*, vol. 27, no. 8, pp. 1334 - 1339, August 2005.

[18] Moon Hwan Kim, Young Hoon Joo, and Jin Bae Park, "Emotion Detection Algorithm Using Frontal Face Image," *ICCAS2005*, pp. 2373-2378, 2015.

[19] Michael Lyons, Shigeru Akamatsu, Miyuki Kamachi, and Jiro Gyoba, "Coding Facial Expressions with GaborWavelets," in *IEEE International Conference on Automatic Face and Gesture Recognition*, Nara, 1998, pp. 200-205.

[20] Douglas S. Shafer and Zhiyi Zhang, *Beginning Statistics*. Charlotte: Creative Commons-licensed, 2012.

This page intentionally left blank.

CASE BASED REASONING: RECOMMENDATIONS OF HIGH SCHOOL STUDENT PROGRAM

Anisa Lutfiyani and Fatchul Arifin
Electronics and Informatics Engineering Education Postgraduate,
Yogyakarta State University, Yogyakarta, Indonesia

Email: an.lutfiyani@gmail.com, fatchul@uny.ac.id

Abstract. Study Program Recommendation System is a computer-based information system that aims to provide an overview or recommendation of study program based on students' interests or character from testing through Case Based Reasoning method. The main purpose of this article is to develop of expert system as study program recommendation for High School student through Case Based Reasoning. This system is made because many parents tend to force their health when choosing a college for their children. It can be impact on the children's developing. Determining the recommendations of a student's study program with a certain character, often not enough for the expert's knowledge, but need to pay attention to cases or events similar to those of the student. A concept held in the field of artificial intelligence that uses case-based or CBR reasons for analysis and decision making. Decisions are determined by calculating the similarities between the new cases and the expert case that has occurred. This study is observation and literature study. The results of this study showed that an engineering inference expertise with the aim of obtaining a decision of study program recommendations can be done using the CBR method based on the characteristics of students.

1. Introduction

Article 1 of the National Education System Law of 2003 explains that Senior High School is a formal education that prioritizes the expansion of knowledge by students to continue to a higher level.

High school / vocational high school students in class XII who plan to continue their education to higher education are sometimes confused to choose the appropriate study program. Often the choice of students does not match their actual interests, so it is unfortunate that a very large desire to enter college, is not mature enough in choosing study programs. Finally it is often heard that quite a lot of new students fail midway or drop out of school when they are accepted into college.

Data from Menristekdikti, there are several higher education institutions whose dropout rates are very large, namely: DKI Jakarta contributed 40,584 students dropping out of school in 2017 or 3.7% students in this province. West Java Province came in second place, with 21,146 dropouts or 2.5% students. Furthermore, North Sumatra, with a total of 12,418 students dropping out of school, 2.9% of the total number of students in the province. Meanwhile, DI Yogyakarta is in tenth position. Of 372.48 students studying in this city in 2017, there were 12,425 or 3.3% students who had to drop out of school.

Meanwhile, according to research conducted by Educational Psychologist from Integrity Development Flexibility (IDF) Irene Guntur, M. Psi., CGA was quoted from okezone.com (Harahap, 2014), as many as 87% of students in Indonesia experienced the phenomenon of taking the wrong study program, its extreme impact was the forerunner of unemployment. Thus, high school / vocational high school students grade XII need to have educational goals in the future and strive to realize real goals in the future. Therefore there is a need for actions that change the way of thinking and encourage prospective students to be more careful in choosing the right study program with their desires and interests and talents.

Based on the above problems, an expert system is needed to provide solutions to high school / vocational high school students who want to go to college without being confused about what study program they will choose based on their talent or characteristics through tests. This system will be built web-based with the method used is Case Base Reasoning.

Case Base Reasoning is a method of reasoning in the field of artificial intelligence (artificial intelligence) which is used to find the best solution of problems related to choosing one solution from many solutions (Daqing & Burrell, 2001). CBR can overcome all the problems mentioned earlier. Case-based CBR is easier to obtain than rules (Salem & HodHod, 2005). CBR can reuse previous results, from the beginning does not require a step by step reason, but their effectiveness will continue to develop by solving new problems (Aamodt & Plaza, 2009). The reasoning used in this study is the character of students while the conclusion is the results of the study program recommendations.

With this research, it is expected to develop an expert system to provide study program recommendation to high school students using case-based reasoning method with similarity measurement method nearest neighbor.

2. Research Method

2.1 Data Collection

This research was conducted in several stages, namely:

2.1.1 *Interviews*, conducted with experts namely psychology, this is to get basic knowledge about the characteristics of students needed in the process of making and developing systems.

2.1.2 *Literature*, carried out by collecting data or written texts in the form of scientific journals, research reports, e-books, and other relevant sources and collecting and studying reference books or literatures that discuss student characteristics and Case Based Reasoning methods.

2.2 Testing Method

2.2.1 Case-based Reasoning (CBR) Process

To determine the results of the study program recommendations, there are several stages of the process carried out, this stage is carried out to find the similarity of new cases with standard cases that are already stored in the database. The stages of the process that occurred at CBR can be seen in Figure 1, as follows:

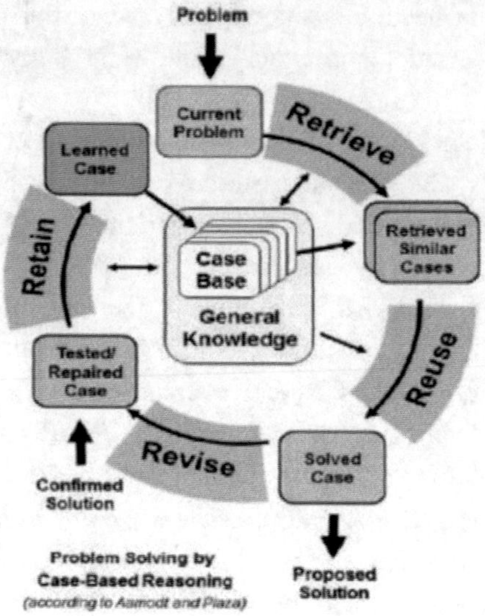

Figure 1. Case-Based Reasoning Cycle (Aamodt & Plaza, 1994).

In general, the case based reasoning method consists of 4 steps, namely:

1. Retrieve
   In this process is a problem recognition process, namely by identifying new problems.
2. Reuse
   In the Reuse process, the system will search for old problems in the database through identifying new problems. Then reuse the old problem information that has similarities to complete the new problem.
3. Revise
   In this process, information will be re-evaluated to overcome problems that occur in new problems. Then the system will issue a new problem solution.
4. Retrain
   In this last process, the system will save new problems into knowledge-based to solve future problems.

Based on the description above, CBR works by comparing case standards with new cases, if case standards are related to new cases, CBR will provide standard case solutions for new cases. Each problem contains a problem and a solution. If there is no match, CBR will divert or revise by inserting new cases into the database of cases, so that knowledge will increase indirectly.

2.2.2 *Inference Engine*

Processes in Case Based Reasoning (CBR) can use various techniques, including the Nearest Neighbor Algorithm. The nearest neighbor algorithm is an approach to look for cases with the closeness between new cases and standard cases, which is based on the weight matching of a number of existing features (Kusrini, 2009). This method looks for distances to the destination of data that has been previously stored. After getting the distance, then the closest distance is searched. The shortest distance is used to find the destination solution.

The formula used in the calculation of closeness (similarity) is as follows (Lubis, 2016):

$$\textbf{\textit{Similarity (problem, case)}} = \frac{s1*w1+s2*w2+\cdots....+sn*wn}{N1+N2+\cdots....+Nn} \quad (1)$$

S = Similarity (similarity value)
W = Weight (weight given)
N = Amount of Input New Entry

3. Results and Discussion

3.1 Flowchart Case Based Reasoning

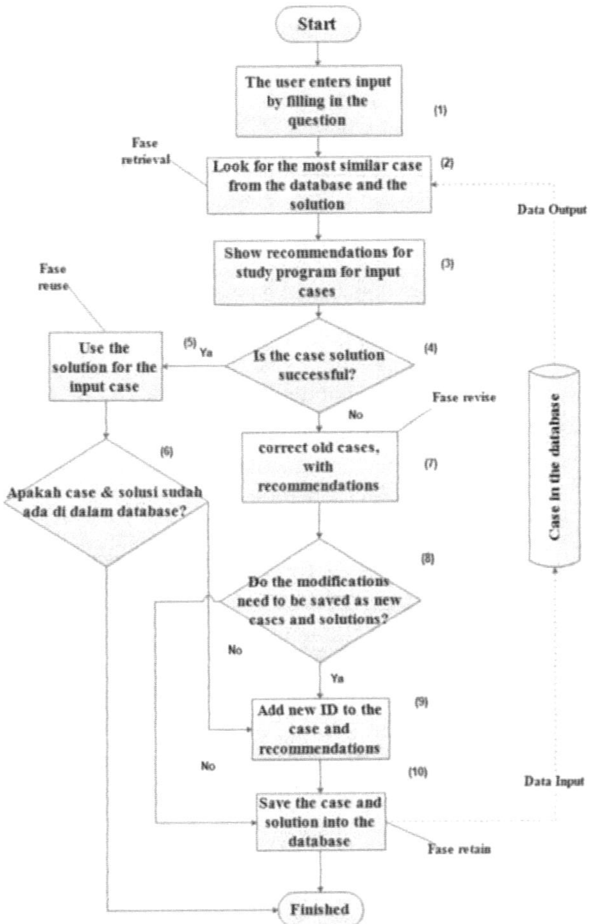

Figure 2. Flowchart Case-Based Reasoning (Yulianti, 2016).

3.2 Determination of Similarity Solutions

The basis of CBR is solving the information that has been stored in the previous case. Based on the stages in the CBR system, three main steps are needed in determining the similarity, (Lubis, 2016) namely:

3.2.1 *Build a case base*

Build a case base that will be used as storage. Each case stored in the case base is divided into 2 parts, namely:

Table 1. Part Factors in Each Case

Part factor in each case
Student character
Study Program Recommendations

This division is done to facilitate the retrieval of data in accordance with the new case, as well as to store new case data into the case base.

3.2.1.1 *Student character*, this section contains the character of students who determine student interest. The examples of some student characters are as follows:

Table 2. Character of students

Code	CHARACTER
K001	Like to play computers and discover how computers work.
K002	Likes to do systematic and thorough work..
K003	Like playing games that use strategies or sharpen the brain, such as chess, mystery games, puzzles, logic, etc.
K004	Easily remember numbers or statistics such as soccer scores, height of the tallest building etc.
K005	Happy to describe visual information in the form of graphics, tables or images.
K006	Like to use technology to solve mathematical problems.
K007	Like to play and find ways to improve electronic equipment.
K008	Like to think critically in analysis.
K009	Like to study or work alone.
K010	Like to do research.
K011	Like to think about the future and goals to be achieved.
K012	Like at home and spend time alone.
K013	Like to contemplate or think about life or myself.
K014	Knowing my strengths and weaknesses.
K015	Like writing diaries to express ideas, memories, Feelings of events that occur.
K016	Like to think or consider the positive or negative side when encountering problems or making decisions.
K017	Like math or counting
K018	Not always using fingers or tools (such as calculators etc.) in counting.
K019	Like to find opportunities that may or can be found based on formulas, formulas that already exist.
K020	Like to read books
K021	....

3.2.1.2 *Recommended Study Program*, this section contains recommendations for study programs that determine student interest. This section has several study programs, namely:

Table 3. Study Program Recommendations

CODE	NAME PRODY
J001	Informatics / Computer Engineering
J002	Electrical Engineering
J003	Psychology
J004	Mathematics
J005	Language
J006	Health
J007	Physics
J008	Economic/accounting
J009	Music Art
J010	Sports
J011	Geography
J012	Law
J013	History
J014	Chemistry
J015	Biology

3.2.2 Determine The Similarity Function

The process of determining the similarity of new cases by adapting existing case standards, there are 4 processes in determining similarity, namely:

3.2.2.1 Retrieve. At the beginning of the process of determining the recommendation of a study program, the user fills in the student biodata, after that it is continued with the user inputting the answer from the questionnaire to be answered. The user fills out the questionnaire consisting of 65 questionnaires in the form of student characters, the user finishes answering the questionnaire, the user will see the results of the study program recommendations that have passed the case-based reasoning process, search for similarities with the previous case standards, in the case memory to find out the recommendations of the study program. Then the system will do the weighting process by matching one by one between the characters that have been entered with those in the knowledge base. After carrying out the weighting process, the highest similarity value is determined for the results in determining the study program. The weighting carried out by the system is displayed in the calculation of the following nearest similar similarity algorithm (Lubis, 2016):

Parameter weight (w):
Selected features (equal) = 1
Selected features (not the same) = 0
Unselected features = 0

$$\textbf{\textit{Similarity (problem, case)}} = \frac{s1*w1+s2*w2+\cdots....+sn*wn}{N1+N2+\cdots....+Nn} \qquad (2)$$

S = Similarity (similarity value)
W = Weight (weight given)

N = Amount of Input New Entry

N is filled with student characters For example:

K001, K004, K011, K014, K020, K028, K040 then n counts as many as 7 new case entries.

3.2.2.1.1 Old Case Tables

In order for the new case to be calculated as close to the old case standard, it is necessary to define the weight of the feature. Defining the weight of CBR features is shown as shown below:

Table 4. Defining the Informatics Engineering Study Program

Code	Character	Similarity	Similarity Value
K001	Like to play computers and discover how computers work.	Same	1
		Different	0
K002	Likes to do systematic and thorough work.	Same	1
		Different	0
K003	Like playing games that use strategies or sharpen the brain, such as chess, mystery games, puzzles, logic, etc.	Same	1
		Different	0
K004	Easily remember numbers or statistics such as soccer scores, height of the tallest building etc.	Same	1
		Different	0
K005	Happy to describe visual information in the form of graphics, tables or images.	Same	1
		Different	0
K006	Like to use technology to solve mathematical problems.	Same	1
		Different	0
K009	Like to study or work alone.	Same	1
		Different	0
Study Programs		**Informatics Engineering/ Computer**	

Table 5. Defining of Electrical Engineering Study Programs

Code	Character	Similarity	Similarity Value
K007	Like to play and find ways to improve electronic equipment.	Same	1
		Different	0
K008	Like to think critically in analysis.	Same	1
		Different	0
K009	Like to study or work alone.	Same	1
		Different	0
K010	Like to do research.	Same	1
		Different	0
K002	Likes to do systematic and thorough work	Same	1
		Different	0
K005	Happy to describe visual information in the form of graphics, tables or images.	Same	1
		Different	0
K006	Like to use technology to solve mathematical problems.	Same	1
		Different	0
Study Programs		**Electrical Engineering**	

Table 6. Defining the Psychology Study Program

Code	Character	Similarity	Similarity Value
K011	Like to think about the future and goals to be achieved.	Same	1
		Different	0
K012	Like at home and spend time alone.	Same	1
		Different	0
K013	Like to contemplate or think about life or myself.	Same	1
		Different	0
K014	Knowing my strengths and weaknesses.	Same	1
		Different	0
K015	Like writing diaries to express ideas, memories, Feelings of events that occur.	Same	1
		Different	0
K016	Like to think or consider the positive or negative side when encountering problems or making decisions.	Same	1
		Different	0
K047	Likes to think of solutions to problems that are happening.	Same	1
		Different	0
Study Programs		**Psychology**	

3.2.2.1.2 New Case Table

An example of a new case table, containing student characters that have not been stored in the case memory. Examples of new cases that will be matched to look for similarities with the standard old case in recommending the study program are:

Table 7. New Case Tables

Code	Character
K001	Like to play computers and discover how computers work.
K004	Easily remember numbers or statistics such as soccer scores, height of the tallest building etc.
K006	Like to use technology to solve mathematical problems.
K014	Knowing my strengths and weaknesses.
K020	Like to read books
K028	Nice to do scientific experiments so that I can train my creativity.
K040	Prefer to move, to learn something that makes it easier for me to remember.

3.2.2.1.3 Calculations using the Nearest Neighbor Retrieval Algorithm

At this stage a case calculation will be carried out in which the calculation of the new case with this old case aims to find similarities, namely using the weighting process using the Nearest Neighbor Retrieval formula.

Journal of Physics: Conference Series **1413** (2019) 012022 doi:10.1088/1742-6596/1413/1/012022

Calculation of Case I

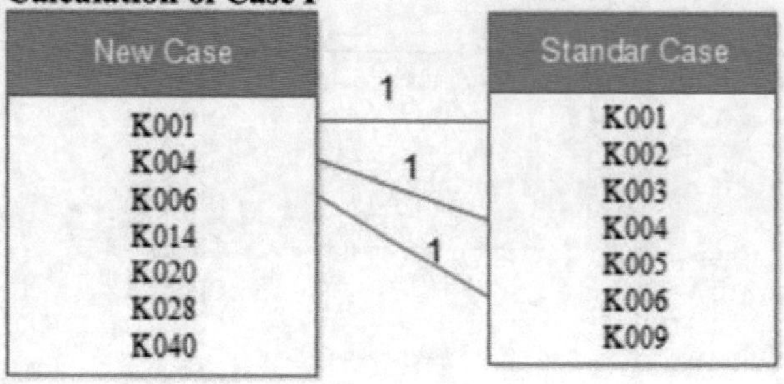

Similarity (x, K1)

$$= \frac{(1*1)+(0*0)+(0*0)+(1*1)+(0*0)+(1*1)+(0*0)}{1+1+1+1+1+1+1}$$

$$= \frac{1+0+0+1+0+1+0}{7}$$

$$= \frac{3}{7} = 0.43$$

Calculation of Case II

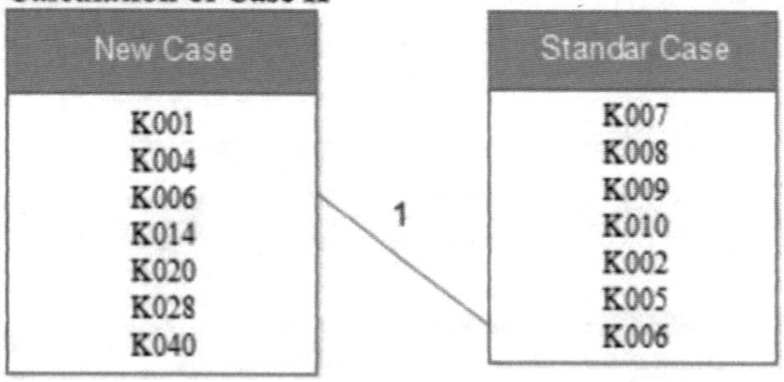

Similarity (x, K2)

$$= \frac{(0*0)+(0*0)+(0*0)+(0*0)+(0*0)+(0*0)+(1*1)}{1+1+1+1+1+1+1}$$

$$= \frac{0+0+0+0+0+0+1}{7}$$

$$= \frac{1}{7} = 0.14$$

Journal of Physics: Conference Series **1413** (2019) 012022 doi:10.1088/1742-6596/1413/1/012022

Calculation of Case III

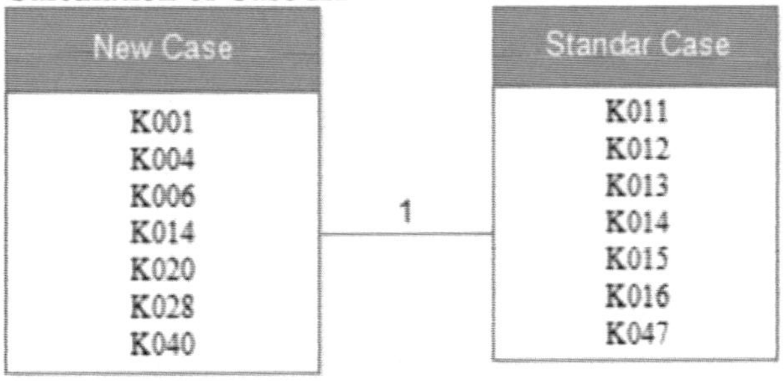

Similarity (x, K3)

$$= \frac{(0*0)+(0*0)+(0*0)+(1*1)+(0*0)+(0*0)+(0*0)}{1+1+1+1+1+1+1}$$

$$= \frac{0+0+0+1+0+0+0}{7}$$

$$= \frac{1}{7} = 0.14$$

After entering the value, the new case will be compared with each standard case as in table 4. The results of the calculation for the similarity of each case stored in the knowledge base with the new case are as follows:

Table 8. Amount of similarity values with new cases

Case Base	Total
K1	3/7 = 0.43
K2	1/7 = 0.14
K3	1/7 = 0.14

3.2.2.1.4 *Data Retrieval*

Criteria for selecting cases are cases that have the highest similarity value with new cases submitted as a solution. The following are the similarity criteria for the division, namely:

Table 9. Similarity Criteria

Similar Decimal Value	Similarity Value
0,8 – 1	High
0.4 – 0.79	Medium
0 – 0.39	Low

4. Conclusion

Based on the table of similarity criteria, each case on the basis of the case has similarity criteria with the new case as follows:

(1) K1 has a similarity of 0.43, the value of similarity of Medium criteria
(2) K2 has a similarity of 0.14, the value of similarity Low criteria
(3) K3 has a similarity of 0.14, the value of the similarity of Low criteria.

Therefore, the results of the criteria for the similarity of new cases will be the suggested solution for new cases, if the level of similarity reaches a minimum of 80%. Because the criteria for the similarity of new cases with the old case K1 are Medium, while K2 and K3 are Low so the solution of each old case is not recommended for the solution of existing new cases, in other words, the new case is probably not the same as the case on the knowledge base. But in this case the value of the K1 Informatics Engineering / computer is the highest value compared to the other cases.

5. References

[1] Aamodt, A., dan Plaza E., (1994), "Case Based Reasoning: Foundational Issues, Methodological Variations, and System Approaches". *Journal of Case-Based Reasoning: Foundational Issues, Methodological Variations, and System Approaches. AI Communications. IOS Press*, Vol. 7: 1, pp. 39-59.

[2] Daqing. C. and Burrerel, P., (2001), *Case Based Reasoning System and Artificilan Neural Network*: A review, Springer-Verlag.

[3] Harahap, R. F., "okezone.com," MNC, 25 February 2014, [Online], Available: http://news.okezone.com/read/2014/02/24/373/945961/duh-87-mahasiswa-indonesia-salah-jurusan. [Diakses 2019].

[4] Kemenristekdikti, (2017), Statistik Perguruan Tinggi *(Higher Education Statistical Year Book),* Jakarta: Pusdatin Iptek Dikti.

[5] Kusrini dan Emha, (2009), *Algoritma Data Mining*. Yogyakarta : C.V Andi Offset.

[6] Law of the Republic of Indonesia Number 20, (2003), Concerning the National Education System.

[7] Lubis,A. (2016), Case-Based Pregnancy Consultation Expert System Using Case Based Reasoning (CBR) Method, Universitas Pasir Pengairan, *Riau Journal Of Computer Science* **Vol. 2**/No.1/2016 : 66 – 75.

[8] Salem . M., Roushdy. M and HodHod. R.A., (2005), A Case Based Expert System for Supporting Diagnosis of Heart Diseases, AIML, *Journal*, Volume (5), Issue 1, March.

[9] Yulianti, W., (2016), Case-Based Reasoning Aptitude Testing in Expert Systems to Determine Interest and Talent of Elementary School Students, *Jurnal Teknologi dan Sistem Informasi* UNIVRAB **Vol. 1** No. 2, Juli 2016.

The Effect of Blended Learning Model to Improve the Conceptual Understanding of Computer and Network Engineering Students

F M Hawi[1] and P Sudira[2]
[1]Electronics and Informatics Engineering Education Department,
Postgraduate Program, Universitas Negeri Yogyakarta, Yogyakarta, Indonesia.
[2]Vocational Technology Education, Postgraduate Program,
Universitas Negeri Yogyakarta, Yogyakarta, Indonesia.

Email: xfmh26hawix@gmail.com

Abstract. Grounded by students' lack of conceptual understanding in identifying and comprehending basic concepts of computer network, this quasi-experimental study aims to identify the effect of applying blended learning model toward the improvement of students' conceptual understanding. It was designed using non-equivalent control group design in which the experimental group and the control group were differently treated using, respectively, blended learning and conventional learning model. There were 58 students selected as the sample of the research; 30 students in experimental class and 28 students in control class. Essay test was chosen to collect the data of the conceptual understanding of the students in both groups. Using MANOVA, the data analysis revealed the significant effect of applying blended learning model toward students' conceptual understanding with $P = 0.000 < 0.05$ of significance level. In conclusion, the result of study identified that 1) the students who taught using blended learning model performed better mastery of basic concepts of computer network with the average score was 74.86, valued in high category; 2) there was a significant difference of conceptual understanding improvement between the group taught by applying blended learning and those by conventional learning model in learning basic computer network.

1. Introduction

The exponential growth of information and communication technology (ICT) in this 21st century has beneficially contributed to many areas. In educational field, an attempt to utilize ICT within instructional methods and strategies has been increasingly developed for the purpose of improving the quality of instruction and expanding students' learning experience. Those who endorse the use of ICT in educational context have grounded their ideas on that ICT has a profound impact which offers new ways of producing, distributing and receiving a learning content (Wang, et. al, 2019). The use of ICT in teaching learning process has also maximized the stage for learning that occurs not only inside a classroom through face-to-face experience but also through the outside-distance learning.

With respect to the fruitfulness of employing ICT in educational field, many believe that learning is better conducted in between the use of ICT and face-to-face experience due to the fact that each extreme has its own particular weaknesses. Kanuka & Anderson (2007), for instance, affirm that using ICT without direct instruction does not allow direct physical interaction among students and students-teacher interactions. This neglects the importance of direct verbal and non-verbal communication which are more able to perceive and sensitize students' different psychological development like feelings, emotions, and motivations that highly support and determine students' learning achievement. Additionally, ICT alone lack the devices and abilities to develop complex cognitive and social skills and may not fully comply with faculty and individual needs for learning (Wright, 2017). On the other side, face-to-face in-class instruction also has weaknesses in many aspects. One most obvious weakness of it is related to the limitation of learning access. This instruction tends to posit teacher as the main, if not the only, source of knowledge which in turn may lead to students' lack of autonomy in learning and accessing learning content. In line with this limitation, face-to-face instruction also ignores the potentialities of exploring myriad and abundant learning resources offered by other than the teacher who is standing and lecturing in front of the class. Still, learning is then constricted by the only resource the teacher has given to the students to learn from.

For this reason, the idea to combine and integrate ICT in traditional learning activity, commonly termed as blended learning (Wright, 2017), has been a growingly interested learning model in many disciplines worldwide, as well as in Indonesia. Bhonk dan Graham (in Thorne, 2003) define blended learning as *"the combination of instruction from two historically separated models of teaching and learning traditional learning systems and distributed learning systems which emphasizes the central role of computer-based technologies"*. It is believed to play a significant role to assist and optimize face-to-face teaching and learning process (Wang, et. al, 2019) by providing additional platforms for learning outside the classrooms where students can self-directedly access learning materials they need. More specifically in practice, blended learning combines the benefits of collaborative, independent, and problem-based learning to reach the broadest range of learning types by involving providing physical and virtual environments and a wide variety of media (Lothridge, 2013).

Studies related to applying blended learning in Indonesian context have proved its significance to enhance learning which in turn enhancing students' conceptual understanding. Conducting an R&D study subjecting physics education college students, Suana et.al (2017) confirmed that schoology-based blended learning media is able to improve students' conceptual understanding and problem-solving ability. They argued that the effectiveness of this approach is contributed by its ability to promote students' autonomy to access learning material. Added to this, Isti'anah (2017) reported that not only does blended learning promote students' conceptual understanding but also their interest and motivation to learn. In her study using Moodle, a kind of online learning management system (LMS), she found that students' interest and motivation in blended learning instruction leads positively to the development of students' cognitive behavior. Meanwhile, Ekayati (2018), in her qualitative study of applying Edmodo for blended learning instruction, proposed that blended learning is widely making possible to escalate learning quality and intensity that may result positively to students' conceptual understanding.

Drawing attention by these abundant potentialities, this study is conducted to investigate the effectiveness of applying blended learning to improve conceptual understanding of computer and network engineering vocational school students. The empirical problems have been identified: 1) the students lacked of conceptual understanding of basic computer material that hampers their performance in designing computer network. It was proved by students learning achievement where their scores only ranged from 50 – 65, far below the passing grade standard (70); 2) the use of conventional lecturing model was not be able to stimulate students' motivation and interest to learn. This ironic situation is caused by teachers' ability and opportunity to provide ICT integrated material; and 3) learning supporting devices and tools were not fully optimized to create effective learning activities. In line with this, the research problem of this study is formulated as follows:
How does blended learning model effect to the improvement of students' conceptual understanding?

2. Research Methods

This research was to investigate the effectiveness of blended learning model in improving conceptual understanding of computer and network students. It was designed quasi-experimental research of nonequivalent control group design. The research subject was 58 students of class XI of Private Vocational School of Informatics Engineering Santo Petrus Ruteng in 2018/2019 academic year. They were divided into two groups: 30 students were in the experimental group and 28 students were in the control group. Blended learning, the investigated learning model, was treated to experimental class with the procedures displayed in table 2. Meanwhile for the control class, conventional lecturing model was applied. The learning topic chosen for both groups was basic computer network materials. The instrument of data collection using in this study was essay test of students' conceptual understanding. After collecting, the data was then analyzed using MANOVA assisted by SPSS 25.

Table 1. The Experimental Research Design

Group	Pretest	Treatment	Posttest
Experiment	O_1	X_A	O_2
Control	O_3	X_B	O_4

(Adaptasi Campbell & Stanley, 1963)

E : Experimental Group
K : Control Group
O_1 : Pretest of Experimental Group
O_2 : Posttest of Experimental Group
O_3 : Pretest of Control Group
O_4 : Posttest of Control Group
X_A : Treatment of Experimental Group
X_B : Treatment of Control Group

Table 2. Design of Blended Learning

Activity	Phases of Blended Learning Model	Steps for Learning the Blended Learning Model	Times	Platform
Introduction	First phase, *seeking of information*	The teacher asks students to download the materials previously uploaded by the teacher in the used learning platform.	10 minutes	*Google Class Room*
Core activities	Second Phase, *acquisition of information*	The teacher explains the learning competence and objectives to initiate students' readiness and prepare them to explore the relevant materials through face-to-face learning as well as technology-assisted blended learning.	5 Minutes	
		The teacher explains and demonstrates the steps of PAN and LAN network simulation, while students are asked to observe the activity. The teacher confronts students' idea or thought after the observation with the result of their interpretation of various-provided sources in the learning platform.	30 Minutes	*Cisco Packet Tracer*
		The teacher assigns students to do exercises about PAN and LAN computer network simulation and asks them to discuss in groups to accomplish the task.	30 minutes	*Cisco Packet Tracer*

		The teacher asks students to send their works via email.		*Google Mail*
Closing	Third Phase, *synthesizing of knowledge*	The teacher helps students synthesizing knowledge into their cognitive structure. The teacher and students make a conclusion about ways of designing computer network. Students reflect their learning mastery by making a record.	15 minutes	

3. Result and Discussion

3.1. Posttest Data of Students' Understanding of Concept

The posttest is provided to know students' understanding of concept after applying a treatment. The posttest scores of understanding of concept is as shown in Table 3.

Table 3. Posttest Result of Students' Understanding of Concept

Class	Sample	Results		Average
		Maximum Score	Minimum Score	
Experimental group	30	84.00	64.00	74.86
Control group	28	78.00	58.00	67.89

The analysis results of students conceptual understanding after implementing blended learning model showed the average mark of 74.86; the highest mark was 84.00 and the lowest one was 64.00. The students were grouped based on the initial ability by referring to the pretest analysis results. The groups were divided into two categories, namely, a. the group of students with the high initial ability; b. the group of students with the low initial ability. This categorization was resulted from initial calculation interval based on the difference between the highest score with the lowest score. The difference was then divided by the sum of two appropriate categories. After the interval specified, then students were classified according category the intervals. The students in the experiment group, with high and low initial ability were 45.00% and 55.00% respectively. While in the control group they were 35.00% and 65.00% respectively.

3.2. The effect of applying blended learning model toward students conceptual understanding.

Before taking into the statistical computation, the hypothesis of this research problem is formulated as follows:

H_0 : There is no significant difference of students' conceptual understanding between those treated using blended learning model and those treated using conventional learning model.

H_a : There is significant difference of students' conceptual understanding between those treated using blended learning model and those treated using conventional learning model.

Table 4. Hotteling's Trace Testing Result

Effect		Value	F	Hypothesis df	Error df	Sig.
Intercept	Pillai's Trace	.967	799.221^b	2.000	2.000	.000
	Wilks' Lambda	.033	799.221^b	2.000	2.000	.000
	Hotelling's Trace	29.063	799.221^b	2.000	2.000	.000
	Roy's Largest Root	29.063	799.221^b	2.000	2.000	.000

Class	Pillai's Trace	.351	14.862^{b}	2.000	55.000	.000
	Wilks' Lambda	.649	14.862^{b}	2.000	55.000	.000
	Hotelling's Trace	.540	14.862^{b}	2.000	55.000	.000
	Roy's Largest Root	.540	14.862^{b}	2.000	55.000	.000

Based on the table 4, the result of Hotteling's Trace testing was significantly different in which the significance level (P) was 0.000 < 0.05. The result of statistical analysis rejects the null hypothesis (H_0) and confirms the alternative hypothesis (H_a). Therefore, the application of applying Blended Learning model effects positively toward conceptual understanding of class X Private Vocational School of Informatics Engineering St. Petrus Ruteng.

Blended learning can improve conceptual understanding because this model helps students to more develop in the learning process, in accordance with learning styles, preferences in learning provides realistic and practical opportunities for teachers and students. That makes the learning be more independent, useful, and grow (Prabowo, 2014). Blended learning model is not monotonous but it is always evolving to improve training. Lessons become interactive sessions that teachers and students look forward to and enjoy. Blended learning enables students to stay focused for longer which brings them academic success. Digital learning mechanisms can sustain the interest of students through graphics, videos and other methods. Teachers have the freedom to create instructional videos and give real-time feedback that promotes engagement with students.

The recognizable impact from blended learning that had been conducted are 1. The students are more enthusiastic with blended learning because the learning process becomes more interesting and not monotonous, (2) by implementing blended learning, teachers can observe students' activities out of school hours and (3) the students can explore other learning resources besides in classes. It is in compliance with Dwiyogo's statement (2010), that blended learning is the best choice to improve effectiveness, efficiency and attractiveness in humans' interaction in various learning environment. Blended learning offers opportunities to learn in groups and to learn individually, as well as in the same or in different time.

The consideration to organize the blended learning composition to be 50/50, 75/25 or 25/75 depends on targeted competence analysis, subject's goal, learner characteristic, face-to-face interaction, online or combination learning strategies, learner's location, teacher's competence and characteristic and available resources (Dwiyogo: 2010). Based on cross analysis toward those considerations, teachers will be able to decide the most suitable learning composition (percentage). Nevertheless, the main consideration in planning learning composition is providing the learning resources which is suitable for various learners 'characteristics in order to learn more effective, efficient and interesting.

3.3. Students' response on the application of blended learning model

To identify students' response on the application of blended learning model, the Likert scale questionnaire was analyzed using descriptive statistics, assisted by Microsoft Excel program. The questionnaire was distributed to the students in experimental group to reveal students' response related to blended learning model in two main aspects: 1) students' response related to learning strategies in blended learning model and 2) students' response regarding the efficacy of applying blended learning model to solve problem in the classroom. The result of the analysis of the questionnaire is displayed in figure 1.

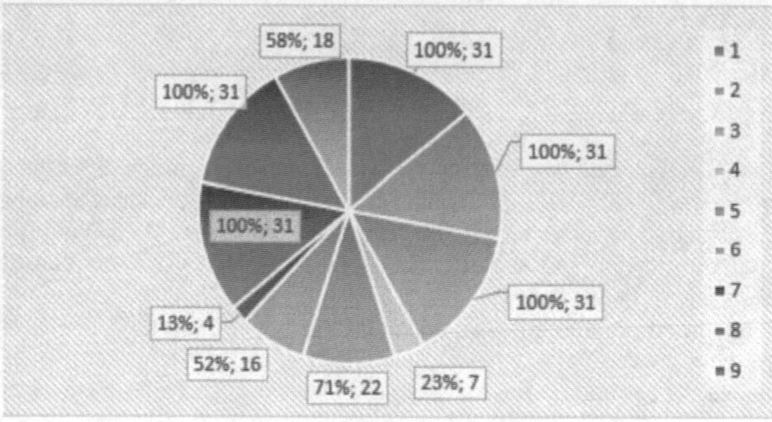

Figure 1. Students' response on the application of blended learning model

The data analysis of the questionnaire shows that most of the students in experimental group have positive response to the application of blended learning model in the classroom. In term of learning strategies of blended learning model, the result of data analysis shows students' high enthusiasm to engage actively in blended learning. This finding endorses the theory saying that blended learning model turns the learning approach in the classroom from teacher-centered to student-centered learning (Park & Ertner, 2007) and results positively to students' active and independent learning (Tarhan & Ayyıldız, 2014; Ramstedt et.al., 2016). Dealing with the efficacy of blended learning to solve students' classroom problem, the data analysis also reveals students' high belief on blended learning model. The students have high belief on blended learning model that can encourage them to be involved in learning real-life problems, enriching problem-solving skills, encouraging multidisciplinary learning, fostering independent learning, facilitating learning to search for information, and enhancing learning of conceptual understanding.

4. Conclusion

In conclusion, as the quality of learning process is characterized by its ability to bridge students' knowledge gap, treating students to involve actively in its process has now become an ideal, yet fruitful learning model. Studies related to applying blended learning in Indonesian context have proved its significance to enhance learning which in turn enhancing students' conceptual understanding. Blended learning is widely making possible to escalate learning quality and intensity that may result positively to students' conceptual understanding. This study confirms that challenging and engaging students' self-directed learning of problem-related concepts is able to accelerate students' conceptual understanding. By its reciprocal relationship, students' conceptual understanding may improve gradually as the result of the precise application and the frequent use of blended learning model in classrooms.

5. Reference

[1] Campbell D T and Stanley J C 1963 *Experimental and Quasi-Experiment Designs for Research* USA: Houghton Mifflin Company.

[2] Dwiyogo W D 2010 *Pembelajaran Blended Learning.* Accessed on 22 July 2018 at http://id.wikibooks.org/wiki/PembelajaranBerbasisBlendedLearning

[3] Ekayati R 2018 *Implementasi metode blended learning berbasis aplikasi edmodo.* Jurnal EduTech, *4(2),* pp *50 – 56.*

[4] Isti'anah A 2017 *The Effect of Blended Learning to the Students' Achievement in Grammar Class.* IJEE (Indonesian Journal of English Education), *4*(1), pp 16-30 doi:10.15408/ijee.v4i1.5697

[5] Kanuka H and Anderson 2007 *Online social interchange, discord, and knowledge construction.* International Journal of E-learning & Distance Education, *13(1)*, pp *57–74*.

[6] Lothridge K, Fox J, and Fynan E 2013. *Blended learning: efficient, timely and cost effective.* Australian Journal of Forensic Sciences, 45(4) pp 407–416. doi:10.1080/00450618.2013.767375

[7] Park S H and Ertmer P A 2007 *Impact Of Problem-Based Learning (Pbl) On Teachers' Beliefs Regarding Technology Use.* Journal of Research on Technology in Education, 40, 247 – 267.

[8] Prabowo T H 2014 *Implementasi model pembelajaran blended learning untuk meningkatkan aktivitas belajar multimedia siswa kelas XI Multimedia 1 SMK Muhammadiyah 2 Klaten Utara tahun ajaran 2013/2014.*

[9] Suana et.al. 2017 *Design and implementation of schoology-based blended learning media for basic physics I course.* Jurnal pendidikan IPA Indonesia, 6(1), 170-178, doi: 1015294/jpii.7205.

[10] Tarhan L and Ayyıldız Y 2014 *The views of undergraduates about problem – based learning applications in a biochemistry course.* Journal of Biological Education, 1 – 11.

[11] Thorne K 2003 *Blended Learning: How to integrate online and traditional learning* (London: Kogan Page)

[12] Wang N, Chen J, Tai M, and Zhang J (2019) *Blended learning for Chinese university EFL learners: learning environment and learner perceptions. Computer Assisted Language Learning,* pp 1–27. (doi:10.1080/09588221.2019.1607881)

[13] Wright B M (2017) *Blended learning: Student perception of face-to-face and online EFL lessons.* Indonesian journal of applied linguistics pp 64-71, (doi: http://dx.doi.org/10.17509/ijal.v7i1.6859)

Learning Media of Electric Lighting Installation Based on *Mobile Augmented Reality*

M Iskandar and M Khairudin
Electrical Engineering Education Department, Universitas Negeri Yogyakarta, Yogyakarta, Indonesia

Email: nandarenergi@yahoo.com, moh_khairudin@uny.ac.id

Abstract. Learning Media of Electric Lighting Installation Based on Mobile Augmented Reality is a technology that can be used as a substitute for conventional learning media to modern virtual learning media. This can be accessed anywhere and anytime by using an Android smartphone. Augmented reality is a technology that helps users to be able to interact directly with virtual components that have been designed to resemble the original. This study applies augmented reality technology to learning aids called learning media at the concentration of household electrical lighting installations. In the operation of learning media, several procedures must be carried out before becoming a product that is ready for use. The results of this study is a media for studying household electric lighting installations assisted by mobile augmented reality.

1. Introduction

Learning media is used as a tool to convey the knowledge of an educator to students in the learning process. Procurement of learning media requires careful planning at the beginning of the semester and the purchase of components and practical tools will issue a large school budget. In a good learning process, the tools in the form of real components and practical tools are very helpful for students to more quickly understand the material presented.

Nowadays one of the most important devices you can use to help with your life is a smartphone. In addition to helping in communication, smartphones are also used for various things and can be accessed anywhere and anytime. Android is an Operating System (OS) that is widely used on smartphone devices. Besides having many applications that can be downloaded for free, the Android OS is chosen by application developers because the facilities are very easy and support their use.

Virtual technology has spread to various lines of life and sections, one of the most widely developed technologies in this era is Augmented Reality and Virtual Reality. The use of virtual technology is widely developed in applications and gaming devices - virtual technology is very familiar to game lovers. One of the famous Augmented Reality Technologies in early 2016 is a game called Pokemon Go - known until June 2019, this game was developed by Niantic. Inc has been able to attract the attention of Android users and has been downloaded by 11,887,394 people on the Google Play Store. Augmented

Reality technology is a technology where a device can display a 3-dimensional visual form that looks like reality by utilizing the camera features on a smartphone device. The use of AR as a learning medium can help streamline budget expenditures for procurement of practical tools and also make it easier to learn the components contained in learning media.

The electric lighting installation is a subject in the competency expertise of Electrical Installation Engineering in Vocational High Schools. In this subject, students learn about introduction of lighting components and types of lighting installations, introduction of components and the types of installation. This is very important for students as a guide so that errors in the installation of lighting can be reduced.

Sudjana, N [1] states that media is one of factors in learning success of students. The activities and processes learned occur, information is transferred from the source to the recipient of information through certain models and media. Learning with models and interesting media will provide convenience for students in achieving learning goals.

Schramm [2] states, in the learning process, it is not enough just to use textbooks or images as learning media so as to make learning less enthusiastic. Teachers must develop their own learning media that are feasible and attract the attention of students, economical, effective and easy in the manufacturing process. The teacher as a facilitator must be able to provide facilities that enable the ease of the learning process.

Persson et al [3] describes mobile technology used in mobile learning. The concept of mobile learning has five characteristics, these are probability, accessibility, personalization, connectivity, and can increase learning motivation. The characteristics described in the statement above are in accordance with the learning media of the lighting installation developed, the media developed are also based on mobile learning.

Triatmaja et al. [4] stated that with the use of smartphones with Android OS in the current era which has become a necessity for every person to communicate without knowing space and time, not least in the world of education, smartphones are a necessity for students to communicate so that media development learning by using a smartphone as its basis will be able to improve learning effectiveness.

Dengzhe Ma et al. [5] describes that AR is the main technology for virtual engineering. AR is the basis for functional virtual prototyping, which allows developers to analyze functional forms and behavior of future products in an immersive and interactive virtual environment. The application of Augmented Reality technology greatly improves communication in product design and production development because this technology helps to identify and avoid design errors in the early stages of the development process and also saves time and money.

Jaya [6] describes the important functions of laboratories and practice tools in technical education because they provide practical knowledge for students. Conventional laboratories are available but limited to a large number of students. Behind its practical functions, laboratories suffer losses, the main disadvantages of laboratories in general are expensive equipment and instruments. The solution to the problem mentioned above is to use a virtual laboratory to help students develop their practical skills.

Odeh et al. [7] explained the advantages of virtual media, which are easy to use, the concept of theory becomes easier to understand, flexible scheduling, the theory of knowledge can be satisfied, a safe environment, advancement of new skills, driven by group cooperation, a comfortable place , more time to experiment.

Jaya [8] explains, there are advantages of Augmented Reality as a solution in practical activities which include: virtual environments can save budgets rather than physical environments, virtual environments are safer than physical environments, interactions through virtual environments provide certain experiences that are not obtained in the physical environment , the practice in the virtual environment can be more developed than the physical environment, the concept of learning takes the form of direct experience, the virtual environment gives a real impression such as the physical environment. Virtual media can increase student motivation in terms of task values and self efficacy [9].

Ogungbenro et al. [10] explained some of the advantages of using emulators on digital techniques, namely native appearance; the nuances and behavior of digital objects remain; initial capital on the production of emulators is greater, but other expenses will be more efficient; reduce laboratory use time; many emulators that have been well developed and licensed; emulators can be used on various device systems; the physical needs of the emulator can be reduced so that it can reduce the cost and time of practice; most emulators have large data memory and can provide many simulated electronic components [11].

Agus Suryanto et al. [12] explained about mobile technology which was developed using augmented reality technology by raising conventional learning problems in lathe learning. the teacher only explained a lathe in front of the class so that not all students understood the characteristics of the lathe so that from these problems a learning media was developed by introducing lathes with augmented reality technology with research and development (R & D) research procedures.

Mohd Erfy Ismail et al. [13] explained in the development of learning media Las Technology assisted by Aumented Reality (AR) in Vocational High Schools needed to create a learning atmosphere that was able to attract students' interest in learning, able to improve student visualization and reduce the cognitive burden experienced by students. In addition, teaching and learning is more interesting because the teacher's teaching style uses theory and practice simultaneously. In addition, the use of AR in education can improve the visualization of students in an abstract manner, which helps them when conducting experience tests and building a safe teaching and learning environment.

2. Learning Media Design

2.1. The Concept of Learning Media

This learning media will use Augmented Reality technology. The interface and learning features in it are made to resemble the actual form of components and work systems. There are three main menus in this learning media, namely Petunjuk, Materi and Profile. A video simulation in the Petunjuk Penggunaan menu explains how to use learning media, so that learning media can be understood and can be used by new users. The Materi menu is the core menu of this learning media. This menu will ask for connection of the camera features on the smartphone. The workings of the Materi menu are like someone who is using a camera, when a smartphone is directed at an object in the form of a marker, certain objects will appear on the smartphone screen. These objects are related to material competencies in the subject of electric lighting installation. The last menu is Profile that displays profiles of researcher and research supervisors.

2.2. Media Design Display

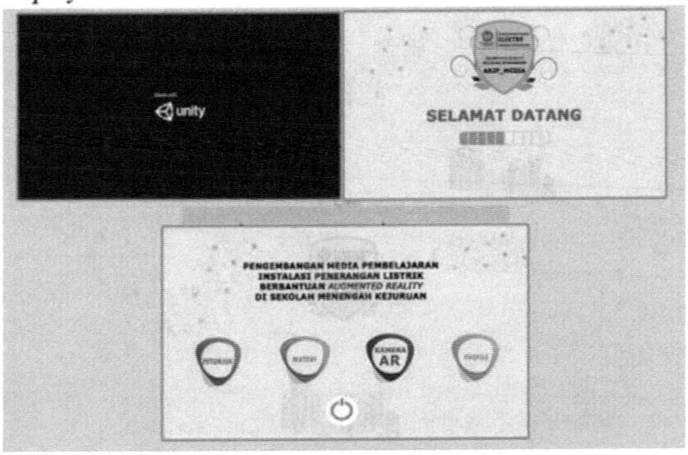

Figure 1. Display of the Main Menu

Figure 1 shows the display of Learning Media of Electric Lighting Installation Based on Mobile Augmented Reality. In the main menu display, there are three menus, namely Petunjuk, Materi and Profile.

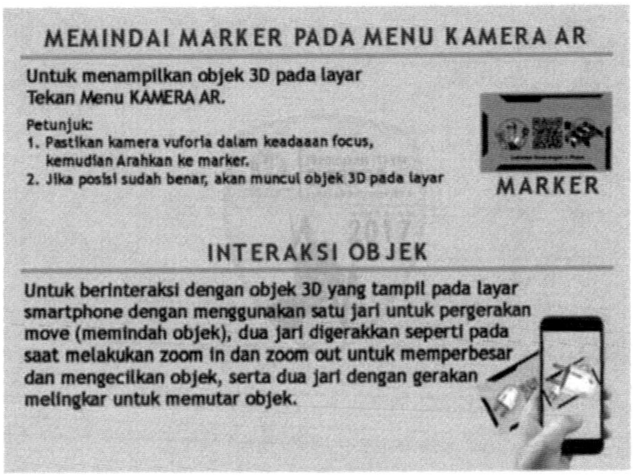

Figure 2. Display of User Manual Menu

Figure 2 shows user manual menu display of the learning media. There are media usage instructions and animation about instructions for use, with hope that media users understand and can use learning media easily.

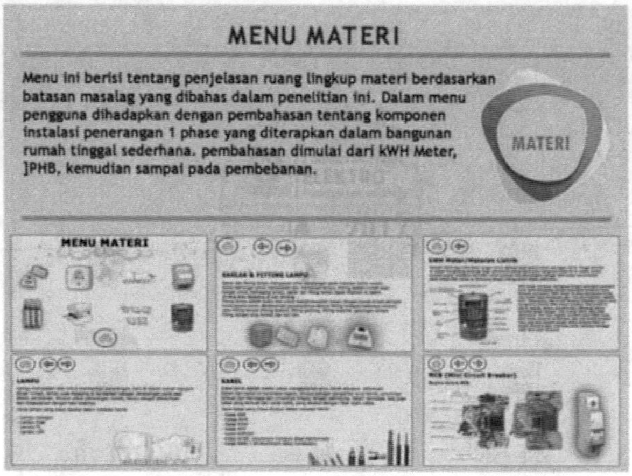

Figure 3. Display of Material Education Manu

Figure 3 Display of Material Education Manu, in the material menu contains the discussion material discussed in this study that is about electric lighting installations and components of 1 phase electric lighting installations

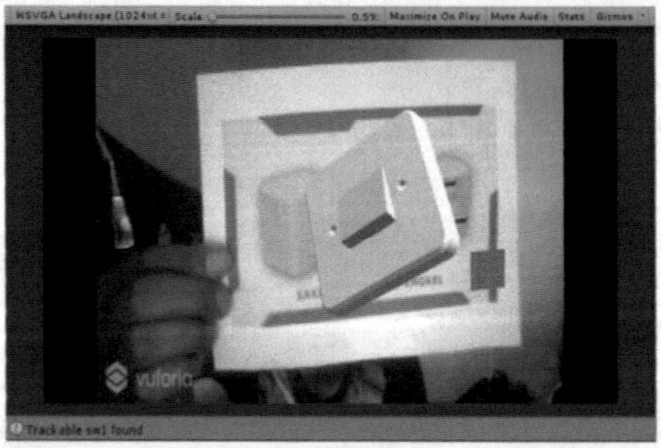

Figure 4. Display of Materi menu 1 "Saklar Tunggal"

Figure 4 shows Materi menu display using augmented reality technology. Materi can be accessed by the user by directing the camera on the smartphone to the marker or directing the marker towards the camera, displays of component and component assembly will appear on the smartphone screen.

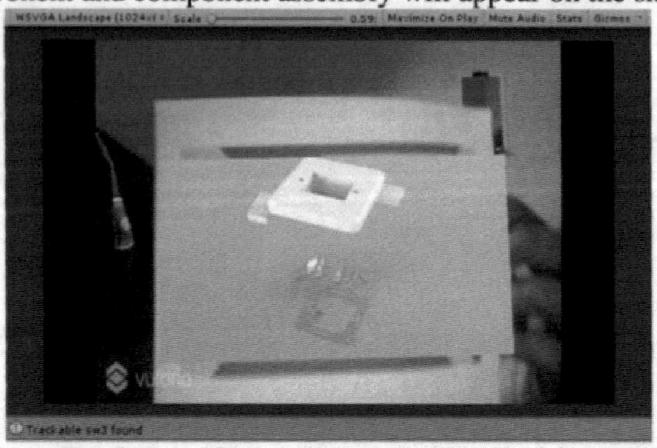

Figure 5. Display of Materi Menu 3 "Simulasi Assembly"

Figure 6. Display of Profile Menu

Figure 6 shows Profile menu display that explains the profile of the researcher.

3. Equipment And Learning Media Devices

Learning Media of Electric Lighting Installation Based on Mobile Augmented Reality Learning media for electric lighting installations are developed using the Unity 3D application on laptops and PCs. Learning media uses augmented reality technology that can run on Android OS-based smartphones. The display design of devices and components was generated from SolidWorks application (.fbx file) and the 3D Blender application (.blend file). Learning media interface design was developed using the Adobe Photoshop CS 6 application. In addition, the Marker was also developed using Adobe Photoshop CS 6. Learning media applications can be installed on an android smartphone by running .apk extension file.

Figure 7. Smartphone Android

4. Features on Learning Media

Some of the features in learning media include:

(1) Interactive Touch of electric lighting components - on the materx menu, when the scaning marker is finished, the components that appear on the smartphone screen can be rotated to see all sides of the component. The result is the compatibility of the characteristics of the components that appear on the screen with the original form of the component can be known by the user. Display of learning media when scanning markers can be seen in Figure 10.

(2) Visual Control and Commands - to get the same learning sensation as when playing games, the visual features and commands on the learning media of electric lighting installation is developed

interactively with augmented reality technology. By utilizing touch sensor on the smartphone's LCD screen, the electrical components are designed to be as detailed as possible.

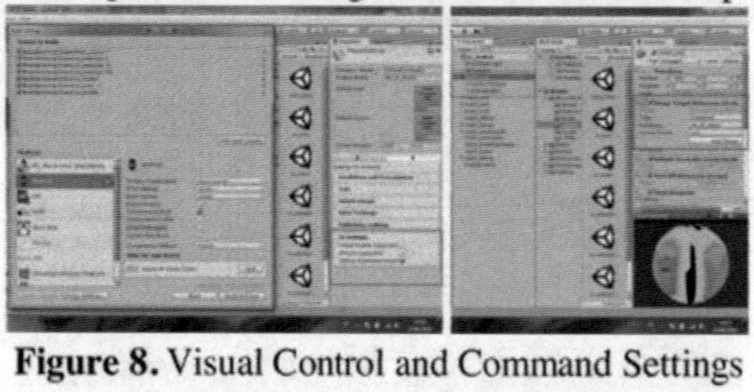

Figure 8. Visual Control and Command Settings

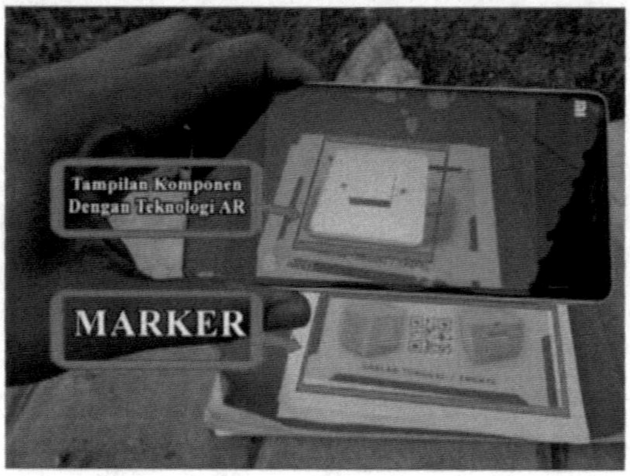

Figure 9. Visual Control and Command settings

Figure 10. The Use of Augmented Reality Display "Scan Marker of Socket Component"

The form of components in Augmented Reality technology can be displayed based on the shape of the marker being developed. Components displayed on this learning media are components used in household lighting installations, namely KWH Meters, Miniature Circuit Breakers (MCB), Cables, Sockets, Single Switches, Double Switches, Three Gang Switches, Lights, Pipes, Fittings, Plug and Pipe. All of these designs are made to resemble the shape of original components and there are also animations of household lighting work system.

In each component displayed, it is also equipped with a video assembly that aims to strengthen the knowledge of students'/users' competence regarding the introduction of the characteristics of lighting components.

In piping simulation, there is a working picture, the working picture will be a reference for the lighting installation to be made. In this simulation, position, function and work system of each component are displayed - starting from the PLN electricity entering the KWH Meter until the loads commonly used in simple house buildings.

5. Research Method

The research model used is the research and development study of Augmented Reality (AR) assisted electric lighting installation media at the Vocational Middle School. The Research Development Model used is the *Waterfall* model

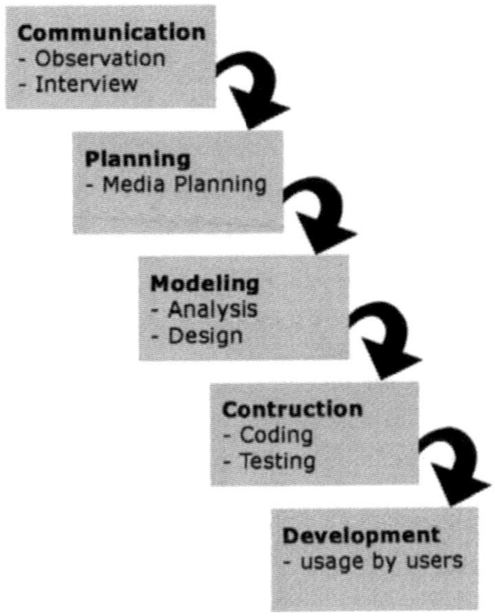

Figure 11. Research Method "*Waterfall*"

5.1. Communication

In this study begins with a communication from researchers to the school that will be used as a place of research, in this communication discusses about the weaknesses that exist in the school, especially in the department of electrical power installation techniques in lighting installation subjects to find a problem that will be raised.

5.2. Planning

In the planning stage the researcher tries to make a solution of the problems found in the school. The solution offered is a learning media that will discuss a particular material in a media package with a touch of the new technology.

5.3. Modeling

In making learning media, researchers try to find technology that can attract students' learning interest to use it so that the mastery of the material raised in the media can be conveyed well to the user.

5.4. Contruction

Researchers use augmented reality technology as a package that will attract users to use, in the construction phase researchers use several development-based applications for media creation, some of which are visual studio, unity 3D, solid works, 3D blanders and Keyshot.

Testing of learning media for lighting installation is done independently before the media is handed over to the user.

5.5. Development

The last process carried out by researchers is the development of instructional media aimed at schools where researchers conduct interviews and observations to determine user responses from augmented reality assisted learning media.

6. Research Result

In this research, testing was conducted on three lecturers as a media validator before carrying out further development to be tested in schools. From the three validators stated that the media is feasible and ready to be implemented in schools.

7. Conclusion

The purpose of developing learning media is to introduce alternative technology in the world of education that creates an atmosphere of learning to be interesting and fun for students without ignoring the elements of learning and competencies that must be achieved. Learners can study in all places and times because all they need is an android smartphone and some markers. This learning media also helps Educators to make time more efficient, because in theoretical subjects students are able to see virtually electrical components along with their characteristics and functions as well as lighting installations in households, without having to go to workshops and be busy by unloading and reloading practical tools. It provides knowledge about the basis of lighting installations in the form of introduction of lighting installation components, introduction to the characteristics of lighting installation components, determining the layout of household lighting installation components and adjusting working picture to actual installation form.

8. References

[1] Sudjana, N., & Rivai, A. (2010) Media Pembelajaran. Jakarta:Sinar Baru

[2] Schramm, 1984. Media Besar Media Kecil, Alat dan Teknologi untuk Pengajaran, *Seri Pustaka Teknologi Pendidikan No. 5. IKIP Semarang.*

[3] Persson, V. & Nouri, J. (2018) A Systematic Review of Second Language Learning with Mobile Technologies. *International Journal of Emerging Technologies in Learning. Vol. 13 (2),* 53-61.

[4] Triatmaja, A. K. & Khairudin, M. (2018).Study on Skill Improve/ment of -Digital Electronics Using Virtual Laboratorium With Mobile Virtual Reality. *Journal of Physics: Conference Series. Vol. 1140, DOI: 10.1088/1742-6596/1140/1/012021*

[5] Dengzhe Ma., Jurgen Gausemeier., Xiumin Fan., & Michael Grafe. (2011). Virtual Reality & Augmented Reality In Industry. The 2nd Sino-German Workshop. *Shanghai Jiao Tong University*

Press, ISBN 978-7-313-06426-4. Springer Heidebery Dordrecht London New York, ISBN 978-3-642-17375-2, e-ISBN 978-3-642-17376-9

[6] Jaya, H. (2010). Laboratorium Virtual Mata Kuliah Praktikum Elektronika Digital: Jurusan Pendidikan Teknik Elektronika Fakultas Teknik Universitas Negeri Makassar. *Jurnal Elektronika Telekomunikasi & Computer, Vol.4(2). 699-710.*

[7] Odeh, S., Shanab, S, A., & Anabtawi, M. (2015). Augmented Reality Internet Labs versus its Traditional and Virtual Equivalence. *International Journal of Emerging Technologies in Learning. Vol. 10 (Issue 3), 4-9.*

[8] Jaya, H. (2010). Laboratorium Virtual Mata Kuliah Praktikum Elektronika Digital: Jurusan Pendidikan Teknik Elektronika Fakultas Teknik Universitas Negeri Makassar. *Jurnal Elektronika Telekomunikasi & Computer, Vol.4(2). 699-710.*

[9] Dyrberg, N, R., Treusch, A, H., & Wiegand, C. (2016). Virtual laboratories in science education: Students Motivation and Experiences in Two Tertiary Biology Courses. *Journal of Biological Education. Vol.51 (4), 358-374.*

[10] Ogungbenro, O, U., Chukwudebe, G, A., Opara, F, K., & Ezeh, G, N. (2017). Design and Implementation of Logic Gate Emulator. 3rd *International Conference on Electro-Technology for National Development, 7-10 November 2017, Owerri, Nigeria*

[11] Khairudin M., Triatmaja A.K., Istanto W.J., Azman M.N.A., (2018). Mobile Virtual Reality to Develop a Virtual Laboratorium for the Subject of Digital Enggineering. *International Journal of Interactive Mobile Technologies. Vol 13, No 04 (2019)*

[12] Agus Suryanto., Diah Ayu Kusumawati., & Ibrahim M. H. Sanhoury. (2018). Development Of Augmented Reality Technology Based Learning Media Of Lathe Machines. *Jurnal Pendidikan Teknologi dan Kejuruan, Vol. 24, No. 1, May 2018, pp. 32-38. ISSN:0854-4735, accredited by KEMENRISTEKDIKTI, Decree No: 51/E/KPT/2017. DOI: 10.21831/jptk.v24i1.18245.*

[13] Mohd Erfy Ismail., Mohd Hasril Amiruddin., Pipit Utami., Irwan Mahazir Ismail., Moh. Khairudin., Badraningsih Lastariwati., & Nopadon Maneetien. (2018). The Effect Of An Augmented Reality Teaching Kit On Visualization, Cognitive Load And Teaching Styles. *Jurnal Pendidikan Teknologi dan Kejuruan, Vol. 24, No. 2, October 2018, 178-184. ISSN:0854-4735, accredited by KEMENRISTEKDIKTI, Decree No: 51/E/KPT/2018. DOI: 10.21831/jptk.v24i2.20031.*

Video-Based Blended Course for Computer Network Learning

S Pambudi, H D Surjono, T Sukardiyono and I Hidayatulloh
Electronics and Informatics Engineering Education Department,
Universitas Negeri Yogyakarta, Yogyakarta, Indonesia

Email: sigitpambudi@uny.ac.id

Abstract. This study aims to develop a blended course for computer network courses with video-based teaching materials. This research refers to the development model of Integrated Learning Design Framework. The development of this learning media product is based on a learning plan for a computer network course for one semester. Based on the results of product testing, this blended course is declared "good" and is suitable for learning. Paired sample t-test results show that the product developed is effective in improving student learning outcomes.

1. Introduction

Electronic learning or abbreviated e-learning is one of the innovations that combines learning activities with the use of electronics. The purpose of the integration is to improve the quality of education [1]. With the existence of internet technology, learning that is packaged in the form of e-learning can be done practically and effectively without being limited by the size of distance or time. Besides this e-learning can be accessed by many people at the same time at once, so that the learning process is also carried out more efficiently.

Internet users in Indonesia are classified as high and continue to increase from year to year. The Indonesian Internet Service Providers Association (APJII), through its official page at https://www.apjii.or.id, released data on internet users in Indonesia in 2018 totaling 171.17 million users from the total population of Indonesia, totaling approximately 264, 16 million people or around 64.8% of the total population [2]. This is a significant increase, where in 2017 Internet users in Indonesia are still at 54.68%. The high number of internet users is actually a potential in developing new learning technologies, for example e-learning. e-learning is a solution to overcome limitations or gaps such as distance, space and time which are often problems in the learning process. The internet makes e-learning have greater reach and benefits.

E-learning is widely used in various higher education institutions in Indonesia as a companion or substitute for conventional learning. One of the institutions that took advantage of electronic-based learning is Universitas Negeri Yogyakarta (UNY). UNY developed a Moodle LMS-based e-learning web called BeSmart intended to increase learning activities and can be accessed at *http://besmart.uny.ac.id/v2*. As one of the study programs in UNY, the Informatics Engineering Education study program also developed electronic learning. However based on observations, it was found that the use of BeSmart is still less than optimal. One of the subjects that is not yet optimal in using BeSmart is the Computer Networking course. These related problems arise when lecturer hours course cannot be present, besides that students also often feel less understanding of the material presented so that it requires more learning duration. Students tend to have initiatives to look for other learning materials, such as learning videos. Based on data obtained from the UPT Puskom UNY, achievement of the semester achievement index for Computer Network courses occupies the fifth lowest position compared with compulsory courses in the Informatics Engineering Education study program, with a cumulative grade point average (GPA) of 3.1.

Based on the results of interviews with supporting lecturers in the Computer Network course, it was found that Problems which hampered the improvement of the quality of learning in the Computer Network course, such as the limited duration of lectures to deliver a number of subject topics, students find it difficult to access learning resource materials, lecture activities are disrupted if the lecturer is unable to attend, and the lack of learning media such as video contains only text and image media, resulting in students lacking in mastering the material specifically related to practical simulations.

This research has the aim to be achieved, namely to produce solutions in the form of video based blended courses for material in Computer Network courses that are effective and efficient for students of the Informatics Engineering Education Study Program, Universitas Negeri Yogyakarta. In addition, the product developed can be used as a means of independent learning and can be used as positive input to further increase the interest and activeness of students or users in the learning process.

2. Literature review

2.1. Blended Course

Some of the literature uses the terminology more blended learning than the blended course. Blended learning is known as learning that combines online learning with face-to-face learning [3]. Through blended learning students can still do online learning outside the classroom. Online learning can be used as a means of interacting between students and other students or teachers directly or indirectly through online discussions [4]. The statement is in accordance with the opinion of Staker & Horn which states that online learning in blended learning can provide students the opportunity to apply concepts that have been obtained with unlimited meeting time in class [5]. Blended learning has a basic framework of concept development based on a constructivist learning perspective by providing experience with investigations that ensure concepts are built and assimilated deeply and meaningfully [6]. Implementation of blended learning can increase interactions between students and the environment, students and teachers, students and students, students with material and students with daily life [7].

The important thing to observe in the design of blended learning is to pay attention to the balance between online learning and face to face [8]. Blended composition that is often used is 50/50, meaning that the allocated time is 50% for face-to-face learning activities and 50% is done online computer-based learning (online and offline) [9]. The Sloan Consortium, the proportion of time division between learning in class and online learning determines what type of course or lesson is used, as explained in Table 1

Table 1. The Prototypical Course Classifications [10]

Proportion of Content Delivered Online	Type of Course	Typical Description
0%	Traditional	Content is delivered orally and in writing face to face
1 to 29%	Web Facilitated	Content is inserted on the web to support face-to-face learning such as for the dissemination of materials or the task collection container.
30 to 79%	Blended/Hybrid	Content delivered online and face-to-face. The technique can be alternately, or a combination of both.
80+%	Online	All content is presented online

2.2. Video for Media Education

Literally, the media has an intermediary or introduction meaning [11]. In the context of learning, the media are all forms of intermediaries whose role is to convey information from the teacher to students. As a conduit of information, the media is important in the learning process. The selection of media should be based on (1) Students' needs, (2) Compliance with learning objectives, (3) Compliance with learning material, and (4) Compliance with learning methods. These four points must be the basis for selecting a learning media, because the media will play an optimal role as an introduction to information if the media is a medium that is liked by students, in accordance with the learning objectives, in accordance with the material to be delivered, and in accordance with the methods determined by the teacher.

As the development of education, the media currently used by teachers varies greatly. Today there are many studies that develop learning media, one of which is video. Video is a type of audio visual media, which means learning media that can be seen using the sense of sight and be heard using the sense of hearing. As a learning media, effective video is used for mass learning processes, individuals and groups [12].

Edgar Dale with Dale's cone of experience theory describes the level of student understanding in a cone [13]. In Figure 1, the video is located in the middle and belongs to the "Television" category. This position means that video is better than image media and audio media.

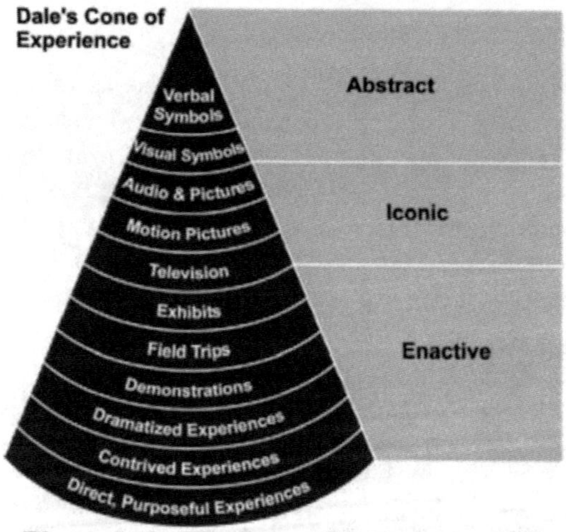

Figure 1. Dale's Cone of Experience [13]

Using video as a solution to overcome the low ability of students to understand a concept. The ease of presenting videos that can be repeated during the learning process makes it easier for students to understand the contents of the video, besides the presentation of a structured material also makes it easier for students to understand the material specifically about concepts [14]. Both of these advantages mean that video is an effective media used to improve the ability of elementary school students to understand concepts.

Many studies have concluded that video can increase student motivation [15] and have a positive contribution to student learning outcomes [16]. The results of the study of Supryadi et.al. stated that learning activities using video can provide a pleasant atmosphere and not boring for students so that students' attention is focused [17]. Besides video can also present events that are not physically possible, so students can find out more about these events. Video can satisfy all students who have different learning characteristics, ranging from audio, visual or audio-visual. This has an impact on improving student-learning outcomes, so it can be said that the use of video as a learning medium is effectively applied to the learning process.

3. Materials and Methods

This research is a Research and Development (R&D) where products developed in the form of video-based blended courses as media teaching courses Computer Networks. The stages of development used in this study adopted the Integrated Learning Design Framework (ILD). ILD Framework, consists of three stages, namely (1) Informed Exploration, (2) Enactment, (3) Local Evaluation, and (4) Broad Evaluation. At each stage the Integrated Learning Design is explained in Figure 2.

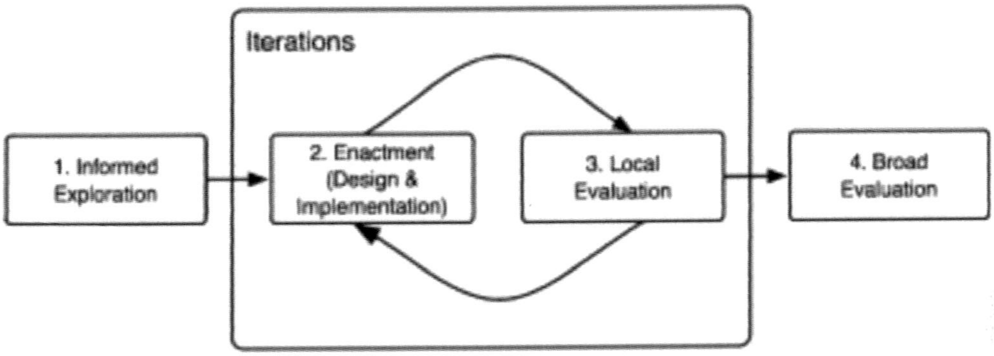

Figure 2. Integrated Learning Design Framework [18]

As the scope of the product development study, an identification process is needed from the Semester Learning Plan of Computer Network Courses in the Informatics Engineering Education Study Program, Universitas Negeri Yogyakarta. Product trials developed in this study include expert validation and trials on a larger scale. Expert tests carried out included material and media validation with the examiner being a material expert lecturer, a computer network lecturer, and a media validator competent in his field. Broad Evaluation was carried out with all respondents, as many as 39, which second semester students of the Informatics Engineering education program in Universitas Negeri Yogyakarta who were conducting a Computer Network course during the research. This study also compared the pretest and posttest values and conducted a t test to measure the effectiveness of a product.

4. Result and Discussion

Video-based Blended Course to support the delivery of Computer Network course material has been completed and successfully installed on Be-Smart. Be-Smart is a Universitas Negeri Yogyakarta e-learning web developed with Moodle LMS. The appearance of the blended course is designed so that students and lecturers as users can easily learn how to use the product. In developing a blended course, the material delivered during lectures is divided into two activities, namely synchronous and asynchronous. Synchronous learning is carried out in real-time, whereas Asynchronous is learning that is carried out online. The order of learning materials is adjusted according to the order of material contained in the Computer Network Semester Learning Plan.

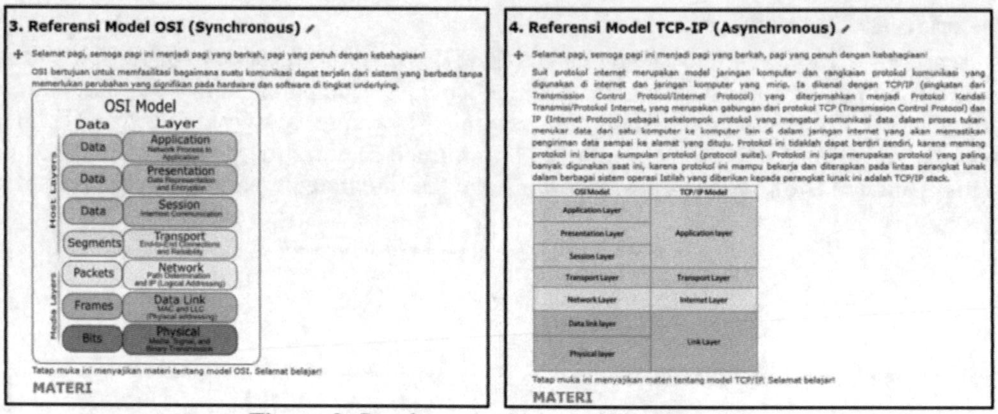

Figure 3. Synchronous & Asynchronous Page

In the learning video developed, there are two techniques presented, namely real shot and animation. With the variations in video presentation, it is expected that students will not get bored easily and will listen to the contents of the material presented..

Figure 4. Real Shot & Animation Video

The results of material expert testing on the quality of video-based blended courses as a product of research results obtained a value of 3.8 in terms of material content. This value is included in the "very good" category. While from the aspect of quality of learning the average value given is 3.6 or included in the category of "very good". Media expert test results on product quality viewed from the aspect of the display interface (interface) get an average score of 3.2 or included in the category of "very good". While from the pedagogy aspect, media experts gave a final score of 3.6 (very good). The results of the assessment in the beta test showed that respondents gave an average value of 2.85 or included in the category of "Good". Noting some of the results of the assessment can be concluded that the product is suitable for use in the learning process.

To measure the effectiveness of a product, it is necessary to do a t test. From the t paired sample t-test, it was found that the t value was -6.089 with a probability of 0.000 because the probability/ 2<0.025, the two population averages were not identical (the mean pretest and posttest values were significantly different). Based on the data, the students' competency improvement towards understanding the concept by comparing the pretest and posttest scores and t test analysis shows the pretest scores are lower when compared with the posttest scores. This shows that the video-based blended learning model has succeeded in increasing students' competence or understanding of Computer Network material or in other words video-based blended learning has proven to be effective in improving student learning outcomes in Computer Network courses.

Table 2. Paired Samples Test

		Paired Differences					t	df	Sig. (2-tailed)
		Mean	Std. Deviation	Std. Error Mean	95% Confidence Interval of the Difference				
					Lower	Upper			
Pair 1	Pretest - Posttest	-14.4872	14.85876	2.37931	-19.3038	-9.67053	-6.089	38	.000

5. Conclusion

Video based Blended Course for learning Computer Network material has been successfully developed using a framework model from Integrated Learning Design (ILD). Alpha test results by material experts and media experts obtained "Very Good" results, while the results obtained from the beta test obtained "Good" results. So that the product is suitable for use in learning Computer Network courses. Based on the pretest and posttest competency test results and paired sample t-test t test analysis shows that the product is effective in improving student-learning outcomes in Computer Networking courses.

Learning innovations in the world of education will always develop following existing technology. At present, the presence of the internet is certainly a breath of fresh air that needs to be utilized as optimal as possible. Blended courses can be used as solutions to problems that often arise. The benefits that can be taken from a blended course make students able to study more independently and efficiently. While the presence of learning videos in the blended course will add to the variety of learning media in lectures, not only for Computer Network courses, but also for other courses. Suggestions for further development related to this research are improvement in terms of interactivity, such as the use of video conferencing and so on.

6. References

[1] Dewi P E D M, Tripalupi L E and Irwansyah M R 2019 The Effect of Blended Learning Method Toward Enterpreneurship Learning Activity and Learning Outcome of Economics Undergraduate Students *International Conference on Tourism, Economics, Accounting, Management, and Social Science (TEAMS 2018)* (Atlantis Press)

[2] APJII 2018 Penetrasi Dan Perilaku Pengguna Internet Indonesia 2018

[3] Baum E 2013 Creating a Blended Cooperative-Learning Classroom *The International HETL Review* 28–36

[4] Sendag S and Duran M 2012 Comparing preservice teachers perceptions of online problem-based learning and online instructor-led learning *Procedia-Social and Behavioral Sciences* **31** 212–7

[5] Staker H and Horn M B 2012 Classifying K-12 blended learning. *Innosight Institute*

[6] Garrison D R and Vaughan N D 2008 *Blended learning in higher education: Framework, principles, and guidelines* (John Wiley & Sons)

[7] Stacey E and Gerbic P 2007 Teaching for blended learning—Research perspectives from on-campus and distance students *Education and information technologies* **12** 165–74

[8] Akkoyunlu B and Yilmaz-Soylu M 2008 A study of student's perceptions in a blended learning environment based on different learning styles. *Educational Technology & Society* **11** 183–93

[9] Dwiyogo W D 2014 Pengembangan Model Rancangan Pembelajaran Berbasis Blended Learning (PBBL) untuk Meningkatkan Hasil Belajar Pemecahan Masalah *ASEAN Comparative Education Research Conference (ACER-N), Institut KWSP Kajang: Universiti Kebangsaan Malaysia*

[10] Allen I E, Seaman J and Garrett R 2007 Blended in: the extent and promise of blended education in the United States *Sloan-Consortium, Needham, MA*

[11] Heinich R, Molenda M, Russell J D and Smaldino S E 2005 Instructional technology and media for learning *New Jersey, Columbus. MULTI MEDIA PEMBELAJARAN* **141**

[12] Daryanto D 2010 Media pembelajaran *Yogyakarta, Gava Media*
[13] Dale E 1946 The cone of experience *Audio-visual methods in teaching* **1** 37–51
[14] Sudiarta I G P and Sadra I W 2016 Pengaruh Model Blended Learning Berbantuan Video Animasi terhadap Kemampuan Pemecahan Masalah dan Pemahaman Konsep Siswa *Jurnal Pendidikan dan Pengajaran* **49** 48–58
[15] Choi H J and Johnson S D 2005 The effect of context-based video instruction on learning and motivation in online courses *The American Journal of Distance Education* **19** 215–27
[16] Sarnoko S, Ruminiati R and Setyosari P 2016 Penerapan Pendekatan Savi Berbantuan Video Pembelajaran Untuk Meningkatkan Aktivitas dan Hasil Belajar IPS Siswa Kelas IV SDN I Sanan Girimarto Wonogiri *Jurnal Pendidikan: Teori, Penelitian, dan Pengembangan* **1** 1235–41
[17] Supryadi P E, Jampel I N and Riastini P N 2013 Penerapan Media Video Pembelajaran Sebagai Aplikasi Pendekatan Contekstual Teaching Learning Untuk Meningkatkan Hasil Belajar IPA Siswa Kelas V *MIMBAR PGSD Undiksha* **1**
[18] Alvarez C, Alarcon R and Nussbaum M 2011 Implementing collaborative learning activities in the classroom supported by one-to-one mobile computing: A design-based process *Journal of Systems and Software* **84** 1961–76

Acknowledgments
This research was Supported by Universitas Negeri Yogyakarta, Indonesia funded by DIPA LPPM UNY.

Optimization of Web-Based Learning Media in SMK PGRI 3 Walikukun

D Iswahyudi and Soenarto
Electronics and informatics Engineering Education Department,
Universitas Negeri Yogyakarta, Indonesia

Email: didikiswahyudi8080@yahoo.com, soenarto@uny.ac.id

Abstract. This study aims to determine: (1) appropriate web-based learning methods to improve student learning outcomes, (2) the teacher's method of using web-based learning media to influence student achievement, and (3) Is the optimization of web-based learning media able to meet standards creative, active, and fun learning at SMK PGRI 3 Walikukun. The method used in this study uses a research and development (R&D) approach. The population in this study were students of SMK PGRI 3 Walikukun Based on data analysis, the results of this study showed that: Optimization of web-based learning media through the PBL approach was more effective in improving ICT competency in SMK PGRI 3 Walikukun students both in the process and learning outcomes which include aspects of competence knowledge, attitudes, skills and participation compared to before are still widely used direct learning models. The results of the conclusions indicate an increase in teacher skills in applying the syntax of the optimization model of web-based learning needs to be improved, by using a problem-based learning system learning activities take place interactively and student-centered, the results can increase student grades.

1. Introduction

The success of students in understanding a learning material is the main goal of every teacher. How the delivery of learning a teacher is very influential on the success of students understanding the material being taught. With the development of technology today many media can be used as infrastructure to deliver material to students, one of which is learning to use website media. According to (Tsai & Machado, 2002) from InkiTiki Corporation the definition of web-based learning (WBL) is web-based learning related to learning resources that are presented through web applications, including if the learning resources are packaged on CD-ROM or other media. The advantages of web-based learning are numerous and varied. Chang (2007) states that researchers have demonstrated advantages of web-based learning such as time saving, cost reduction and space saving, and increased learning opportunities for non-school students. The use of instructional media using website media can reduce the static atmosphere and can create an effective, interesting, interactive learning process and can also arouse student learning motivation (Darussalam, 2012). The same thing was also stated by Gilakjani (2011), the use of digital technology in learning provides learning conditions with the opportunity to create a conducive learning environment for students, rich in information and learning resources, and can be inserted with various multimedia-based learning elements. SMK PGRI 3 Walikukun already has a media website, but the teacher has not been able to use it well, in conveying learning the teacher uses website media only to display the theory as in the text and teach it theoretically. This makes students feel bored and do not understand the material presented by the teacher. The results obtained by students are not satisfactory, so the results are far from Teacher's expectations. For this reason, this research is important in conducting research on web-based learning media optimization in SMK PGRI 3 Walikukun.

2. Study Review

2.1 Learning

Wina Sanjaya (2008: 51) learning is an activity aimed at student learning. Every teacher is important to understand the learning system, because by understanding this system, every teacher will understand about the learning objectives or expected results, the process of learning activities that must be carried out, the use of each component in the activity process to achieve the goals to be achieved and how to know the success of achieving the. Sanjaya (2008: 88) learning objectives must contain elements of ABCD, namely Hearings (which must have abilities), Behavior (behaviors that are expected to have), Conditions (in conditions and situations where the subject can show ability as a result of learning that has been obtained), and Title (quality or quantity of behavior expected to be reached as a minimum).

2.2 Understanding Learning

According to Hilgard and Bower in Purwanto (1987) learning relates to changes in a person's behavior towards certain situations caused by repeated experiences in the situation. Winkel (2009) mental or psychological activity, which occurs in active interactions with the environment, which results in a number of changes in knowledge, understanding, skills and value-attitudes. Slameto (2010) learning as a business process carried out by someone to get a whole new behavior change, as a result of his own experience in interactions with his environment.

2.3 Understanding Learning Outcomes

Learning outcomes according to Winkel, quoted by Sudjana (2004: 142) as follows::

2.3.1 Learning outcomes as an indicator of the quality and quantity of knowledge students have mastered;
2.3.2 Learning outcomes as a symbol of concentration of curiosity;
2.3.3 Learning outcomes as information material in educational innovations;
2.3.4 Learning outcomes as internal and external indicators of the situation of educational institutions;
2.3.5 Learning outcomes can be used as indicators of student intelligence.

Bloom (in Irwanti, 2011) divides learning outcomes into three domains, namely the cognitive, affective and psychomotor domains.

a. Cognitive domain

This domain relates to intellectual learning outcomes consisting of six aspects, namely knowledge, understanding, application, analysis, synthesis and evaluation;

b. Affective Domain

The affective domain is related to attitudes and values. Types of affective learning outcomes arise in students in a variety of behaviors such as attention to learning, discipline, learning motivation, respect for teachers, study habits, and social relations;

c. Psychomotor domain

Psychomotor learning outcomes appear in the form of individual skills and abilities.

2.4 Types of Learning Outcomes

According to Sukmadinata (2005), "Achievement or learning outcomes are the realization or expansion of a person's potential skills or capacities". Learning achievement is said to be perfect if it meets three aspects, namely: cognitive, affective, and psychomotor.

2.5 Factors That Influence Learning Achievement

According to Purwanto (2006) learning factors are divided into 2 (two) groups: (1) individual factors, including maturity / growth factors, intelligence, training, motivation, and personal factors; and (2) factors outside the individual that we call social factors, including family / household factors, teachers and their teaching methods, tools used in teaching and learning, the environment and opportunities available, and social motivation.

2.6 Understanding Learning Media

The word media is a plural form of the word medium which literally means an intermediary (Djamarah, 2006) in the world of education, teachers often use media to teach what is called learning media (Rusman, (2012) and Daryanto (2010)). Learning media are a number of assistive devices, simulation materials or programs used in learning activities (Erlina, 2009), messenger technology that can be used for learning purposes and are a physical means of delivering subject matter (Rusman, (2012) and Sukiman (2012) from the sender to the recipient so that they can stimulate the thoughts, feelings, attention, and interests and desires of students in such a way that the learning process occurs to achieve learning objectives effectively.

2.7 Types of Learning Media

Learning media that are often used in learning activities are very diverse and some are deliberately designed according to the needs of learning activities (Setiawan, 2013). Can be used to stimulate the thoughts, feelings, attention and abilities or skills of students so that it can encourage the learning process (Febrian, 2016). Asyhar (2012), there are 4 (four) types of learning media, namely: "(1) Visual media, are types of media that are only used to rely solely on the student's sense of sight, for example: visual media that is not projected (objects of reality, protective models, and graphics), and projection media (power point, paint and automatic cad), (2) audio media, is the type of media used in the learning process by relying solely on students' sense of hearing, for example: radio, vocal cords, and recordings vinyl (3) Audio-visual media, is a type of media used in learning activities that involve hearing and vision and in a process or activity, for example: video tapes and film frames, and (4) Multimedia, is media that involves several types media and equipment are integrated in the process or learning activities, for example: TV and power point.

2.8 Learning Media Function

Learning media function to convey learning messages (Ouda Teda Ena, 2001). According to Asyhar (2011), learning media is divided into seven parts, namely: (1) media as a source of learning; (2) semantic functions; (3) manipulative functions; (4) fixative function; (5) distribution function; (6) psychological functions and (7) socio-cultural functions. According to Usman and Asnawir (2002) including: (1) helping facilitate learning for students and teachers (2) providing more tangible experiences; (3) attract more student attention; (4) all students' senses can be activated, weaknesses of one sense can be balanced with other sensory strengths; (5) attract the attention and interest of students in learning; and (6) can bring the world of theory into reality.

2.9 Definition of Optimization
Optimization is an action, process, or methodology to make something (as a design, system, or decision) to be more / fully perfect, functional, or more effective (Balai Pustaka, 1994: 800). According to Machfud Sidik related to the optimization of an action / activity to improve and optimize. (Sidik, 2001).

3. Research Methods
This type of research is conducted using quantitative descriptive. The subjects used in class XI were two classes with 70 students. In this study using data analysis techniques using surveys

3.1 Understanding Problem Based Learning (PBL)
Broadly speaking there are two approaches to learning, namely teacher centered learning (TCL) and student centered learning (SCL). Teacher centered learning (TCL) is teacher-centered learning, so this approach is often called the conventional approach. Whereas student centered learning (SCL) is student-centered learning. This method is a relatively new learning paradigm in developing countries.
MacVaough (1988: 37) states that: Based on this opinion it is clearly stated that problem based learning there are 4 (four) of these activities in problem based learning (PBL), namely:

3.1.1 Problematisation
Problematisation (Problematisation) requires students to consider what they already know about the problem, what is the problem of that knowledge, and the most important is what is not known. This process is very good for continuing education because it refers to the ability of learners, asking for problematic issues for themselves, and making changes in attitude towards an action.

3.1.2 Problem Investigation
Problem Investigation requires learners to consider the elements of the problem that are currently understood and seek knowledge to understand the dimensions, content, location, and solutions that are known for the problem, it is important for learners to look for knowledge outside their scientific discipline , placing problems into new understanding frameworks, and looking for interdisciplinary knowledge solutions that are not yet known.

3.1.3 Problem Solving
Problem solving gives learners the opportunity to take action, and shows that their learning is more than contemplation on an infinite idea. This is perhaps the most important element for continuing education because success can only be realized through significant behavioral change and not just an understanding element.

3.1.4 Critical Reflection
Critical reflection requires students to evaluate the process used in problem solving and evaluate the benefits of the recommended actions to enlighten current understanding. At the same time, learners might know that the solutions they make cannot solve problems. Thus learners must create new concepts again for the continuation of solving the problem.

4. Results and Discussion

4.1 TKJ learning Basic competence Installing a Graphical User Interface (GUI) and Command Line Interface (CLI) based Operating System

The preliminary study was carried out by collecting data and analyzing various matters related to the analysis of TKJ learning needs of basic competencies. Installing an Operating System based on Graphical User Interface (GUI) and Command Line Interface (CLI) in aspects of the TKJ curriculum and the factual conditions of TKJ learning in schools. TKJ learning for students of SMK PGRI 3 Walikukun which specifically includes basic competencies. Installing a Graphical User Interface (GUI) and Command Line Interface (CLI) based Operating System is about Preparing a GUI-based operating system installation and Implementing a GUI-based operating system installation according to the manual intallation. Specifically the factual conditions of TKJ learning in SMK PGRI 3 Walikukun are described as follows,

4.1.1 Teacher

Preliminary study activities are carried out by collecting data and analyzing various matters relating to the analysis of the needs and factual conditions of TKJ learning in schools. TKJ learning as a study in the preliminary study includes: TKJ curriculum especially relating to basic competencies Installing Operating Systems based on Graphical User Interface (GUI) and Command Line Interface (CLI), Semester Learning Plan (RPS), materials, methods and strategies, media , interactions, evaluations, and tools used. This condition factually reflects TKJ's basic competency in Installing an Operating System based on Graphical User Interface (GUI) and Command Line Interface (CLI) in schools. Specifically the results of the preliminary study in the teacher's perspective are described as follows

Table 1. Results of Preliminary Study Analysis conducted at SMK PGRI 3 Walikukun

Number	Component	Answer	
		Yes	No
1.	Conduct analysis and mapping of basic competency curriculum Installing Operating Systems based on Graphical User Interface (GUI) and Command Line Interface (CLI)	70%	30%
2.	Design and create a Semester Learning Plan (RPS) TKJ Learning basic competencies Install Operating Systems based on Graphical User Interface (GUI) and Command Line Interface (CLI)	80%	20%
3	Developing basic competency TKJ material Installing an Operating System based on Graphical User Interface (GUI) and Command Line Interface (CLI) according to needs, level of student development, and problems in practice in the field	80%	20%
4	Using TKJ learning methods basic competency Installing Operating Systems based on Graphical User Interface (GUI) and Command Line Interface (CLI) that is varied and student-centered	80%	20%
5	Using diverse and innovative media in TKJ learning basic competencies Installing Operating Systems based on Graphical User Interface (GUI) and Command Line Interface (CLI)	70%	30%
6	Developing TKJ learning processes and interactions of basic competencies Installing a student-centered Graphical User Interface (GUI) and Command Line Interface (CLI) based Operating System	80%	20%
7	Assessing TKJ learning basic competencies Installing an	80%	20%

	Operating System based on Graphical User Interface (GUI) and Command Line Interface (CLI) in accordance with the objectives and includes all aspects (knowledge, attitudes, and psychomotor)		
8	Utilizing books, learning resources, and technology that supports TKJ learning basic competencies Installing Operating Systems based on Graphical User Interface (GUI) and Command Line Interface (CLI)	60%	40%
9	Develop learning by honing the ability to solve social problems and desicion making in learning TKJ basic competencies Installing Operating Systems based on Graphical User Interface (GUI) and Command Line Interface (CLI)	20%	80%
10	Developing TKJ learning interactions is more group than individual	20%	80%
11	Develop inquiry and group investigations in TKJ learning basic competencies Install Operating Systems based on Graphical User Interface (GUI) and Command Line Interface (CLI)	20%	80%
12	Develop learning that hones independence, self-confidence, and skills in learning TKJ basic competencies Installing Operating Systems based on Graphical User Interface (GUI) and Command Line Interface (CLI)	20%	80%

Based on the table above shows that most teachers do an analysis of the basic competency curriculum Installing an Operating System based on Graphical User Interface (GUI) and Command Line Interface (CLI). The analysis conducted in relation to the TKJ curriculum is curriculum analysis in preparing the Semester Learning Plan (RPS), mapping the level of difficulty, breadth, scope of material, and integration of TKJ. In general, teachers are of the view that mapping is done to facilitate the preparation and implementation of effective learning in achieving learning objectives.

4.1.2 Student

Analysis of needs in terms of students shows TKJ learning basic competencies Installing an Operating System based on Graphical User Interface (GUI) and Command Line Interface (CLI) has not been able to develop the ability to provide solutions to solve social problems in everyday life. More learning emphasizes cognitive aspects, aspects of attitude and skills are less developed optimally.

As part of the work environment, teachers should be directly involved in solving various problems that occur. Because in general the learning process is not fully student-centered, the teacher is still in control of the learning activities. Student independence in learning is not optimal. Students in learning more carried out individually, both through the tasks given and the learning approaches given by students. This condition causes students to lack the courage to actively participate in solving various problems in the field.

In students' view, the learning approach taken emphasizes more on the individual approach. The group approach is only on specific tasks and is not routinely carried out by students in each learning. In learning interactions, students are generally dominated. Students are still the main source in learning by using the lecture method, questions and answers, and class discussion.

Students still use textbooks as the main handle in students' views. Additional books relating to learning material are rarely used. In certain materials, the teacher gives assignments using books as a primary learning source. While web-based in schools has not been used as a source and the main learning media in the learning process carried out by students so far.

TKJ material developed by the teacher according to the teacher, has not led to field-based problems that occur in the work environment around students, it can even be said to tend to be ignored. During this time TKJ learning is sourced more from handbooks.

4.1.3 Material
TKJ material becomes a study at every level of education with increasingly complex Information System-based problems both in local and national contexts. Analysis of TKJ material so far has been:

Basic competency TKJ material Installing an Operating System based on Graphical User Interface (GUI) and Command Line Interface (CLI) has not been fully developed by the teacher in learning. TKJ RPS mapping has not been carried out by all teachers. Material of basic competency TKJ Installing an Operating System based on Graphical User Interface (GUI) and Command Line Interface (CLI) so far is still based on textbooks as the main source in learning. TKJ material basic competency Installing an Operating System based on Graphical User Interface (GUI) and Command Line Interface (CLI) is very challenging to use various sources and media available in the work environment. Along with the complexity of social problems that exist in the life of the work environment and the importance of efforts to preserve sustainable diversity, through learning TKJ basic competencies Installing the Operating System based on Graphical User Interface (GUI) and Command Line Interface (CLI) becomes very important to be implemented. Learning TKJ basic competencies Installing the Operating System based on Graphical User Interface (GUI) and Command Line Interface (CLI) has not fully utilized the problem of field practice as a source and learning medium. Though the essence of TKJ is nothing but the study of the work environment with its environment both physically and non-physically

4.2 Early Learning Design Optimization of Web Based Learning
Web-based learning optimization model Problem Based Installing a Graphical User Interface (GUI) and Command Line Interface (CLI) based system departs from problems in the daily work environment, by raising issues and problems Installing a Graphical User Interface (GUI) based Operating System) and Command Line Interface (CLI). The main objective of this development is to improve the competence of Installing Graphical User Interface (GUI) and Command Line Interface (CLI) based systems, both in terms of knowledge, attitude / values, skills, and participation in Installing Graphical User Interface (GUI) based Operating Systems and Command Line Interface (CLI). In the initial design of problem-based web based learning optimization Installing an Operating System based on Graphical User Interface (GUI) and Command Line Interface (CLI) the construction consists of: (1) syntax, (2) operating system, (3) working principles, (4) support systems, (5) instructional impacts and companion impacts.

4.2.1 Syntax
The syntax in the design of problem based web based learning optimization Installing the Operating System based on Graphical User Interface (GUI) and Command Line Interface (CLI) was developed into 6 syntaxes, namely: (1) student orientation to the problem, (2) mapping and determining the priority of the problem , (3) investigation of problems Installing Graphical User Interface (GUI) and Command Line Interface (CLI) based Operating Systems, (4) interdisciplinary and crossdisciplinary analysis, (5) group and classical discussions, (6) developing and presenting the work, and (7) analyze and evaluate the problem solving process. Optimization of problem based web based learning Installing Operating System based on Graphical User Interface (GUI) and Command Line Interface (CLI) in TKJ learning consists of 6 main steps that begin by orienting students to a problem situation and ending with the presentation and analysis of student work. The following is the development of

the Problem Based Learning syntax with the development of a problem-based web based learning optimization syntax. Installing an Operating System based on Graphical User Interface (GUI) and Command Line Interface (CLI).

Table 2. Syntax Model Problem Based Learning and Development of Problem Based Optimization Installing Operating Systems based on Graphical User Interface (GUI) and Command Line Interface (CLI)

Sintaks *Problem Based Learning*		**Syntax of Web Based Learning Optimization**	
Stage	**Behavioral**	**Stage**	**Behavioral**
Stage 1 Student orientation to problems	- Explain the learning objectives - File a phenomenon - raises a problem - Motivate students to be involved in problem solving	*Stage 1* Student orientation to the problem	- Raising the phenomenon of problems Installing Operating Systems based on Graphical User Interface (GUI) and Command Line Interface (CLI) - Evidence and examples of problems Installing Operating Systems based on Graphical User Interface (GUI) and Command Line Interface (CLI) - Motivate students to solve problems
Stage 2 Organizing students for learning	- Organizing learning tasks related to problems - Setting up the device in problem investigation	*Stage 2* Mapping and prioritizing problems	- Identifying the factual conditions of the problem Installing the Operating System based on Graphical User Interface (GUI) and Command Line Interface (CLI) - Mapping issues Installing a Graphical User Interface (GUI) and Command Line Interface (CLI) based Operating System - Determine 5 main problems Installing Operating Systems based on Graphical User Interface (GUI) and Command Line Interface (CLI)
Stage 3 Guide individual and group investigatio ns	- Encourage students to gather information accordingly - Carry out investigations / experiments to get explanations and problem solving	*Stage 3* Investigating problems	- Conduct an investigation - Develops critical thinking & analysis skills - Description and condition of the problem Installing the Operating System based on Graphical User Interface (GUI) and Command Line Interface (CLI) - Analysis of data and facts, causes, impacts, and solutions to problems

Stage 4 Develop and present the work	- Assist students in preparing and planning and preparing appropriate work such as: reports, videos, and models	*Tahap 4* Performing Multidisciplinary and Crossdisciplinary Analysis	- Specifically conducting Multidisciplinary and Crossdisciplinary studies on issues Installing Operating Systems based on Graphical User Interface (GUI) and Command Line Interface (CLI)
Stage 5 Analyze and evaluate the problem solving process	- Helping students to reflect or evaluate the investigation and the processes carried out - Overall student reflection	*Stage 5* Guiding group and classical discussions	- Conduct presentations and discussions - Analyzing, criticizing, and responding to various data, facts, and solutions in the discussion
		Stage 6 Develop and present the work	- Present the work in the form of reports, drawings, photos, pamphlets, and posters - Design and conduct practicum Installing Operating Systems based on Graphical User Interface (GUI) and Command Line Interface (CLI) and presented in social media in various forms
		Stage 7 Analyze and evaluate the problem solving process	- Reflecting or evaluating the process and results - Fix weaknesses - Perform or provide reinforcement

Source: Development based on PBL Arends syntax, 1997; Eggen & Kauchak, 2012; Ibrahim & Nur, 2000)

4.3 Test Result

Tabel 3. Pretest Recapitulation of Competency Installing Operating Systems based on Graphical User Interface (GUI) and Command Line Interface (CLI) Students In Limited Test 1

Aspect	**Pretest Value**		**Posttest Value**		**Average Increase**	
	Control	Experiment	Control	Experiment	Control	Experiment
Knowledge	62.40	64.32	60.84	67.56	-1.56	3.24
Attitude	63.78	63.61	60.98	65.71	-2.80	2.10
The skills	66.23	68.13	67.30	69.87	1.07	1.74
Participation	66.53	62.20	63.17	65.80	3.36	3.60

Tabel 4. Recapitulation of Competency Pretest Value Installing Operating Systems based on Graphical User Interface (GUI) and Command Line Interface (CLI) Students in Limited Trials 2

Aspect	Pretest Value		Posttest Value		Average Increase	
	Control	Experiment	Control	Experiment	Control	Experiment
Knowledge	66.57	70.77	70.66	76.54	4.09	5.77
Attitude	56.37	60.73	56.63	70.17	0.26	9.44
The skills	56.54	59	59.60	73.67	3.06	14.67
Participation	64.90	69.48	65.50	74.52	0.6	5.04

5. Conclusion

The results of research on optimizing web-based learning media through the PBL approach in ICT learning for students at SMK PGRI 3 Walikukun can be concluded, that:

Optimization of web-based learning media through the PBL approach is more effective in increasing ICT competency in students in SMK PGRI 3 Walikukun both in the process and learning outcomes which include competencies in aspects of knowledge, attitudes, skills, and participation compared to before which there are still many using direct learning models (direct instruction). Improvement of teacher skills in applying the syntax of web-based learning optimization models, learning systems and reaction principles takes place interactively and is student-centered.

The advantages of optimizing Web-Based Learning make the community environment as a source and medium of learning, making the knowledge possessed by students as centers of learning centered on students, shaping students' thinking critically, logically, and able to solve problems. The application of Web-Based Learning optimization also fosters positive attitudes towards ICT subjects. While the weaknesses are efficiency and time constraints in implementing the model, the syntactic process series of optimization models Web-Based Learning is a long process, requires special skills for teachers to implement the model, and incomplete teacher understanding of the model, especially in the application of the syntactic model can influence the success Web Based Learning optimization application.

6. References:

[1] Brown, J.S., Collins, A., and Duguid, P. (1989). *Situated Cognition and the Culture of Learning*. Educational Researcher 18, no. 1 (January-February 1989): 32-41.

[2] Adhikary, P.K. (2005). Educational Reform For Linking Skills Development With Employment In Nepal. In M. SINGH (Eds.), Meeting Basic Learning Needs in the Informal Sector Integrating Education and Training for Decent Work, Empowerment and Citizenship (pp. 215-228). Hamburg, Germany: UNESCO Institute for Education.

[3] Brown, J.S., Collins, A., and Duguid, P. (1989). *Situated Cognition and the Culture of Learning*. Educational Researcher 18, no. 1 (January-February 1989): 32-41.

[4] Bukit, Masriam. (2002). Beberapa Masalah Dalam Implementasi Pendidikan Sistem Ganda di SMK. Dalam Dedi Supriadi (Ed.), Sejarah Pendidikan Teknik dan Kejuruan di Indonesia, Membangun Manusia Produktif, Dirdikmenjur Depdiknas, (p. 527-542). Bandung: Rosdakarya.

[5] Bungin, Burhan. (2005). *Analisis Data Penelitian Kualitatif*. Jakarta: PT Raja Grafindo.

[6] Chinien, C. and Singh, M. (2009). *Overview: Adult Education for the Sustainability of*

Human Kind. In R. Maclean, D. Wilson, & C. Chinien (Eds.), International Handbook of Education for the Changing World of Work, Bridging Academic and Vocational Learning (pp. 2521-2536). Germany: Springer.

[7] Cunningham, I., Dawes, G., & Bennett, B. (2004). *The Handbook of Work Based Learning*. Burlington: Gower Publishing Limited.

[8] Depdikbud. *(1995). Garis-garis Besar Program Pengajaran (GBPP). Jakarta: Pusat*

[9] Depdiknas. *(2005). Peraturan Pemerintah RI Nomor 19, Tahun 2005, tentang Standar* Nasional *Pendidikan*

[10] Depdiknas. *(2006). Peraturan Menteri Pendidikan Nasional RI Nomor 23, Tahun 2006, tentang Standar Kompetensi Lulusan untuk Satuan Pendidikan Dasar dan Menengah.*

[11] Depdiknas. *(2009). Rencana Strategis Departemen Pendiikan Nasional, Tahun 2010–2014,17 September 2009,Jakarta.*

[12] Depdiknas. *(2003). Undang-Undang RI Nomor 20, Tahun 2003, tentang Sistem Pendidikan Nasional.*

[13] Depnakertrans. *(2003). Undang-Undang RI Nomor 13, tahun 2003, tentang Ketenagakerjaan*

[14] Deseco. (2005). *Defining and Selecting Key Competencies: Executive Summary*. Diambil pada tanggal 19 September 2014 dari: http://www.oecd.org/pisa/35070367.pdf

[15] Dikmenjur. (1999). *Keterampilan Menjelang 2020 untuk Era Global: laporan satuan tugas pengembangan Pendidikan dan Pelatihan Kejuruan di Indonesia*. Depdikbud, Jakarta.

[16] Djojonegoro, Wardiman. (1998). *Pengembangan Sumberdaya Manusia Melalui Sekolah Menengah Kejuruan (SMK)*. Jakarta: PT Balai Pustaka.

[17] Finlay, I., Niven, S., & Young, S. (1998). *Changing Vocational Education and Training an International Comparative Perspective*. London: Routledge.

[18] Gray, K., Bae, S. (2009). *Skills Shortages, Over-Education and Unemployed Youth: An International Dilemma*. In R. Maclean, D. Wilson, & C. Chinien (Eds.), International Handbook of Education for the Changing World of Work, Bridging Academic and Vocational Learning (pp. 175- 187). Germany: Springer.

[19] Harris, I.B. (2011). *Conceptions and Theories of Learning for Workplace Education*. In Janet P. Hafler (Eds.), Extraordinary Learning in the Workplace, Innovation and Change in Professional Education 6, DOI 10.1007/978-94-007-0271-4_3. (pp. 39-62) New York: Springer + Business Media B.V.

[20] Hemsley-Brown, J.V., & Sharp, C. (2004). *The use of research to improve professional practice: a systematic review of the literature*, Oxford Review of Education. (fourth coming).

[21] Mason, J. (2002). *Qualitative Researching*, London: SAGE Publications Ltd.

[22] Matveeva, N. and Lapp, J. (2009). *Overview: Participation in Formal Programmes of Learning and Skills Development*. In R. Maclean, D. Wilson, & C. Chinien (Eds.), International Handbook of Education for the Changing World of Work, Bridging Academic and Vocational Learning (pp. 2025- 2037). Germany: Springer.

[23] McNamara, O., Murray, J., Jones, M.(2013). *Framing Workplace Learning*. In *Workplace Learning in Teacher Education*, ed. McNamara, O., Murray, J., Jones, M, 1-27. New York: Springer, 2013. eScholarID:196034

[24] Mendagri. *(2013). Peraturan Menteri Dalam Negeri Republik Indonesia Nomor 2 Tahun 2003, Tentang Pedoman Pengembangan Sistem Pendidikan Dan Pelatihan Berbasis* Kompetensi *Di Lingkungan Kementerian Dalam Negeri Dan Pemerintahan Daerah.*

[25] Merriam, S.B. (1998). *Qualitative Research and Case Study Applications in Education*. San Fransisco: Jossey-Bass.

[26] Miles, M.B., & Huberman, A.M. (2014). *Qualitative Data Analysis: A Methods Sourcebook. 3th edition*. USA: SAGE Publications, Inc.

[27] Miswardi, Y., & Pardjono. (2013). *Proses Dan Hasil Belajar Pada Prakerind Bidang Keahlian Kendaraan Ringan: Studi Kasus Pada Industri Pasangan SMKN 3 Yogyakarta*. Jurnal Pendidikan Vokasi UNY, Vol 3, Nomor 2, Juni 2013 pp.268-281.

[28] Moleong, Lexy J. (2002). *Metodologi Penelitian Kualitatif*. Bandung: Remaja Rosdakarya

[29] Munadi, Sudji. (2012). *Penerapan Assesmen Kinerja Dalam Praktik Proses Permesinan Untuk Pengembangan Budaya Kerja Peserta Didik*. Pidato Pengukuhan Guru Besar. Di Universitas Negeri Yogyakarta. Maret. 2012.

[30] OECD. (2010). *Innovative Workplaces: Making Better Use of Skills within Organisations*. OECD Publishing.

[31] Pakpahan, J. (2002). *Perkembangan Pendidikan Kejuruan Pada Pelita VI*. Dalam Dedi Supriadi (Ed.), Sejarah Pendidikan Teknik dan Kejuruan di Indonesia, Membangun Manusia Produktif, Dirdikmenjur Depdiknas, (p. 221-272). Bandung: Rosdakarya.

[32] Pardjono. (2011, Februari). *Peran Industri Dalam Pengembangan SMK*. Makalah disampaikan pada workshop Peran Industri dalam Pengembangan SMK di SMKN 2 Kasihan Bantul

[33] Poell, R.F. and Woerkom, M. V.(2011). *Chapter 1 Introduction: Supporting Workplace Learning*. Dalam *Supporting Workplace: Learning Towards Evidence-based Practice*. Rob F. Poell dan Marianne van Woerkom Eds. Netherlands: Springer. e-ISBN 978-90-481-9109-3

[34] Rau, D. (1998). *Transformation And Reform of Vocational Education And Training In Taiwan, Republic of China*. In I. Finlay, S. Niven, and Stephanie. *Changing Vocational Education and Training: An International Comparative Perspective*. (p.61-74). USA: Routledge.

[35] Setyabudi, Ismanto. (2014). *Kemunduran Pendidikan Sistem Ganda (PSG) di Indonesia*. Diakses pada tanggal 20 September 2014 dari http://www.vedcmalang.com/pppptkboemlg/index.php/menuutama/otomotif/944-kemunduran-pendidikan-sistem-ganda-psg-di-indonesia

[36] Slamet PH. (2013). *Pengembangan SMK Model untuk Masa Depan*. Jurnal Cakrawala Pendidikan, Februari 2013, Th. XXXII, No. 1. pp. 14-26.

[37] Soeharto. (1988). *Desain Instruksional sebuah Pendekatan Praktis untuk Pendidikan Teknologi dan Kejuruan*. Jakarta: Departemen Pendidikan dan Kebudayaan, Direktorat Jenderal Pendidkan Tinggi, Proyek Pengembangan Lembaga Pendidikan Tenaga Kependidikan.

[38] Starr, P. (1984). *The social transformation of American medicine: The rise of a sovereign profession and the making of a vast industry*. New York: Basic Books, dalam Definition of Professionalism-Social Sciences Literature. diakses tanggal 19 September 2014 dari https://www.afmc.ca/pdf/SocScsLit.pdf

[39] STEMNET. *Employability Skills*. Diakses pada 19 September 2014 dari: http://www.exeter.ac.uk/ambassadors/HESTEM/resources/General/STEMNET%20Employability%20skills%20guide.pdf

[40] Sudira, Putu.(2011). *Praksis Ideologi Tri Hita Karana Dalam Kebudayaan Kompetensi Pada SMK Di Bali*. Disertasi. Universitas Negeri Yogyakarta. Diakses tanggal 12 Juli 2014. Tersedia : http://staff.uny.ac.id/sites/default/files/131655274/DISERTASI-DR-PUTU-SUDIRA-UNY-2011.pdf

[41] Sudrajat, A.A., dkk. (2002). *Pendidikan Pertanian Dari Masa Ke Masa*. Dalam Dedi Supriadi (Ed.), Sejarah Pendidikan Teknik dan Kejuruan di Indonesia, Membangun Manusia Produktif, Dirdikmenjur Depdiknas, (p. 379-436). Bandung: Rosdakarya.

[42] Sudrajat, A.A. (2002). *Perkembangan SMK Bidang Teknik/Teknologi*. Dalam Dedi Supriadi (Ed.), Sejarah Pendidikan Teknik dan Kejuruan di Indonesia, Membangun Manusia Produktif, Dirdikmenjur Depdiknas, (p. 329-358). Bandung: Rosdakarya.

[43] ___________.(2011). *Pendidikan Vokasional Sebagai Investasi*. Paper Competition ASC

2011.

[44] Sukardi, Th. (2011). *Peranan Bimbingan Kejuruan Terhadap Pembentukan Karakter Kerja Siswa Di Jurusan Mesin SMKN 2 Wonosari*. Jurnal Cakrawala Pendidikan, Mei 2011, Th. XXX, Edisi Khusus Dies Natalis UNY.

[45] Supriadi, D. (2002). *Pendahuluan: Satu Setengah Abad Pendidikan Kejuruan di Indonesia*. Dalam Dedi Supriadi (Ed.), Sejarah Pendidikan Teknik dan Kejuruan di Indonesia, Membangun Manusia Produktif, Dirdikmenjur Depdiknas, (p. 1 – 32). Bandung: Rosdakarya.

[46] Sutiyono, Agus. (2010). *Pengaruh Pendidikan Dan Pelatihan Berbasis Kompetensi Competence Base Education And Training) Dan Motivasi Kerja Terhadap Kinerja Petugas Satuan Polisi Pamong Praja*. Disertasi. Universitas Negeri Jakarta. Diakses pada tanggal 10 Oktober 2014 pada http://unj.ac.id/fip/sites/default/files/Agus%20Sutiyono%20-%20Pengaruh%20Pendidikan%20dan%20Pelatihan%20Berbasis%20Kompetensi%20dan%20Motivasi%20Kerja%20terhadap%20Kinerja.pdf

Website-Based Personnel Information System Design in Vocational High School

J Widodo, C H A Dirgatama and A W Wirawan
Office Administration Education, Universitas Sebelas Maret, Indonesia

Email: kangfikri@staff.uns.ac.id, chairulhudaatmadirgatama@gmail.com, id.arifwahyu@gmail.com

Abstract. Education is a means of creating high-quality human resource. Science and Technology development, in addition to affecting industrial world, also affects education realm. A website-based personnel information system simulation application used for the learning of personnel administration management automation course in supporting learning media in Vocational High Schools throughout Ex-Surakarta Residency is designed using FAST (*Framework for the Application Of System Thinking*) model with object-oriented method and UML (*Unified modeling language*) tools. The use of FAST model made the website-based personnel information system simulation system designed more structured and more organized.
Keywords: Information System, Personnel, Website

1. Introduction

Education is a means of creating high-quality human resource. A high-quality human resource can build state and improve a nation's quality. Education becomes an important asset to life and inseparable from life, as without education an individual will not know any thing existing in this world, so that education is absolute in nature in individual, family, nation, and state's life through science and technology development.

Science and technology development in addition to affecting industrial world also affects education world in Indonesia. Technology affects positively not only social life but also education, because technology becomes more prevalent in education organization, and educators are expected to utilize digital tools to support teaching-learning process in the class (Akyuz & Yavuz, 2015; Gan, Menkhoff & Smith, 2015). It can be seen from the infrastructures and learning media used by educators in delivering material to students. Technology advance enable educators to innovate in order to create a learning media that can support the students' process of understanding the teaching material, either theoretically or practically (Kingsley, 2007; Norman & Furnes, 2015). The use of learning media, in addition to facilitate the delivery of material from teachers to students, can also improve students' interest in a course (subject).

The learning appropriate to apply to the 2013 curriculum is the one emphasizing on learning by doing as an attempt of practicing the students' competency. The competency is expected to improve problem solving and critical thinking abilities of students in the learning process. During learning process, the competencies that can be implemented through scientific approach are, among others: observing, questioning, presenting, reasoning, and trying. Those competencies can be optimized by applying student-centered learning, so that students can learn independently to solve the problems given during learning process.

Considering Focus Group Discussion (FGD) of studies in 2018 along with Office Administration teachers, it can be seen that teachers want the availability of learning source in the form of 2013 curriculum-based learning media. Office administration specialty teachers still lack of learning media to support theory through the 2013 curriculum-based learning media. Educators are expected to have an ability of improving their professionalism through the learning media development itself, so that the development conducted can answer the students' problem in the learning process. Learning media is expected to facilitate the students in learning, so that they can learn according to their ability and their competency can be accomplished.

The objective of research is to provide a website-based personnel information system design in Office Administration Management Automation Course in Vocational High School. The presence of website-based personnel information system design is expected to help students implement learning media and to facilitate them to understand the material and the procedure of personnel administration delivered by educators corresponding to business and industrial world.

2. Methodology

The research conducted was Research and Development (R&D). The finished product expected from research and development should meet the existing criteria: a new product or a product different from the preexisting one and ready to apply and develop in the field. The product is produced through evaluation and perfecting process in order to met the criteria specified in the beginning of study.

The research procedures carried out are based on the stages proposed by Gall, Gall & Borg (2003: 570-571), in this study namely: (1) The Exploration Stage, (2) The Draft Model Development Stage, (3) The Model Testing Stage, And (4) Product Dissemination. The place of research was conducted at the Vocational School of Business and Management which has the area of office administration expertise in the Ex-Residency of Surakarta (1) Exploration, (2) Model Draft Development, (3) Model Trial, and (4) Product Dissemination. The research was conducted in Vocational High School in Business and Management field having Office Administration specialty throughout Ex-Surakarta Residency.

Data validation was conducted using source and method triangulations. Source triangulation was used to examine data validity from different source, while method triangulation was used to examine data through different method. Technique of analyzing data used in this study was mix method. Qualitative data was analyzed using interactive model of analysis (Miles & Huberman, 1992: 20), while quantitative data was analyzed using statistic descriptive technique and variance test. Mix method strategy was used in this study in order to obtain a complete analysis corresponding to the characteristics of data obtained and the need.

3. Result And Discussion

3.1 *Use Case Diagram*

Use Case diagram depicts how the users interact with the system used, use case was used to find out the functions contained inside an information system and who are entitled to use the functions (A.S, Rosa & Shalahuddin, 2013). In a personnel information system simulation application, the users can interact with, among others: Dashboard, Management, Report and Logout.

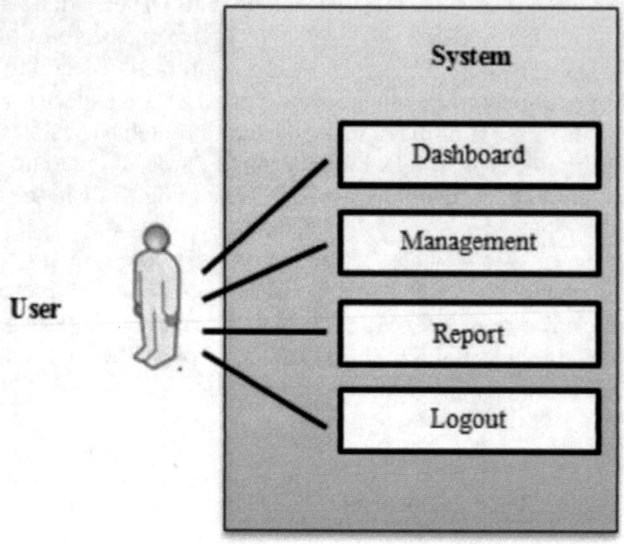

Figure 1. Use Case Diagram

3.2 *State Chart Diagram*

State Chart Diagram depicts a design of personnel information system simulation and activity application in corresponding Company/Institution. The design of personnel information system simulation is developed as follows.

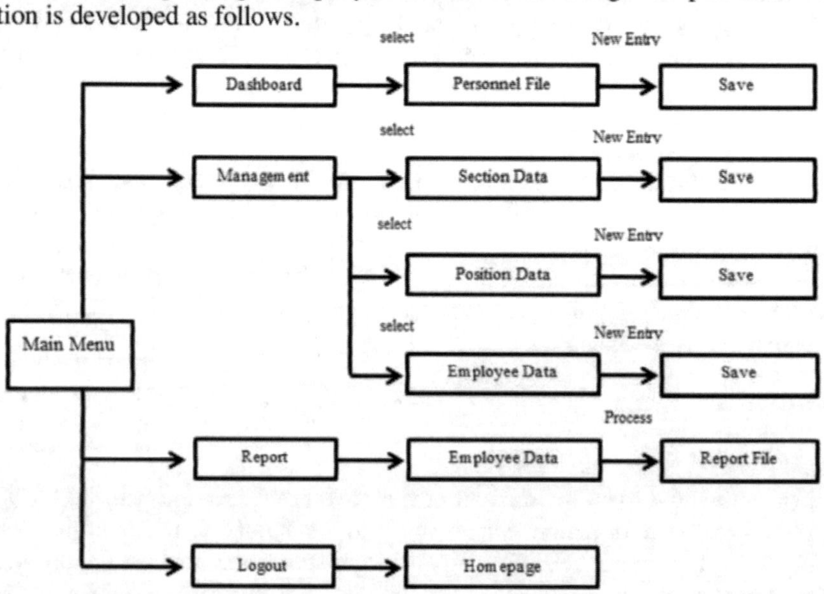

Figure 2. State Chart Diagram

3.3 *Activity Diagram*

In the model ran, Activity Diagram can be used to elaborate the operational working procedure of components existing in the system step by step. Activity Diagram depicts entire control path. The Activity Diagram of personnel information system simulation made and designed is shown in the figure below.

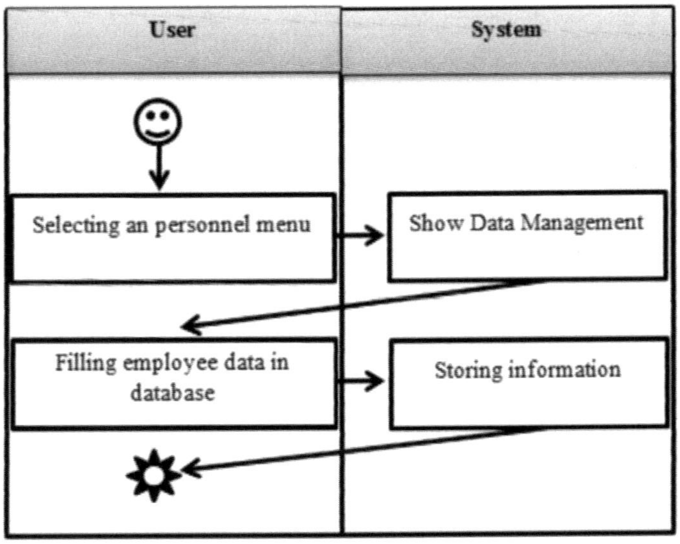

Figure 3. Activity Diagram of Personnel Information

Figure 3 it is explained that the user selects the personnel menu, then the system will display employee data management. In addition, users can also fill employee data in the database in managing employee data needed. In this case, the system will store the data in the database as database information on the personnel information system and if the data will be searched, then the data sought will be found on the system and will display information relating to the employee, whereas if no word is found it will not be found and the system will inform the user to repeat by entering the employee's name again.

3.4 *Application Display*

This desktop application display is the front display; having been by media expert, material expert, and practitioners, this will be developed into android-based online application in playstore.

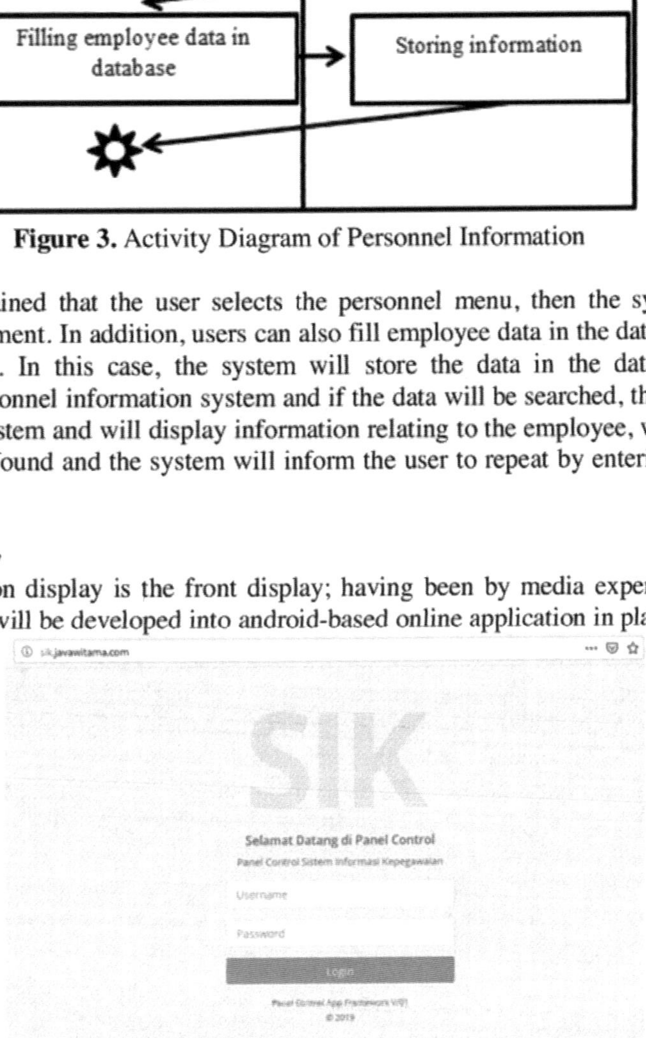

Figure 4. Desktop Display of Personnel Information System Simulation

3.5 *Display of Personnel Administration Material*

This display is Personnel Administration Management Automation teaching material to facilitate and to enable the students to understand concept and practice of personnel information system simulation application.

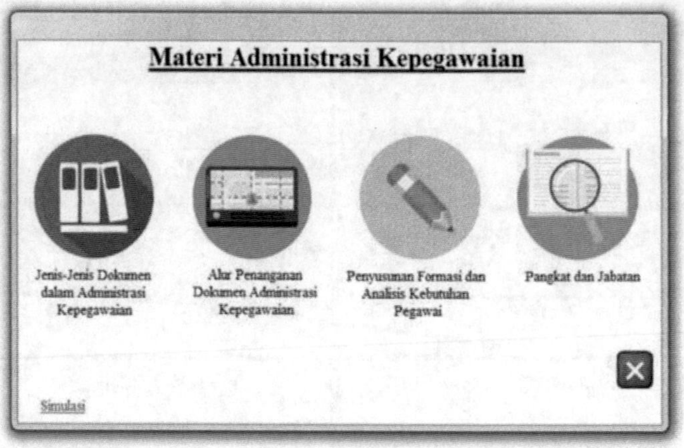

Figure 5. Desktop Display of Personnel Administration Material

System is a set of interrelated, interacting, and interdependent elements or variables to achieve the objective, while information is a data that has been processed in such a way that is more meaningful to its users (Tohari, 2014).

Personnel information system is a systematic procedure to collect, to store, to maintain, to attract, and to validate data needed by an organization concerning human resource, personnel activities, and organizational unit Simamora, 2001; Septyana, 2011; Tantra, 2012). To develop a more dynamic design, website approach that is dynamic in nature is used to facilitate the design and the development of web-based personnel information system application.

FAST (Framework for the Application of System Thinking) defines the procedure used to identify and to evaluate expected problems, opportunities, constraints occurring and needs so that some corrective measures can be recommended. The procedure of FAST builds on problems and opportunities encountered with the improvement expected from the system to be developed (Whitten, 2004).

4. Conclusion

A web-based personnel information system simulation application used for the learning the Personnel Administration Management Automation course is designed using FAST (Framework for the Application of System Thinking) with object-oriented method with UML (Unified modeling language) tools. The use of FAST model makes the web-based personnel information system simulation application designed and made more structured and organized. The web-based personnel information system simulation is used in Personnel administration management automation to enable students and educators to simulate personnel job more easily and simply involving: inputting, searching for personnel data, managing personnel data, storing data in database more effectively and efficient, and preventing the inputted data from being lost. This application is used to simulate web-based personnel information system to support the learning media in Vocational High Schools throughout Ex-Surakarta Residency.

5. References

[1] Akyuz & Yavuz. 2015. Digital Learning in EFL Classrooms, Procedia - Social and Behavioral Sciences.Vol.197, pp 766-769.

[2] Gall, D, Meredith., Gall, P, Joyce., Borg, Walter, R. 2003. Educational Research: An Introduction. Boston: Allyn and Bacon. Human Behavior. Vol.54, pp 301–309.

[3] Gan, B., Menkhoff, T., & Smith, R.2015. Enhancing students' learning process through interactive digital media: New opportunities for collaborative learning, Computers in Human Behavior. pp1-12, http://dx.doi.org/10.1016/j.chb.2014.12.048.

[4] Kingsley, K. V. 2007. Empower diverse learners with educational technology and digital media. Intervention in School and Clinic.Vol.43, No.1, pp 52–56.
[5] Miles, Matthew dan Huberman, A. Michael.1992.Analisis Data Kualitatif: Buku Sumber Tantang Metode-Metode Baru. Jakarta:UI Press.
[6] Norman, E., & Furnes, B.2016.The relationship between metacognitive experiences and learning: Is there a difference between digital and non digital study media?, Computers in Human Behavior.Vol.54. pp 301–309.
[7] Septyana,. Nanda. 2011.Sistem Informasi Kepegawaian Pada Dinas Perhubungan Komunikasi dan Informatika Kabupaten Cilacap, Yogyakarta.
[8] Simamora, Henry.2001.Manajemen Sumber Daya Manusia. Cetakan 3. STIE YKPN: Yogyakarta.
[9] Tantra, Rudy. 2012.Manajemen Proyek Sistem Informasi. Andi Offset: Yogyakarta.
[10] Tohari, Hamim.2014.Analisis Serta Perancangan Informasi Melalui Pendekatan UML.Andi Offset: Yogyakarta.
[11] Whitten, L.J. et al.2004.Metode Design dan Analisa Sistem. Edisi 6. Andi Offset: Yogyakarta.

Acknowledgement
Thank to Research institutions and community service of Universitas Sebelas Maret and Subject Teacher Discussion (MGMP) of Office Administration teachers throughout Surakarta Ex-Residency.

Design and Analysis of Database Clustering System of Micro, Small and Medium Enterprises (MSME) in Bantul Regency

Rr. Y Rachmawati, M Sholeh, and Suraya
Department of Informatics Engineering, Faculty of Industrial Technology, Institut Sains & Teknologi AKPRIND Yogyakarta, Yogyakarta, Indonesia

Email: muhash@akprind.ac.id

Abstract Information system support is really important for the existence of MSME in economic competition today. The implementation of information system for MSME is expected to develop the competitiveness through the added value for products and services. One of the examples of the utilization of information system implemented for MSME is the geographical information system which informs the location and the potential MSME. In order to present the information of the potential of MSME optimally in the geographical information system, analysis and design should be made as complete as possible. In this research, the geographic-based information system is developed for the MSME businessmen in Bantul. The system design is started by building a database. The purpose of this research is to design a database that is designed to avoid duplication of data and simplify the storage process and when displaying data. The process of geographical information system database of MSME is detailed into tables. The database process of the MSMES geographic Information system is detailed into tables. By being detailed into some of these tables the duplication can be avoided. Table designed between MSMES table that stores SME master data such as latitude and longitude position, MSME detail table, video, gallery, Medkom and tables that store the location of the area, such as sub-district table, Village table, Dusun table and Clustered tables from MSMES. The implementation of database design is developed into geographical information system using PHP language and utilizing google map to display the geographical position of MSME businessmen. The next application development can be developed based on Android.

1 . Introduction

The current era of disruption must be faced by an adaptation process, especially for business people. In the disruption era, the use of information technology becomes an important factor to attract the customers. The disruption era facilitates the customers and procedures. Facilitating the customers and procedures during the transaction becomes a challenge that must be faced by business people.

MSME is one of economic agents should be improved and be ready to face the disruption era. The development of Information and Communication Technology (ICT) gives new opportunity which overcomes some of the MSME problems. Although they are not the main problems, the opportunity is really huge and based on the field condition, meaning that the MSME has not fully adopted the Information and Communication Technology compared to other big companies. The ICT can be utilized for doing promotion, introducing the location/centres of MSME in a location, and obtaining other advantages such as dialogue, discussion, and online consultation with the customers, so the customers can be involved proactively and interactively in designing, developing, marketing, and marketing product.

According to the Indonesian Constitution No.20 of 2008 concerning on Micro, Small, and Medium Enterprise (MSME), it is said that the criteria of Micro Enterprise is having net worth of at most Rp50,000,000, excluding land and building of business premises; or having annual sales of at most Rp300,000,000.00 (three hundred million rupiahs). The criteria of small-scale enterprise is having net worth of more than Rp50,000,000 - Rp500,000,000 (five hundred million rupiahs) excluding land and building for business; or having annual sales of more than R300,000,000 - Rp2,500,000,000. Meanwhile, the criteria of medium enterprise is having net worth of more than Rp500,000,000 - Rp10,000,000,000, excluding land and buildings of business premises; or having annual sales of more than Rp2,500,000,000 - Rp50,000,000,000 (fifty billion rupiah).

Specifically in the Special Region of Yogyakarta, Small Micro Industry production growth in the second quarter of 2016 compared to the first quarter of 2016 experienced positive growth, which amounted up to 4.03%. The growth in the second quarter of 2016 was in line with the growth rate of the production of small micro industries at the national level, which experienced positive growth of 5.74%. [1]

Some related studies on the use of ICT for MSME businessmen have been conducted by researchers with different hypotheses in each research. The problems of MSME in utilizing ICT are not only problems in Indonesia, but also other countries. The research conducted by Olise to Peel SMES in Nigeria [2] .In Nigeria, MSME businessmen have not fully used ICT; MSME businessmen are recommended to use ICT and must arrange policies to facilitate the implementation of ICT. This facility is really essential due to its high potential in developing performance of MSME growth; MSME owners should try to develop their productivity and competitiveness; to develop the productivity and competitiveness, SME owners should invest in ICT as well as its components since it has been proven that ICT influences organization performance significantly.

Based on the research conducted by Valdez-Juárez [3], it is apparent that ICT implementation in SMES carries a competitive advantage., it is shown that the implementation of ICT in MSME brings competitive advantage. The analysis result of the knowledge management helps MSME businessmen to understand the condition where MSME might be more effective in developing innovation and performance. One of the important findings of this research focuses on the positive correlation between strategy and policy of innovation management in MSME through the knowledge (ICT). The research conducted is limited to the use of ICT for MSME, and has not examined the specific application used to track the location of MSME.

A paper written by Rulhudana, [4] focuses on the pattern of empowerment or the use of ICT hardware and software in the business management process of MSME which forms two main patterns, namely the technology is able to give advantages (usefulness) and easy to use. It is seen in MSME Buana Property which utilizes/empowers ICT during the process of business management using ICT. This research more focuses on the use of ICT to empower an MSME and has not discussed the use of ICT for promotion and distribution process.

The optimal use of ICT does not only support the administration but also marketing process. Therefore, the users except those involved in MSME are able to know the potential managed by MSME. The research conducted by Susanto [5] examines the system development method used for marketing, selling, and payment process.

Setiawan in this research [6], it is discussed about business process in batik industry. In batik processing, the process covers not only the process of making batik, but also several other processes that can be carried out by implementing technology innovation, such as the research and model design, administrative process, inventory control process, marketing process, communication process, and integration with other external parties. Although the processes have not implemented technology maximally, especially Information and Communication Technology (ICT), the Laweyan Community has started to use ICT in its business processes even though its penetration is still very low; out of 60 (sixty) businessmen becoming the members of the Laweyan community, only 10% have implemented ICT in their business activities.

2 . Research Methodology

The development of information system is started by designing database. The purpose of designing database is to obtain information containing the users' needs specifically and its implementation. Database is a set of data which is interconnected one another, and certain software is used to manipulate it. Database is one of the important components in information system since it contains data which provides information for the users.

The Description of diagram of research flow on figure 1

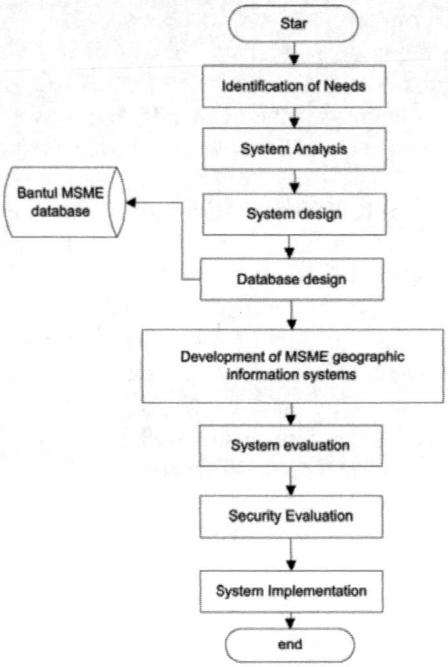

Figure 1. Diagram of Research Flow

Generally, diagram of research flow is:

2.1 Need Identification
- Conducting a survey, especially on what information will be presented. The information must be adjusted to the wishes of the user (MSME).
- Identifying MSME in Bantul
- Clustering MSME
- Mapping the position of latitude and longitude of the location of MSME

2.2 System Analysis
- Doing primary data inventory through interview and questionnaires. Interview and questionnaires is conducted with MSME staff, especially in Bantul.

2.3 System Design
- System design, proposing logical problem solving.
- Design detail, conducting system design of problem solving in detail.

2.4 Database Design
- Designing database structure used to store the data of MSME
- Evaluating database design to avoid data duplication

2.5 Information System Coding
Program coding is the translation step of system design made in the form of commands understood by computer. Programing language of PHP, HTML, CSS, and Jquery is used to create software in We category. Database uses MySQL.

2.6 Program Testing
Program testing is the testing step of system which has been successfully translated in listing program. In this step, all processes of input and output are tested, so the possible error and bug can be immediately known and the reparation on program coding can be performed. Trial is carried out locally on computer.

2.7 Implementation
The implementation of program is a step in which the developer team implements software, that has been created and tested, and provides training to the users. When conducting training, it is important to make sure that the community can use without any difficulties.

3 . Result and Discussion

3.1. Clustering ICT-based MSME

Clustering ICT-based MSME which is developed is the process of classifying MSME based on some classifications, such as clustering based on location, types of business of MSME businessmen, types of commodity, and Standard Classification of Indonesian Business Field. Types of MSME are included in the classification of micro, small, and medium enterprise. Detail of clustering process is as follow:

a. Clustering based on area is differentiated by MSME area. The area will be detailed from district, sub-district, village, and RT/RW. Clustering based on area is expected to ease the process of searching of MSME businessmen in an area.
b. Clustering based on the types of business is classified based on the capital or expense of MSME, such as micro, small, and medium enterprise. The clustering process aims to ease the data collection of the expenses of MSME in certain area or all areas in Bantul.
c. Clustering based on the types of commodity of MSME, such as food business like meatball, soto, chicken noodles, clothing business, convection business, or others.
d. Clustering based on Standard Classification of Indonesian Business Field [7].

3.2. Identification of Users' Need

Database design process is started by identifying the users' need and purpose of the application which is developed. To make the process of designing database is in accordance with the need, and to keep the database design away from data duplication, all needed data are classified. According to the purposes of the application which is developed, such as clustering MSME based on either location or types of MSME and illustration of database system should represent the whole illustration of the system. Database design is classified into the main data and supporting data. The main data is a master file of which data growth is static, such as data related to the location in urban to rural areas. Supporting data is data from MSME, and the data is dynamic and grows. MSME data becomes one of the data used in clustering process.

3.3. Database Implementation

Database implementation process is carried out through normalization process. The purpose of normalization process is to eliminate and to reduce data redundancy and to make sure data dependency, as well as data stored in the right table. During normalization process, all tables designed is analyzed whether the table contains deviations, especially in the process of adding data, removing data, and editing data.

Normalization process obtains table that has been avoided from:

3.3.1 INSERT Anomaly: a situation in which it is not possible to include several types of data directly to the table before putting the main file. One of the examples of analysis process is adding data of MSME businessmen on the table of SME should be conducted first. If the data of MSME businessmen has already existed in the MSME tables, the process of adding data related to MSME can be done on the supporting tables. The process of adding data directly to the tables, such as produk_ukm table and galeri-foto table cannot be carried out if the data of MSME businessmen have not included on SME table.

3.3.2 DELETE Anomaly: Deletion of unexpected data is only done on certain tables, while the same data in other tables is not deleted, meaning that the data that should not be deleted may unintentionally be deleted or data that should have been deleted is not deleted. The analysis process carried out includes the deletion of MSME data which cannot be immediately deleted in MSME tables, deletion can be done if UMKM data deleted in the produk_ukm table and galeri-foto table has been deleted first.

3.3.3 UPDATE Anomaly: this is a situation where changed values cause database inconsistencies; data that is changed is not as ordered or desired. The analysis process carried out includes changes in data which can only be done on general fields. Changes to the field that becomes the primary key cannot be made; data changes are made to the table according to the desired data changes.

The results of the database design and after implementing analysis using mysql there are some table-making processes, namely:

- Making Regency table:

```
CREATE TABLE `alamat_kabupaten` (
  `kabupaten_id` int(11) NOT NULL AUTO_INCREMENT,
  `kabupaten_nama` varchar(100) DEFAULT '',
  `kabupaten_latitude` varchar(200) DEFAULT NULL,
  `kabupaten_longitude` varchar(200) DEFAULT NULL,
  `date_created` datetime DEFAULT current_timestamp() ON UPDATE current_timestamp(),
  `date_updated` datetime DEFAULT NULL,
  `user_created` int(11) DEFAULT NULL,
  `user_updated` int(11) DEFAULT NULL,
  PRIMARY KEY (`kabupaten_id`)
) ENGINE=MyISAM AUTO_INCREMENT=10 DEFAULT CHARSET=latin1 ROW_FORMAT=DYNAMIC;
```

- Making UMKM table

```
CREATE TABLE `alamat_ukm` (
  `alamat_ukm_id` int(11) NOT NULL AUTO_INCREMENT,
  `ukm_id` int(11) NOT NULL,
  `kabupaten_id` int(11) NOT NULL,
  `kecamatan_id` int(11) NOT NULL,
  `kelurahan_id` int(11) NOT NULL,
  `dusun_id` int(11) NOT NULL,
  `rw_id` int(11) DEFAULT NULL,
  `rt_id` int(11) DEFAULT NULL,
  `alamat_detail` varchar(200) NOT NULL,
  `alamat_jenis` varchar(10) DEFAULT 'cabang',
  `alamat_latitude` varchar(200) DEFAULT NULL,
  `alamat_longitude` varchar(200) DEFAULT NULL,
  `date_created` datetime DEFAULT current_timestamp() ON UPDATE current_timestamp(),
  `date_updated` datetime DEFAULT NULL,
  `user_created` int(11) DEFAULT NULL,
  `user_updated` int(11) DEFAULT NULL,
  PRIMARY KEY (`alamat_ukm_id`) USING BTREE
) ENGINE=MyISAM AUTO_INCREMENT=15 DEFAULT CHARSET=latin1;
```

3.4. Relations between Tables

Relations between tables become the main key in the implementation of database. In programming, the needed information during the process of displaying data is the result of query of interrelated tables. Figure 4 is the result of relation between tables in the database. The relation will connect all existing tables. The main relation is done by relating alamat_kabupaten table, alamat_kecamatan table, alamat_kelurahan table, RW table, and RT table. The relation aims to display the location of MSME. Meanwhile, other relations connect UKM table to other supporting tables for MSME, such as produk_ukm table and galeri-foto table.

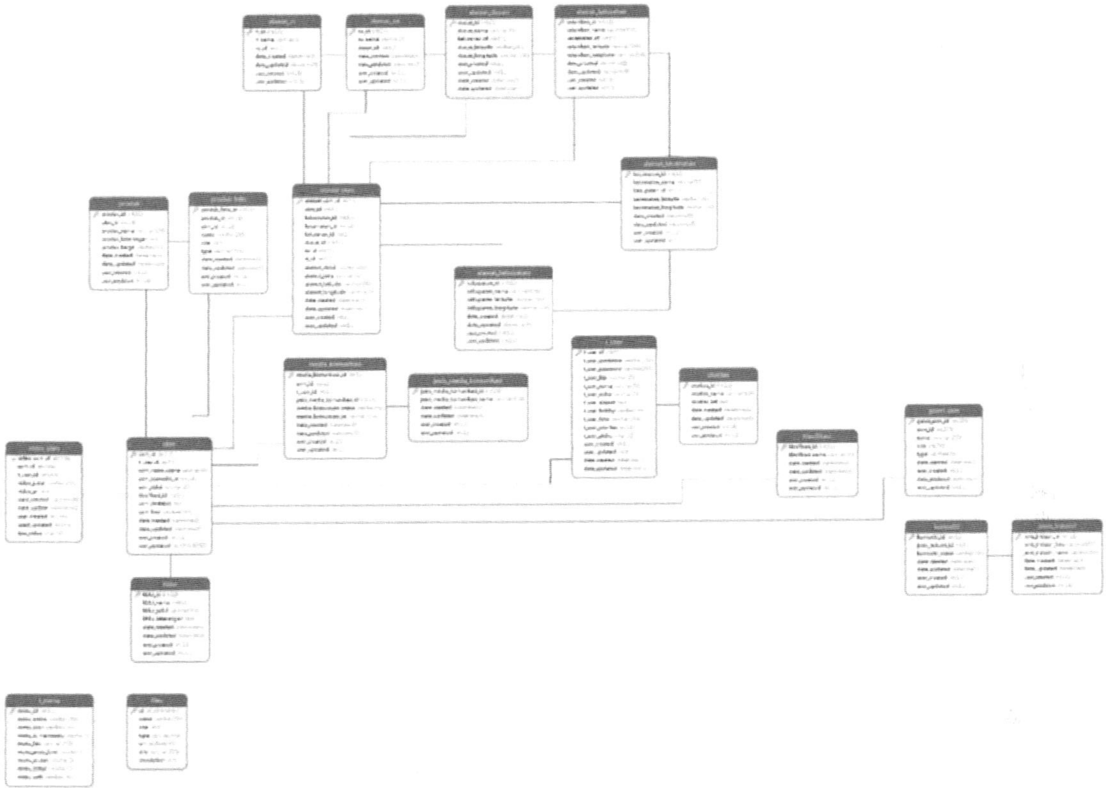

Figure 2. Relationship between MSME database tables

Database design consisting of some small tables aims to avoid duplication during the process of storing data. Although the data is stored in the tables, the process of creating information does not become a problem. By using query, the process of displaying information can be done easily. Some of the results of query are as follow:

- Displaying information of the location of MSME

```
SELECT
` alamat_ukm_id `, ` ukm_nama_usaha `, ` kabupaten_nama `, ` kecamatan_nama `, ` kelurahan_nama `, `
dusun_nama `, ` rw_nama `, ` rt_nama `, ` alamat_detail `, ` alamat_jenis `, ` alamat_latitude `, `
alamat_longitude `
FROM ` alamat_ukm `
JOIN ` ukm ` ON ` ukm `.` ukm_id `=` alamat_ukm `.` ukm_id `
JOIN ` alamat_kabupaten ` ON ` alamat_kabupaten `.` kabupaten_id `=` alamat_ukm `.` kabupaten_id `
JOIN ` alamat_kecamatan ` ON ` alamat_kecamatan `.` kecamatan_id `=` alamat_ukm `.` kecamatan_id `
JOIN ` alamat_kelurahan ` ON ` alamat_kelurahan `.` kelurahan_id `=` alamat_ukm `.` kelurahan_id `
JOIN ` alamat_dusun ` ON ` alamat_dusun `.` dusun_id `=` alamat_ukm `.` dusun_id `
LEFT JOIN ` alamat_rw ` ON ` alamat_rw `.` rw_id `=` alamat_ukm `.` rw_id `
LEFT JOIN ` alamat_rt ` ON ` alamat_rt `.` rt_id `=` alamat_ukm `.` rt_id `
```

- Displaying potential of MSME

```
SELECT
` ukm `.` ukm_id `, ` ukm `.` ukm_nama_usaha `, ` ukm `.` ukm_foto `, ` ukm `.` ukm_deskripsi `, `
alamat_ukm `.` alamat_detail `, ` alamat_ukm `.` alamat_latitude `, ` alamat_ukm `.` alamat_longitude `, `
jenis_industri_nama `, ` t_user_nama `, ` t_user_foto `
```

```
FROM ` ukm `
LEFT JOIN ` t_user ` ON ` t_user `.` t_user_id `=` ukm `.` t_user_id `
LEFT JOIN ` komoditi ` ON ` komoditi `.` komoditi_id `=` ukm `.` ukm_komoditi_id `
LEFT JOIN ` jenis_industri ` ON ` jenis_industri `.` jenis_industri_id `=` komoditi `.` jenis_industri_id `
LEFT JOIN ` klasifikasi ` ON ` klasifikasi `.` klasifikasi_id `=` ukm `.` klasifikasi_id `
LEFT JOIN ` kblui ` ON ` kblui `.` kblui_id `=` ukm `.` ukm_kblui `
LEFT JOIN ` alamat_ukm ` ON ` alamat_ukm `.` ukm_id `=` ukm `.` ukm_id `
WHERE
` ukm `.` ukm_id ` = `ukm_id`
```

3.5. Analysis of database constraints

Another important analysis in database management is data integrity or constraints in database. Database constraints are basically the limitation of values which are allowed to be included in a field in a table. By implementing the limitation in a table, only data that matches these limitations can be put in the table. The main function of constraints in a table is to ensure the consistency and integrity of data included in a table. The result of the analysis is:

```
mysql> ALTER TABLE alamat_ukm ADD FOREIGN KEY (kabupaten_id) REFERENCES alamat_kabupaten (kabupaten_id) ON UPDATE CASCADE ON DELETE CASCADE;
```

These restrictions aim to maintain data consistency when deleting data. With this limitation, if deletion is carried out on district data, all data in the address table will be deleted, especially the same data.

Compliment :

```
mysql> SELECT kabupaten_id, kabupaten_nama from alamat_kabupaten ORDER BY kabupaten_id DESC LIMIT 2;
+--------------+-----------------+
| kabupaten_id | kabupaten_nama  |
+--------------+-----------------+
|           13 | klaten          |
|            8 | Kota Yogyakarta |
+--------------+-----------------+
2 rows in set (0.69 sec)

mysql> select alamat_ukm_id, kabupaten_id, kecamatan_id from alamat_ukm order by alamat_ukm_id desc limit 2;
+---------------+--------------+--------------+
| alamat_ukm_id | kabupaten_id | kecamatan_id |
+---------------+--------------+--------------+
|            20 |           13 |         NULL |
|            14 |            8 |           22 |
+---------------+--------------+--------------+
2 rows in set (0.18 sec)

mysql> DELETE FROM alamat_kabupaten WHERE kabupaten_id = '13';
Query OK, 1 row affected (1.55 sec)

mysql> SELECT kabupaten_id, kabupaten_nama from alamat_kabupaten ORDER BY kabupaten_id DESC LIMIT 2;
+--------------+-----------------+
| kabupaten_id | kabupaten_nama  |
+--------------+-----------------+
|            8 | Kota Yogyakarta |
|            7 | Gunung Kidul    |
+--------------+-----------------+
2 rows in set (0.19 sec)

mysql> select alamat_ukm_id, kabupaten_id, kecamatan_id from alamat_ukm order by alamat_ukm_id desc limit 2;
+---------------+--------------+--------------+
| alamat_ukm_id | kabupaten_id | kecamatan_id |
+---------------+--------------+--------------+
|            14 |            8 |           22 |
|            12 |            6 |           18 |
+---------------+--------------+--------------+
2 rows in set (0.19 sec)
```

4 . CONCLUSION

The results of database design need to be done to examine the tables that have been designed whether they have been avoided from the problems during the process of adding, deleting, and editing data. Limitations in a database must be defined in database design. With this restriction, it is certainly influential in making information system. In this study, the database design has been analyzed, so all boundaries have been defined in the database implementation. Database implementation using MYSQL and all restrictions like setting data deletion conditions can only be done if all data in the relation table has been deleted. The process of deleting data from one of the MSMEs in alamat_ukm, table, then the MSME data in the galeri_ukm table and foto_ukm table must be deleted first. After the data in the galeri_ukm table and foto_ukm table is deleted, the deletion process of the desired UMKM data can only be deleted in the parent table (alamat_ukm table).

5. References

[1] BPS, "Perkembangan UMKM pada Periode 1997 -2013," BPS, Jakarta, 2014.

[2] M. C. Olise, "aktor Penentu Adopsi TIK untuk Peningkatan Kinerja UKM di Anambra Negara, Nigeria," *American International Journal of Contemporary Research,* vol. 4, no. 7, pp. 163-176, 2014.

[3] L. E. Valdez-Juárez, "Management of Knowledge, Innovation and Performance in SMEs," *Journal of Information, Knowledge, and Management,* pp. 142-176, 2016.

[4] F. Rulhudana, "Pola Penggunaan TIK dI Manajemen Bisnis UKM (Studi Kasus pada UKM Buana Property di Provinsi Bali)," *Jurnal Penelitian Komunikasi dan Opini Publik Vol. 19 No. 2, Agustus 2015,* pp. 113-126, 2015.

[5] A. Susanto, "Rekayasa M-Market (Mobile Market) untuk Kelompok Usaha Pemuda Binaan Dinas Pemuda dan Olahraha PropinsiI Jawa Tengah Sebagai Upaya Peningkatan Penjuala Produk UMKM," *Techno.COM, Vol. 14, No. 1, Februari 2015,* pp. 8-12, 2015.

[6] A. B. Setiawan, "Studi untuk Pemanfaataan Teknologi dan Komunikasi dan Komunikasi untuk Aktivitasn dan Strategi Bisnis pada UMKM Batik (Studi Kasus pada Komunitas Batik Laweyan, Kota Solo," Pusat Litbang Aplikasi Telematika dan Sarana Komunikasi dan Diseminasi Informasi, Badan Litbang SDM, Kementerian Komunikasi dan Informatika, Jakarta, 2013.

[7] Anonim, "Klasifikasi Baku Lapangan Usaha Indonesia 2015," Badan Pusat Statistik, Jakarta, 2015.

The Design of Relational Database for Multipurpose WebGIS Applications

E Sutanta and E K Nurnawati

Department of Informatics Engineering, Institut Sains &Teknologi AKPRIND Yogyakarta
Kalisahak Street #28, Yogyakarta, Indonesia

E-mail: edhy_sst@akprind.ac.id, ernakumala@akprind.ac.id

Abstract. GIS concepts and applications evolved from a desktop-based into web-based and mobile. These changes require a new approach in the GIS database, from spatial data separation and attribute data into a database structure approach. The problem is the database for GIS is generally designed just for specific users, objects, and territory scopes, so it can't be applied to other applications. This paper proposed a relational database model (RDBM) design that can accommodate new changes and needs in the user's aspect, the objects that are processed and displayed, and the scope of the region in GIS Web Applications (WebGIS). The study was conducted in 3 stages, namely a preliminary study to obtain an overview of the concept, design, and application of RDBM on various WebGIS. Next, identified potential problems in database design for WebGIS, and finally proposed an alternative solution in the form of a new database design template that could accommodate the possibility of new changes and needs. Database design is done using MySQL. The database design process for WebGIS applications needs to anticipate the possible growth and development of different user group data, object categorization to subcategories, and the scope of the object area. Designs that meet these criteria can be done from the beginning of the design process, that is by changing two things. First, change the view of the database designer from the user view level to the conceptual view level so that the database design results can be used by several different applications. Second, expand the view of the database designer from the specific scope into a broader scope in the user aspect, object categorization, and territory coverage. Thus, the database design obtained will be in accordance with the RDBM concept.

1. Introduction

GIS concepts and applications continue to evolve. GIS is a system designed to work with spatially-referenced data or geographic coordinates, having the ability to receive, manage, analyze, and display data [1]. The first computer-based GIS application was CGIS (Canadian GIS) implemented by the Department of Energy, Mining and Resources, Ottawa, Ontario, Canada in 1967 [2]. The development of microcomputer has spurred ESRI, CARIS, MapInfo, and other vendors incorporating spatial data separation approaches and attributes on GIS applications into database structures. Industrial developments in the 1980s and 1990s enabled GIS to be applied to UNIX workstations and personal computers, and since then GIS applications have been developed for diverse fields and objects [3]. ICT alterations have had an impact on GIS applications, from desktop-based, changed to web-based applications (WebGIS), and are currently mobile-based. In general, GIS applications can be divided into two groups, namely desktop-based and distributed GIS. Distributed GIS consists of WebGIS and Mobile GIS applications [2].

The availability of easily accessible online maps with no geographical restrictions and a wide range of attractive display options is one of the key drivers of WebGIS application development. For example, the results of fast-tracking on the internet using the keyword "WebGIS application" found as many as 66,400 posts. If keyword replaced using "WebGIS application example" found 17,000 posts, on the other hand, changed using the keyword "WebGIS application research" found 13,400 words. Furthermore, specifically if using the keyword "WebGIS application thesis" found 6,330 posts. The result shows that WebGIS application has become an interesting topic for many developers. Some examples of GIS applications for college have been developed by[3] and [4]; a tourist attraction or culture found on [5], [6], [7], [8], [9], [10], and [11]; layout by [12]; industry by [13] and [14]; health services on [15] and [16]; Other objects such as houses of worship, boarding houses, cargo agents are found on [17], [18], [19], and [20]. Most of these applications are WebGIS and other mobile-based, mostly developed using open source software and a few other paid apps. WebGIS applications are also widely developed by government agencies in Indonesia, i.e., found at http://gis.pusair-pu.go.id/, http://kewilayahan.bantulkab.go.id/, and https://wartacara.com/ukm/.

GIS applications outline has the same model with computer-based processing, including input, process, and output. The sources of input data on GIS can come from tables, reports, field measurements, digital data, maps, satellite images, aerial photographs, and others. The entire data entered is stored in the GIS database. GIS database processing in principle is to store the inputted data into the database, manipulate the data as needed, and retrieve data to be displayed as the outcome of the application. There are several form of GIS application, i.e., maps, tables, reports, digital information, and other forms according to user needs.

Like the GIS desktop application, WebGIS requires database support that contains graphical data (vector and raster) and attributes (contains tables of tables that relate to graphical databases). Currently, database management for GIS applications is mostly done by utilizing DBMS (Database Management System) software. The main reason for the convenience store (store) and retrieve (retrieve) data. While other reasons are the innate nature of the DBMS, including related to the problem of physical data independency, data redundancy, logical data independency, access speed, data standardization, and data integrity [21]. The database approach in GIS applications also facilitates management and avoids errors due to data loss or duplication.

The most important support of the database in a GIS application is to provide the attribute data query function to obtain a spatial position or vice versa, then display it on the map. In the Relational Data Base Model (RDBM) approach, attribute data in GIS is stored in tables (called relation), each table is composed of some columns with unique names (called attributes) [22]. Each row of data in a RDBM is called a record that can be identified by using a unique primary key (PK). Thus, the query function to obtain spatially or otherwise, the position can be fulfilled by connecting between tables in GIS database using foreign key (FK). Furthermore, the relationship between spatial data and attributes is performed using the unique code provided in each of these data.

A common problem with WebGIS database design in those examples is that databases are developed for the scope of a particular area or region of a specific nature and the type and a limited number of users. As a result, the developed databases will be difficult to accommodate changes in new needs, such as expanding fields or coverage areas to be processed and displayed or adding users to WebGIS applications. T the developed databases can only be overcome since the early stages of WebGIS application development, especially in the database design phase. Attempts to resolve the issue were once proposed on [23] through the integration of several database designs, which include tourist attractions, schools, cultures, hotels, and inns, as well as government offices, into a new database design which is then tested on mobile applications. In the integrated database design, all objects are stored in the same table in the database and are differentiated by object code. The proposed database for mobile applications is composed of ten classes, i.e., province, regency, district, category, object, object_detail, user, user_level, log_activity, and change_password. The classes are then implemented into 13 database tables, i.e., province, regency, district, pos_code, user, user_level, pict, object, object_detail, category, sub_category, log_activity, and temp_login. Relationships between tables are shown in FK definitions in the regency, district, and category, object, and user tables.

Proposed on [24] emphasizes on how to obtain a flexible database design that can accommodate the growth and development of data so that it can be implemented for various GIS applications. The proposal is a database design consisting of 11 tables, i.e., village, district, regency, province, sub_category, category, user, user_level, user_status, object, and object_status Relationships between tables are shown in FK definitions in user tables, objects, sub_category, village, district, and regency. The integrity of the referential database is implemented through ONUPDATECASCADE and ONDELETECASCADE constraints. Approach on [23] and [24] is an integration solution that is done at the back end level [25]. The integration process begins with a schema matching stage that can be done with many different strategies, e.g., hybrid as done on [26].

The novelty of the approach in this study lies in the expansion of user views from one specific WebGIS application into a wider range of WebGIS applications and the expansion of the designer's view. From design to only one specific WebGIS application becomes more widespread on the merger of multiple WebGIS applications. Thus the database design can be used for a wider scale in the aspect of the user, object category, and region.

2. Methods

The study was conducted in 3 stages, namely a preliminary study to obtain an overview of the concept, design, and application of RDBM on various GIS Web applications. We then identified potential problems in database design for WebGIS, and finally proposed an alternative solution in the form of a new database design template that could accommodate the possibility of new changes and needs. The steps in this study are as follows:

2.1 Preliminary study to obtain an overview of the conceot, design, and applications of RDBM on various WebGIS applications.
2.2 Identified potential problems in database design on various WebGIS applications.
2.3 Proposed a new database design template for various WebGIS applications.
2.4 Comparing the old and new database designs for data growth on user groups, object categories, and scope of the object areas.

3. Result & Discussion

3.1. Minimum Database Requirement in WebGIS Applications

In theory, database needs for WebGIS applications in the form of two types of data namely spatial and attributes data. Spatial data is represented by the coordinate position data of each stored object, in which the location of each object is identified and associated with the attribute data. The coordinate position data of each object shows the position of latitude and longitude. In RDBM, attribute data is represented by the detailed values of data stored in linked tables in the database. The simplest database design for WebGIS applications is to store spatial data for one particular object type and attribute data consisting of administrator user data and detailed object-related data in a particular region. The simplest representation of the needs of the database can be defined, e.g., given the name *webgis01* as shown in the following schema:

a. Table structure for *object*:
```
CREATE TABLE IF NOT EXISTS `object` (`obj_code` char(10) NOT NULL DEFAULT '',
`obj_latitude` int(11) NOT NULL, `obj_longitude` int(11) NOT NULL, `user_name`
varchar(10) NOT NULL);
```
b. Table structure for *object_attribute*:
```
CREATE TABLE IF NOT EXISTS `object_attribute` (`obj_code` char(10) NOT NULL
DEFAULT '', `obj_name` char(50) NOT NULL, `obj_description` varchar(100) NOT
NULL);
```
c. Table structure for *user*:
```
CREATE TABLE IF NOT EXISTS `user` (`user_name` varchar(10) NOT NULL DEFAULT
'',`user_password` varchar(10) NOT NULL,`user_email` varchar(50) NOT NULL);
```

Journal of Physics: Conference Series 1413 (2019) 012029 doi:10.1088/1742-6596/1413/1/012029

3.2. Problems on webgis1 Database

There are many problems on *webgis1* database i.e.:

1. User data can only be increased on the record number.
2. Data object can only be increased on the record number.
3. Data object can only be increased on the record number.
4. Data objects represent a category of objects, it is difficult to adding of new categories.
5. Data object is not possible to do a track by area/region certain.

It is necessary to design an appropriate database table structure so that the detailed values stored can meet the needs of users, both current and future. Such needs changes can occur in the type and number of users, the scope of the application, and/or the scope of its territory.

3.3. The webgis2 Database Design Improvement

The proposed WebGIS database design improvement called *webgis2* is made by changing the approach of design that is specific to the user, the object category, and the particular region. *webgis2* being a more general design that allows for use by multiple groups of users, for various categories of objects, and for wider territory/different. The improvement can be achieved in two ways at a time. The first is to change the database designer's viewpoint to a conceptual view position so that the database design results can be used by several different applications. Both extend the view of the database designer so that the database is designed to be used for a wider scale in the user, object, and region aspects. Figure 1 shows a database design approach often used in applications such as webgis1, while Figure 2 shows a new database design approach for proposed *webgis2*. In Figure 1 it appears that each WebGIS application uses separate files with other applications, while in Figure 2 shows a database used by some WebGIS applications. This approach fits in with the stable database concept as found on [27], [22].

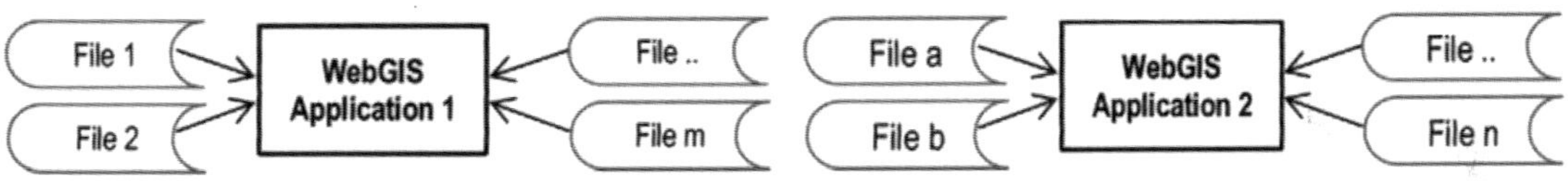

Figure 1. The database design approach on *webgis1*

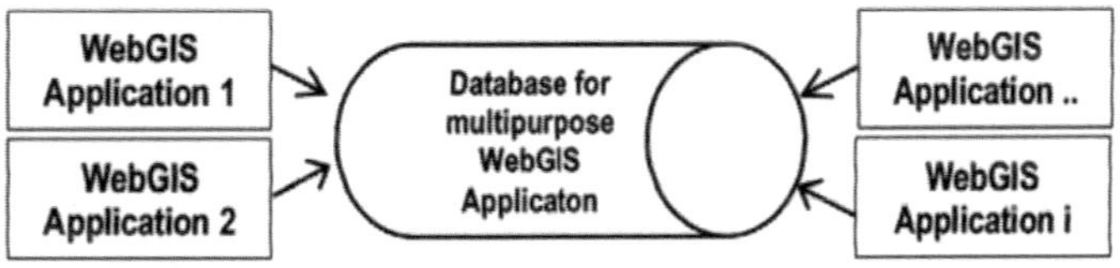

Figure 2. The database design approach on *webgis2*

Some of the considerations required when drawing up new database design improvement proposals for WebGIS applications are as follows:

1. Need to add status data on user data to record user status. The type of status should be anticipated to not only contain "active" and "not active" because there is a possibility of adding a new user but not yet verified "not active yet". For this change to be done more easily, it is necessary to add a reference table to record the status code and its description.
2. Need to add object category data. Each data object is associated with the appropriate subcategory or category to facilitate the search. For this change to be done more easily, we need to add a reference table to record the category code and its description. If required, sub-category data can be broken down into sub-categories, and so on.
3. Need to add data object area. The division and encoding of the object area can be done by referring to the division of existing and commonly used areas, eg provinces, districts, subdistricts, and villages.

In order to make the area change easier, you need to add the reference table of the region code and its description.

For WebGIS applications covering areas in Indonesia, territorial codes have been formally established by the Government of the Republic of Indonesia in the Minister of Home Affairs Regulation number 66 of 2011, on the Code and Regional Data in the Information System of Population Administration (ISPA) which applies nationally. In the regulation the regions is encoded as `province: char[2]`, `regency: char[4]`, `district: char[6]`, and `village : char[10]`. Examples of the use of area code in ISPA are 34 (DIY Province), 3401 (Kulon Progo District), 340106 (Sentolo regency), and 34010620008 (Banguncipto villages). The *webgis02* database design is displayed on the following schema:

a. Table structure for *district*:
```
CREATE TABLE IF NOT EXISTS `district` (`district_code` char(6) NOT NULL,
`district_name` varchar(50) NOT NULL,`regency_code` char(4) NOT NULL
);
```
b. Table structure for *object*:
```
CREATE TABLE IF NOT EXISTS `object` (`object_code` char(10) NOT NULL DEFAULT
'',`object_name` varchar(50) NOT NULL,`object_address` varchar(50) NOT
NULL,`object_contact` varchar(50) NOT NULL,`object_description` varchar(100)
NOT NULL,`object_latitude` decimal(9,6) NOT NULL,`object_longitude`
decimal(9,6) NOT NULL,`object_picture` blob NOT NULL,`village_code` char(10)
NOT NULL,`sub_sub_category_code` char(4) NOT NULL,`user_name` varchar(10) NOT
NULL);
```
c. Table structure for *object_status*:
```
CREATE TABLE IF NOT EXISTS `object_status` (`object_status_code` char(1) NOT
NULL,`object_status_name` varchar(12) NOT NULL);
```
d. Table structure for *province*:
```
CREATE TABLE IF NOT EXISTS `province` (`province_code` char(2) NOT
NULL,`province_name` varchar(50) NOT NULL);
```
e. Table structure for *regency*:
```
CREATE TABLE IF NOT EXISTS `regency` (`regency_code` char(4) NOT NULL,
`regency_name` varchar(50) NOT NULL,`province_code` char(2) NOT NULL);
```
f. Table structure for *sub_category*:
```
CREATE TABLE IF NOT EXISTS `sub_category` (`sub_category_code` char(4) NOT
NULL,`sub_category_name` varchar(20) NOT NULL,`category_code` char(2) NOT
NULL);
```
g. Table structure for *sub_sub_category*:
```
CREATE TABLE IF NOT EXISTS `sub_sub_category` (`sub_sub_category_code` char(6)
NOT NULL,`sub_category_code` char(4) NOT NULL,`sub_sub_category_name`
varchar(20) NOT NULL);
```
h. Table structure for *user*:
```
CREATE TABLE IF NOT EXISTS `user` (`user_name` varchar(10) NOT NULL DEFAULT
'',`usser_password` varchar(10) NOT NULL,`user_email` varchar(50) NOT NULL,
`user_phone` varchar(12) NOT NULL,`user_level_code` char(1) NOT
NULL,`user_status_code` char(1) NOT NULL);
```
i. Table structure for *user_level*:
```
CREATE TABLE IF NOT EXISTS `user_level` (`user_level_code` char(1) NOT NULL,
`user_level_name` varchar(18) NOT NULL);
```
j. Table structure for *user_status*:
```
CREATE TABLE IF NOT EXISTS `user_status` (`user_status_code` char(1) NOT NULL,
`user_status` varchar(12) NOT NULL);
```
k. Table structure for *village*:
```
CREATE TABLE IF NOT EXISTS `village` (`village_code` char(10) NOT NULL,
`village_name` varchar(50) NOT NULL,`district_kode` char(6) NOT NULL);
```

3.4. The webgis2 and webgis2 Comparison

Design on *webgis2* allows to accommodate the growth and development of different user group data, object categorization to sub-categories, and the scope of the object area. User data can grow and expand from only one predetermined level category (in *webgis1*) can be expanded with reference tables user_status and user_level (in *webgis2*) so that it can be applied to the status and level of the wider user. Data objects can grow and expand from only one category of objects (in *webgis1*) can be expanded with reference tables district, object_status, province, regency, sub_category, sub_sub_category, user_level, user_status, and village (in webgis2) so that it can be applied to wider region. In some cases, attribute data in the WebGIS database is also equipped with photos, videos, and other supporting data. This can be done by adding it to the object. Please follow these instructions as carefully as possible so all articles within a conference have the same style to the title page. This paragraph follows a section title so it should not be indented.

4. Conclusion

The database design process for WebGIS applications needs to anticipate the possible growth and development of different user group data, object categorization to subcategories, and the scope of the object area. Designs that meet these criteria can be done from the beginning of the design process, that is by changing two things. First, change the view of the database designer from the user view level to the conceptual view level so that the database design results can be used by several different applications. Second, expand the view of the database designer from the specific scope into a broader scope in the user aspect, object categorization, and territory coverage. Thus, the database design obtained will be in accordance with the RDBM concept.

5. Acknowledgments

The authors would like to thanks the Directorate of Research and Community Service, Directorate General of Research and Development Strengthening at the Ministry of Research, Technology and Higher Education of the Republic of Indonesia under Fundamental Research Grant Contract No.: B/1435.21/L5/RA.00/2019 and Institut Sains & Teknologi AKPRIND Yogyakarta for supporting facilities for this research.

6. References

[1] Charter D 2016 *Desain dan Applikasi GIS* (Jakarta: PT. Elex Media Komputindo)
[2] Irwansyah E 2016 *Sistem Informasi Geografis: Prinsip Dasar dan Pengembangan Aplikasi* (Yogyakarta: DigiBooks)
[3] Sutanta E 2010 Membangun SIG Perguruan Tinggi DIY dengan CMS Joomla dan Google Map *Information Management and Technology (IMATEC)* **1** 37
[4] Suratmo D Widyastuti N and Suraya 2014 Membangun Sistem Informasi Geografis Perguruan Tinggi Daerah Istimewa Yogyakarta Berbasis Web *Script* **2** 68
[5] Prabowo E T Sholeh M and Iswahyudi C 2013 Sistem Informasi Geografis dalam Pencarian Lokasi Museum di Daerah Istimewa Yogyakarta *Script* **1** 13
[6] Sholeh M and Haryanto R 2014 Sistem Informasi Berbasis Geografi untuk Pendukung Promosi Wisata Budaya di Kota Yogyakarta *Jarlit* **10** 64
[7] Zuliyanto Hamzah A and Suraya 2014 Sistem Informasi Lokasi Wisata Candi di DIY Berbasis GIS *Script* **1** 184
[8] Wijayanto P Sutanta E and Nurnawati E K 2014 Aplikasi Sistem Informasi Lokasi Hotel Berbintang di Yogyakarta Berbasis Webgis *Script* **2** 50
[9] Setiadi T and Yuliansyah H 2015 Kampung Wisata Online Berbasis SIG sebagai Upaya Meningkatkan Partisipasi Warga dalam Mengelola dan Mempromosikan Pariwisata Kota, *Jarlit* **11** 85
[10] Suraya and Sholeh M 2015 e-Museum Sebagai Media Memperkenalkan Cagar Budaya di Kalangan Masyarakat *Jarlit* **11** 24
[11] Kusumaningsih R Y R Kurniawan T and Pradhityo S 2016 Aplikasi Pengembangan Sistem Informasi Bangunan Cagar Budaya di Kota Yogyakarta Berbasis Web Mobile dan Location Based Service *Jarlit* **12** 53
[12] Andika L A A and Ariandi M 2015 Sistem Informasi Geografis Ruang Terbuka Hijau

Kawasan Perkotaan (RTHKP) Palembang, *Student Colloquium Sistem Informasi & Teknik Informatika (SCSITI)* Universitas Bina Dharma Palembang
[13] Ariandono Y Hamzah A and Nurnawati E K 2014 Sistem Informasi Lokasi Industri di Kabupaten Serang Berbasis Geographic Information System (GIS) *Script* **1** 176
[14] Wibowo H Lestari U and Triyono J 2014 Sistem Informasi Potensi Industri di Kabupaten Bantul Berbasis Geographic Information System dan Location Based Service *Script* **1** 72
[15] Hege Y B L Lestari U and Nurnawati E K 2014 Sistem Informasi Geografis (SIG) Pelayanan Kesehatan di Kotamadya Yogyakarta Berbasis Web *Script* **1**
[16] Nurnawati E K Muryanto J 2014 Aplikasi Mobile Berbasis Lokasi untuk Penyedia Lokasi Layanan Kesehatan di Yogyakarta *Proceeding of Seminar Nasional Aplikasi Sains & Teknologi (SNAST)* (Yogyakarta: IST AKPRIND Yogyakarta) A-293
[17] Sari W E Sholeh M and Hamzah A 2013 Penerapan Jquery Mobile dan PhP Data Object pada Aplikasi Pencarian Lokasi Tempat Ibadah di Yogyakarta *Script* **1** 69
[18] Pratikto H S Suraya and Sutanta E 2014 Sistem Pencarian dan Pemesanan Rumah Kos Menggunakan Sistem Informasi Geografi (SIG) *Script* **1** 110
[19] Suharto U H Triyono J and Sutanta E 2014 Aplikasi Pencarian Lokasi Agen TIKI (Titipan Kilat) di Yogyakarta untuk Android, *Final Project* (Yogyakarta: IST AKPRIND)
[20] Ardianto J Lestari U and Triyono J 2016 Sistem Informasi Geografis Pemetaan Area Persawahan Desa Gantung Kabupaten Belitung Timur *Script* **3**
[21] Sutanta E 2010 *Basis Data Dalam Tinjauan Konseptual* (Yogyakarta: Andi)
[22] Silberschatz A Korth H F and Sudarshan S 2011 *Database System Concepts* (*6th ed.)* (New York: The McGraw-Hill Companies, Inc.)
[23] Nurnawati E K and Ermawati 2018 Design of Integrated Database on Mobile Information System: A Study of Yogyakarta Smart City App *IOP Conf. Series: Materials Science and Engineering* **306** 1-11
[24] Sutanta E and Kumalasanti R A 2017 Desain Database yang Skalabel untuk Berbagai Aplikasi Berbasis GIS, Proceeding *of Seminar Nasional Science dan Teknologi Indonesia (SENSEI)* (Jember: Universitas Muhammadiyah Jember) 1-9
[25] Sutanta E Wardoyo R Mustofa K and Winarko E 2016 Survey: Models and Prototypes of Schema Matching *International Journal of Electrical & Computer Engineering (IJECE)* **6** 1011
[26] Sutanta E Wardoyo R Mustofa K and Winarko E 2016 A Hybrid Model Schema Matching Using Constraint-Based and Instance-Based *International Journal of Electrical and Computer Engineering (IJECE)* **6** 1048
[27] Elmasri R and Navathe S B 2011 *Fundamentas of Database Systems (6th ed.)* (Massachusetts: Addison-Wesley)

Early Detection of Leaks on Gas Cylinders Using Arduino Based MQ-6 Sensors

E Fatkiyah, D Persada and D Andayati
Department of Informatic Engineering, Institut Sains & Teknologi AKPRIND, Yogyakarta, Indonesia

erfanti@akprind.ac.id, dias.ist.akprind@gmail.com, dina asnawi@yahoo.com

ABSTRACT: Liquid Petroleum Gas (LPG) which has components in the form of propane and butane currently has an important role for people in the household and industry. However, there are things that must be considered when using LPG in terms of safety. This is because the gas is volatile and there is a possibility of leaking which will be very prone to cause explosion or fire hazard. The gas leak early detection device is a very appropriate effort in dealing with early detection of LPG leak in order to minimize the occurrence. This study was designed using a prototype system that functions as a simulation if there is a gas leak in a room by using the MQ-6 sensor which detects LPG content in the room. The way this tool works is that, if there is a gas leak, the MQ-6 sensor will detect it and then send the data to the microcontroller on Arduino in the form of analog data. In the MQ-6 sensor-based LPG leak early detection system using Arduino it has been successfully running according to an algorithm that has been designed and installed before in the test results, and send a warning message to the specified number. This system can detect gas faster in a closed room, on the contrary if in an open room the sensor can detect longer because the level of contaminated gas will be immediately wasted into the air.
Keywords: Arduino, LPG gas, MQ-6 sensor, and microcontroller.

1. Introduction

LPG (liquified petroleum gas) is a liquefied petroleum gas by adding pressure and lowering the temperature of the gas so that the gas becomes liquid, LPG plays an important role for the needs of society now with the use of LPG kerosene which was very high and even in 2017 land is very rarely used. Quoted from Republika.co.id "Consumption of LPG in Indonesia has increased.in 2007 only around one million metric tons per year almost reached 7 million metric tons in 2016. The figure grew by 700 percent for nine years. Pertamina's Deputy President Director Ahmad Bambang said, the increase in LPG consumption was due to the program to convert BBM to LPG conducted the government for household needs since 2007 and for fishing boat engines which are carried out starting in 2016, the plan will be continued in 2017 " [1].

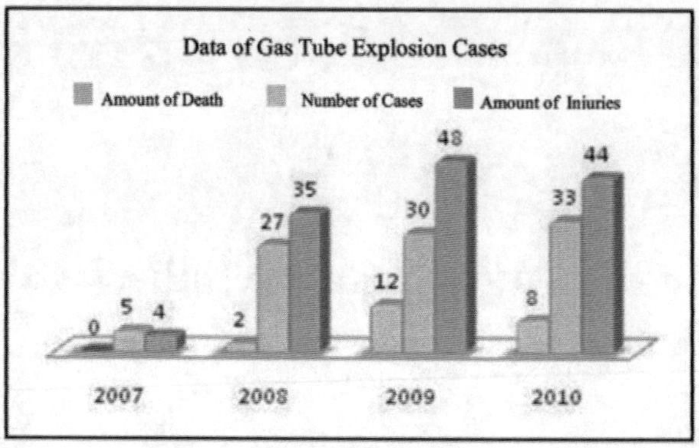

Figure 1. Graph of Data on LPG Gas Tubes Explosion Cases According to BPKN June 2010 (https://fahmizaleeits.com/ accessed on August 16, 2018)[2]

From the above data, the increase in LPG use is very significant, but in terms of the conversion of BBM to LPG there are problems that have arisen, such as image 1 data obtained from the BKPN (Investment Coordinating Board) institution about LPG gas explosions from 2007 to 2010[2]. Based on the above problems, the idea arises to make "LPG GAS Cylinders Leak Detectors Using Arduino Based MQ-6 Sensors" expected by making this system can reduce the number of cases of LPG GAS explosions.

The first research design a telemetry system from the ammonia gas concentration measurement unit to the data collector to monitor of ammonia gas in realtime and continuously. The data collected is integrated with software, so it can serve as an early warning of changes in the level of concentration of ammonia gas through the deployment of short message system (SMS Gateway) to the public or the industry responsible[3].

The second research produced an LPG gas cylinderleak detector design using the MQ-6 sensor as gas sensor, and an ethernet shield as a module on the arduino uno microcontroller to connect arduino to the internet network[4]. The other research produced a design og LPG leakdetection equipment using MQ-6 sensor as gas sensor, and GSM module as sender of SMS notification to user's mobile phone [5].

The research discussed the prototype of room refresher system from cigarette smoke and LPG gas based on ATMega 8535. The design of air refresher devices from cigarette smoke and LPG based microcontroller ATMega 8535 with MQ-9 sensors through the stage of making hard traps (hardware) and software (software). Hardware consists of a box that is a room prototype and there is a MQ-9 sensor to detect cigarette smoke and LPG gas. ATMega 8535 controller as a controller, buzzer as an early warning that there is detected cigarette smoke and LPG gas and fans that will be active to remove cigarette smoke and LPG gas from the room to the outside[6].

The research design of methane gas detection devices in Arduino Uno-based coal mines. This final project is capable of detecting methane gas using the MQ-4 gas sensor. So when the sensor detects methane gas, the system activates the buzzer and indicator lights (LED). In addition, this tool is also equipped with an LCD to provide information on methane gas leakage or as a monitoring so that users can always observe it [7]. From a number of references above, it can be concluded that technology is now increasingly advanced with the existence of intelligent systems so that all the hazards that exist in the surrounding environment can be minimized by making detection devices such as those made above.

The research produced LPG detection system because most of LPG explosions are caused by undetected gas leakage in the pre-detection condition. The purpose of this system is to detect gas leakage, neutralize it, and prevent the explosion. Gas leakage could happen due to improper regulator installation or the hose is broken. This detection should not work in just one location because gas can leak at the gas regulator and its hose. Therefore, Wireless Sensor Network (WSN) is one of the methods that suitable for detecting gas leakage in the wider area. This method uses two or more gas sensors to detect leakage in two or more locations around the gas tube and its distribution line. WSN system works based on gas sensor MQ-6 and wireless module Bluetooth HC-05. Explosion prevention system works based on alarm/buzzer, exhaust fan, and automatic gas regulator. If the gas leaks, the sensor will send its data wirelessly to Arduino. Then, explosion prevention system will be activated. The system will turn the alarm/buzzer on, automatically releases gas regulator, and neutralizes the air with the exhaust fan. Both systems will be fully controlled by Arduino platform[8].

This research developed system device monitoring and early detection of gas leaks by utilizing gas sensors MQ-4 and AVR microcontroller family as control devices. The system is also equipped with XBee PRO S2B nirkable devices as the interface to the wireless networking system that is used to transmit sensor data from the detection point to the monitoring center equipped with a PC and software integrated Visual Basic. Gas leak alert is sent with the message formats through social networking Gtalk[9].

This research deals with the development of an advance technology gas sensor for detection, monitoring and control system of LPG leakage. Using DC motor the stove knob is automatically controlled[10]. The research procused gas leak sensor is such a device which detects the gas leaks at initial levels and warns the people of the same. This paper basically deals with the development of a simple gas leak detector at the initial stage and then transforming this simple device into a most advanced gas detector system in the future[11].

This research deals with the detection, monitoring and control system of LPG leakage. Using relay DC motor the stove knob is automatically controlled. Along with safety measures the system has additional advantage of automatic rebooking of cylinder when the level of gas goes below the normal weight of cylinder[12].

2. Research Method

2.1. Program in Arduino IDE

In this study, Arduino IDE software for Windows is used, with serials, namely series 1.8.5.0. This software is built-in Arduino which is useful in compiling the source code for programming the Arduino board. This study uses the GSM SIM800l Module which functions to send messages, SIM800L GSM Module can be accessed by adding a library to the program source code that will be created, as shown in Figure 3 which displays the addition of the GPRS SHIELD library. sim card on the SIM800L GSM Module, how to add a library in Arduino that is by using hastag included with the name of the library that will be used.

2.2. Display Prototype

In the prototype section the gas leak detection system consists of several components connected including:

2.2.1 Arduino Uno board that functions as a controller of all existing components.
2.2.2 Project board is a place to connect between components one to another component.
2.2.3 Sensor MQ-6 is a sensor that detects the intensity of gas in the room.
2.2.4 SIM800L GSM module is a hardware that functions as a medium for sending SMS gateways to gas leak warnings.
2.2.5 The fan functions as a gas suction device in the room.
2.2.6 channnel relay functions as a regulator of electric current from outside.
2.2.7 Led lights as a warning for a gas leak.
2.2.8 Jumper cable serves as a liaison between components

2.3. SMS Delivery Scheme

From the schema in Figure 4, it can be seen that the pranti in the system is 3, namely there are inputs, processes and outputs. In the hardware input process that is used is the MQ-6 sensor which functions as a gas detector sent with data in the form of analog data to the Arduino microcontroller,

Journal of Physics: Conference Series **1413** (2019) 012030 doi:10.1088/1742-6596/1413/1/012030

after input is available, then enter the data processing to be entered into conditioning in the microcontroller that has been programmed, in this process the Arduino microcontroller acts as a process device and that will control all output devices. On the hardware output device that is used is an LED light that serves to indicate a leak in the gas cylinder, a fan that functions as a device that sucks gas in the room and the last is a SIM800L module that functions as a message sending media if a gas leak is detected in the room this condition SIM800L module has a weakness, namely in the network search section which is often unreadable in the network process, it is not readable, the sms gateway that has been conditioned is not delivered, for the scheme flow can be seen in the picture 2.

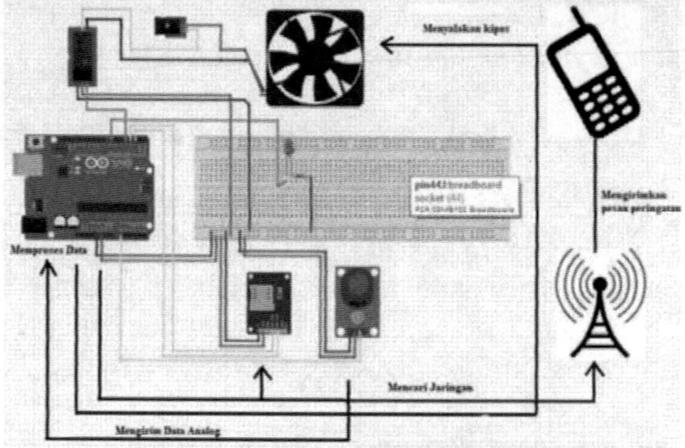

Figure 2. SMS sending scheme

3. Result and Analysis

3.1. Block Program

In the block diagram there are 3 blocks consisting of input, process and output. The input block contains the detection of the gas sensor MQ-6, then there is a block process in which there is an Arduino microcontroller component which acts to receive the value sent by the gas sensor MQ-6 and finally the output block displays the output in the form of LED lights SMS gateway.

The input block here is the MQ-6 gas sensor that reads the analog data input value, in this condition the sensor will turn on if the Arduino microcontroller has power, the sensor will then start reading and sending analog data to the microcontroller and then enter the block process.

The block process acts as the recipient of data that is sent from the input block and then processed to do the conditioning that was previously programmed using the Arduino IDE software. After conditioning, the output will be continued to the output block.

The output block consists of a led light as a warning of a gas leak in the room then the fan functions as a gas suction in the fan room will automatically turn on if the block process sends a command that is when there is a gas leak in the room, the fan will automatically turn off if the analog value is the input block has returned to normal and the last one is the GSM SIM800l module which functions as a message sending media in the event of a gas leak in the room, sending messages will occur repeatedly with a delay of 5 seconds if the gas conditions in the room are still high intensity. Block diagram can be seen in the figure 3.

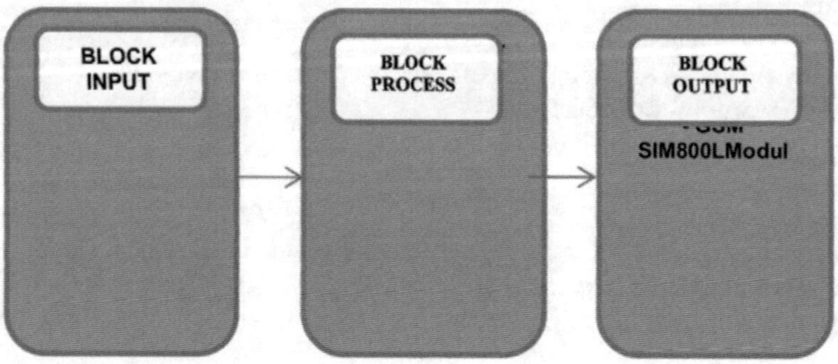

Figure 3. Block diagram

3.2. Black Box Testing

Black box testing based on prototype systems for early detection of gas leaks using Arduino Microcontrollers can be seen in Table 1. Based on black box testing in table 1 it appears that all the processes that have been tested can run well. The first scenario of testing is by flowing electric current on the Arduino microcontroller board with the results of all components running well including the SIM800l module which automatically installs and automatically searches for the GSM network then there is the MQ-6 sensor that reads the gas intensity in the room. The second test is to bring the gas closer to the MQ-6 sensor with the result that it can detect a gas leak then turn on the led as a sign of danger and turn on the fan in the room, and give a warning by sending an sms gateway.

Table 1. Testing Black Boxes

Nu	Scenario	Test Result	Security	Conclusion
1.	Provides a power source	Able to install all components Able to turn on LEDs and fans and give SMS alerts	Read automatically the value of gas content in the room	*Valid*
2.	Bring the gas source closer to the MQ-6 sensor	Able to turn on LEDs and fans and give SMS alerts	The LED and Fan Lights up as a warning and sends a message that there is a gas leak	*Valid*

3.3. Testing the GAS Leak Detection System Algorithm

Algorithms in designing tools are very important in order to obtain accuracy in terms of control. The algorithm functions as a reference for making programs where the algorithm contains important points contained in the programming logic that will be made. As in Table 2 below, how the detector system works. Range ADC is the value of a small amount of gas in the air that is read by the sensor. The condition is an order from the microcontroller to the output, based on the program logic given in this test it can be concluded that the algorithm implanted in the microcontroller tool is appropriate.

Table 2. Testing of the gas leak detection system algorithm

Analog data	Condition (Output)			explanation
	Red Led	Fan	SMS	
<=199	Off	Off	There is no message sended	Secure
>=200 &<=400	On	On	"Gas content increases, Warning !!!"	Standby
>=400	On	On	"Indicated Gas Leaks !! Danger !! "	Danger

3.4. GAS Sensor Testing in MQ-6

The gas sensor testing in MQ-6 is done by using a small size gas cylinder (the size of a perfume can) by spraying the gas at a predetermined distance carried out in the room. It can be concluded in the results of table 3 this test is an analog value read by the MQ-6 sensor depending on distance and wind conditions in the room. The distance gas source in the range of 15 cm down the readable value is high, namely 300 to 400 and at a distance of 15 to above the analog value that is read is getting smaller because the gas source does not go directly to the MQ-6 sensor / gas source exposed to wind. Table testing of Gas Sensor MQ-6 can be seen in the Table 3.

Tabel 3. Testing of GasSensor MQ-6

Testing to-	Distan-ce (cm)	Value of Analog Sensor MQ-6 Data at Seconds To- 5(s)	10(s)	15(s)	Testing Scenario	Test Result
1	-	50 ppm	54 ppm	54 ppm	There is no gas source	The led and fan do not turn on, do not send SMS alerts
2	2	443 ppm	431 ppm	435 ppm	Bring the gas source closer to The MQ-6 sensor	The led and fan are on, sending a warning SMS
3	5	388 ppm	370 ppm	332 ppm	Bring the gas source closer to The MQ-6 sensor	The led and fan are on, sending a warning SMS
4	10	287 ppm	330 ppm	276 ppm	Bring the gas source closer to The MQ-6 sensor	10th seconds Led and the fan light up, sending a warning SMS
5	15	222 ppm	230 ppm	199 ppm	Bring the gas source closer to The MQ-6 sensor	The led and fan do ot turn on, do not send SMS alerts
6	20	221 ppm	221 ppm	218 ppm	Bring the gas source closer to The MQ-6 sensor	The led and fan do ot turn on, do not send SMS alerts
7	25	192 ppm	221 ppm	211 ppm	Bring the gas source closer to The MQ-6 sensor	The led and fan do ot turn on, do not send SMS alerts
8	30	112 ppm	100 ppm	102 ppm	Bring the gas source closer to The MQ-6 sensor	The led and fan do ot turn on, do not send SMS alerts
9	35	98 ppm	98 ppm	93 ppm	Bring the gas source closer to The MQ-6 sensor	The led and fan do ot turn on, do not send SMS alerts
10	40	93 ppm	80 ppm	81 ppm	Bring the gas source closer to The MQ-6 sensor	The led and fan do ot turn on, do not send SMS alerts

3.5. Testing the led indicator circuit

Led testing is done by conneting the anode on the Arduino output pin that passes through the resistor first and the cathode is connected to ground on Arduino. Then give the program logic the On-Off control program (Highlogic or low logic). Led will be On if given High logic and Offif given Low logic.

3.6. Fan Testing

Tests carried out by giving a voltage to the fan Table 4 are the test result. In this circuit a relay component is added to disconnect and connect an electric current to the fan and serves to adjust the voltage needed by the fan.

Tabel 4. Fan Testing

Voltage is given	Fan Condition
0 Volt DC	Off
4,9 Volt DC	On

3.7. Software Testing

Software testing this mean is program testing. The finished program is created using Arduino sketch, then uploaded to Arduino using a USB cable. After that, testing the program whether it is in accordance with good result or not by observing the outputsof the device such as a LED, fan, and GSM module.

3.8. Implementation of Tools

In the implementation of this tools, additional components are used, as figure 4, namely the addition of a relay to 2 channels which serves to regulate electrical power and control the flow of electricity connected to fan and light. The principle and method of operation of the circuit below is the same as the circuit in the prototype above the difference only the fan is replaced from the USB fan to the exhaust fan which is commonly used in the room, the plug-in is replaced by a plug-in that is directly connected to PLN electricity. Similiar to the fan, the lamp used in the circuit below is replaced by a large wattage lamp and the plug-in used is directly connected to PLN electricity.

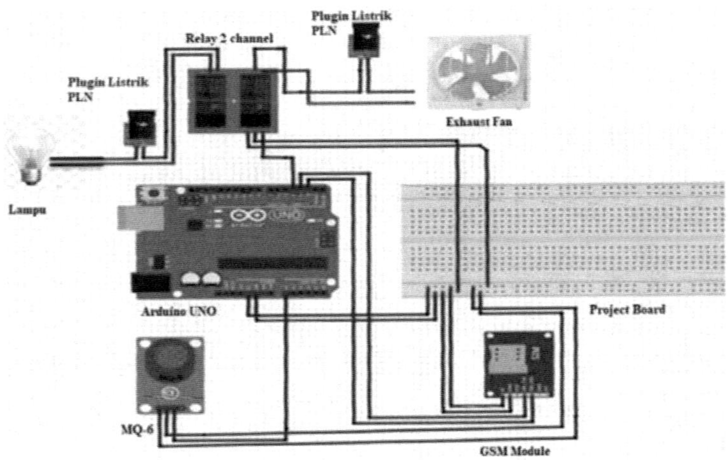

Figure 4. Series of Implementation

4. Conclusion

With early detection of leaks on gas cylinders system using arduino based MQ-6 sensors, we can conclude:

4.1 In the MQ-6 sensor-based LPG gas leak early detection system using Arduino it has been successfully running according to an algorithm that has been designed and installed before in the test results and send a warning message to the specified number.

4.2 This system can detect gas faster in a closed room, on the contrary if in an open room the sensor can detect longer because the level of contaminated gas will be immediately wasted into the air.

4.3 This application system is only a prototype and has not yet been tested for accidental leakage of LPG gas cylinders in relevant agencies which can cause explosions and fires.

4.4 In designing and making a prototype system, there are various kinds of weaknesses, both from the system planning and the equipment that has been made, it needs development including adding buzzer output that can alert the surrounding environment if there is a gas leak then adding a controller feature electrical switch, so that if a gas leak occurs it will automatically turn off and the last one will make a website or android based system to be able to see data in more detail and realtime.

5. References

[1] Republika. (2018, January 17). *Republika.* Retrieved Agustus 3, 2018, from Republika: https://www.republika.co.id/berita/koran/ekonomi-koran/17/01/18/ojz2k5-konsumsi-elpiji-meningkat

[2] Https://fahmizaleeits.com/ accessed on August 16, 2018

[3] C I Parwati, H P Suseno, E Fatkhiyah, 2015, *Modeling telemetry system"Multi-Node" early warning ammonia gas leak based SMS gateway,* Jurnal Teknologi, volume 8 number 2, December 2015, p 151-160

[4] M F Putra, A H Kridalaksana, Z Arifin, 2017, *Designing an LPG gas leak detector with MQ-6 sensor microcontroller-based via android smartphone as an information media,* Jurnal Informatika Mulawarman, volume 12 number 1, February 2017

[5] I Hidayat, 2018, *Gas leak system using MQ-6 based on wireless sensor network,* Techno.COM, volume 17 number 4, November 2018, p 355-364

[6] D Akhwandi, A Yudhana, 2017, *Sistem penyegaran ruangan dari asap rokok dan gas LPG berbasis mikrokontroler A2tmega 8535,* Yogyakarta: Universitas Ahmad Dahlan.

[7] A Nofriyeldi, 2017, *Rancangan alat pendeteksi gas metan dalam tambang batu bara berbasis arduino UNO,* Padang: Politeknik Negeri Padang.

[8] D Laksmita & Somantri Y., 2018, Wireless Sensor Network on LPG Gas Leak Detection and Automatic Gas Regulator System Using Arduino. IOP Conference Series: Materials Science and Engineering. 384. 012064. 10.1088/1757-899X/384/1/012064. https://www.researchgate.net/publication/326350308_Wireless_Sensor_Network_on_LPG_Gas_Leak_Detection_and_Automatic_Gas_Regulator_System_Using_Arduino/ accessed on August 26, 2019.

[9] K Medilla & Firdaus, Firdaus & Yulianto, Andik & K Syakban, 2018, Early detection of LPG gas leakage based Wireless Sensor Networking, https://www.researchgate.net/publication/323451555_Early_detection_of_LPG_gas_leakage_based_Wireless_Sensor_Networking/ accessed on August 26, 2019.

[10] Girish Karikatti, Sameer Sattur, roopa lokeshwar, 2019, Advance Technology Gas Leakage Detection And Control System https://www.academia.edu/16157642/ADVANCE_TECHNOLOGY_GAS_LEAKAGE_DETECTION_AND_CONTROL_SYSTEM/ accessed on August 26, 2019.

[11] Mr. Akshay D. Prabhu , Mr. Ashwin D. Pathak, 2017, Gas Leak Detector using Arduino UNO Microcontroller, International Journal for Research in Applied Science & Engineering Technology (IJRASET) ISSN: 2321-9653; IC Value: 45.98; SJ Impact Factor: 6.887 Volume 5 Issue VII, July 2017. https://www.ijraset.com/fileserve.php?FID=9021 accessed on August 26, 2019.

[12] T.Soundarya, J.V. Anchitaalagammai , G. Deepa Priya, S.S. Karthick kumar, 2014, C-Leakage: Cylinder LPG Gas Leakage Detection for Home Safety, IOSR Journal of Electronics and Communication Engineering (IOSR-JECE) e-ISSN: 2278-2834,p- ISSN: 2278-8735.Volume 9, Issue 1, Ver. VI (Feb. 2014), PP 53-58, http://www.iosrjournals.org/iosr-jece/papers/Vol9-Issue1/Version-6/J09165358.pdf/ accessed on August 26, 2019.

Visualization of Multi-Storey Building Visitor Guide Animation By Optimizing Macromedia Flash MX of Action Script Functions and 3D Max Studio (Study: Engineering Faculty Building, Muhammadiyah University of Yogyakarta)

M Yamin[1], A Faslih[2], and H Yuliansyah[3]
[1] Halu Oleo University, Informatics Engineering, Kendari, Southeast Sulawesi, Indonesia
[2] Halu Oleo University, Architecture, Kendari, Southeast Sulawesi, Indonesia
[3] Universiti Kebangsaan Malaysia, Faculti Teknologi & Sains Maklumat, Bangi Selangor, Malaysia

Email: muh_yamin@uho.ac.id, arman.faslih@uho.ac.id, herman.yuliansyah@tif.uad.ac.id.

Abstract. Yogyakarta Muhammadiyah University (UMY) has 7 faculties, divided into 5 zones, namely the north zone, the west zone, the south zone, the east zone, and the middle zone. The Faculty of Engineering (FT-UMY) occupies the north zone. The problem is how one knows the location, position, and function of the room in the building, because every educational institution has many buildings, the shape of the building the same / similar, each building is occupied by many faculties, and has a large number of rooms. The concept of Visualization or Visual Communication Design was used to solve the problem, Visualization or Visual Communication Design is the study of communication concepts and creative expressions, techniques and media to convey messages and ideas visually, such as pictures, audio, video by processing graphic design elements . The Software / Tools used are Macromedia Flash MX and 3D Max Studio by utilizing and creating animation and visualization data base information systems with simple action script commands, generating visitor guide software has input data, namely: 1 faculty button and has 4 search buttons , i.e., find the lecturer button, find a room, search for majors, and find employee button. Has 4 output data consisting of: View (side view, rear view, and front view), path, floor plan, and building.

1. Introduction

Yogyakarta Muhammadiyah University (UMY) is one of the many private tertiary institutions in the City of D.I. Yogyakarta which has many lecture buildings, because UMY has 7 faculties with 18 strata-1 study programs and 2 strata-2 study programs and the total number of active students in the academic year 2007/2008 was 8,332 people. In the 2007/2008 academic year UMY received 4130 students from 6974 registrants (UMY Student Affairs Bureau, 2008) which had to be balanced with facilities and infrastructure such as lecture buildings, laboratory buildings, library buildings, etc.

UMY was officially established on March 1, 1981. The UMY building occupies ± 25 hectares of land located on the ring road west ring, Tamantirto sub-district, Kasihan sub-district, Bantul district, Yogyakarta. (UMY Guide, 2010). From the development team of the UMY Campus in 2010, the UMY buildings are divided into 5 zones (area), namely:

a. West Zone, consists of: Masjid
b. East Zone, consist of: building A. R Fahrudin, and Ahmad Dahlan building

c. Centre, consists of: building stars
d. North zone, consists of: construction of the Faculty of Medicine, construction of the Faculty of Engineering, construction of the Faculty of Agriculture and building of an Eksakta laboratory.
e. South zone, consists of: the Faculty of Law Building, the Faculty of Economics building, the Faculty of Islamic Religion building, the Isipol Faculty building, the library building, and the non-exsacta.

The Faculty of Engineering at Muhammadiyah Yogyakarta University (FT-UMY) occupies the northern area of the Maeshorah Hillal building. FT-UMY existed and stood in the first year of established UMY, which is dated 01 March 1981, initially there is only one department of civil engineering, then developed so far there are three departments, namely civil engineering courses, engineering courses and mechanical engineering courses.

If we see from the building side, FT-UMY have a lot of room such as:

a. FT Conference Room.
b. Board room (TU) FT.
c. Board room course (TU).
d. Speaker Room D.
e. Reference space E. FT.
f. FT Zone.
g. Seminar Room G. FT.
h. Major seminar rooms.
i. FT. Teaching room.
j. General FT Stadium Hall.
k. Space SPMU-TPSDP.
l. Space Information Technology Reseach Group (ITRG-FT).
m. Space Electrical English Club (EEC).
n. PUSPER space, and
o. Laboratory Room.

How does one know the location, position and function of the room in the FT-UMY building? This is because UMY has many buildings, the shape of the building is the same/similar, and each building is occupied by many faculties, and has a large number of rooms. Initial research was conducted in all faculties of the UMY and non-UMY campus areas randomly as many as 35 questionnaires.

Tabel 1. First Respondents

No	Major	Amount
1	Communication/UMY	13
2	Fisipol/UMY	1
3	Management/UMY	3
4	HI/UMY	5
5	Accounting Economics/UGM	1
6	Agriculture/UMY	1
7	Government Sciences/UMY	3
8	LAWI/UMY	4
9	FAI/UMY	1
10	Economics/UMY	2
11	Electro/UII	1
	Total	**35**

From the results of research conducted obtained data

Table 2. Visitor Problems

No	Question	Percentage (%) ≈ Respondent	
		Ya	Tidak
1	Are you confused about finding a room in FT	82.86 % ≈ 29 orang	17.14 % ≈ 6 orang
2	Do you need a long time to find the room you are looking for at FT	85.71 % ≈ 30 orang	14.29 % ≈ 5 orang

2. Research Method

According to Sunardi Purwosuwito, visualization or design of visual communication is a science that learns the concept of communication and creative expressions, techniques and media to convey messages and ideas visually, such as images, Audio, Video by processing Graphic design elements form images, fonts and colors, and layout them, so that messages and ideas can be accepted by the lens.

2.1. Visualization Term

The term visualization is developed from graphic design, namely:

a. Graphic Art/Graphic Arts, including in the fine arts group.
b. Chart/Chart, is a question related to a text or image that contains meaning to convey a message or information.
c. Graphic design, the term used before using the term Visual Communication Design, derived from the Greek word "Graphos" which means "writing/drawing".
d. To anticipate the development of visual communication and its broader role, he used the term visual communication design visualization or design.

2.2. Principles and Visualization Elements

Visual messages must be creative (original, innovative and fluid), communicative, effective and effective, yet beautiful/aesthetically pleasing. The elements of graphic design are just like the basic elements in other design disciplines. Elements (including shape, shape, texture, line, value and colour) form the basic principles of visual design. Principles, such as balance, rhythm, pressure (emphasis) and unity, then form a broader structural aspect of the composition.

2.3. Visualization Tools

The equipment used for the design visualization of an object is resourceful (mind), eye, hand, traditional tools (such as pencil or ink), and computer. A concept or idea is generally not considered a drawing until it is realized or expressed in visual form.

The software/tools that can be used to design the visualization of an object that more attractive and vivid are as follows:

2.3.1. Design

- Adobe Photoshop: Point-based design (bitmap).
- Adobe Image Ready: Reframe images in HTML format.
- Adobe Illustrator: vector design.
- CorelDraw: Vector design.

2.3.2. Animation

- 3D Studio Max and SketchUp 5: To create objects and 3D animations.
- GIF Building Set: Create an animated GIF file.
- Macromedia Flash: Shows small vector animations.
- Microsoft Gif Animator: Create an animated GIF file.
- Swift 3D: Design 3D animations with FLASH file formats.
- Swish: Create a wide range of text effects with the FLASH file format.
- Ulead Cool 3D: Create a 3D text effect animation.

2.4. *The Need for Knowledge in the Design of Visual Communication*

a. Know the concept of visual communication design as the basis of the design/design and communication strategy.
b. Familiarize with graphic design (visual communication design) and language as a visual information data processor.
c. Know the technical of principles, information technology processes and information systems management.
d. Understand the elements of graphic design as an effective, effective, communicative and aesthetically creative messaging tool in the context of conceptual policy/planning/strategy and implementation and evaluation.
e. Understand communication strategy, psychology and social/cultural anthropology.
f. Understand some new media, especially the world of media/cyberspace and technology:
   - Animation - Visual, audio-visual video (Mix Media).
   - Interactive media and web/website commonly used to complement online media and mix media/multimedia.
g. Master the concept of visual communication design and universal global marketing. Master process and design techniques that can anticipate the development of the entrepreneurial/enterprenuership world.

2.5. *Flow Chart*

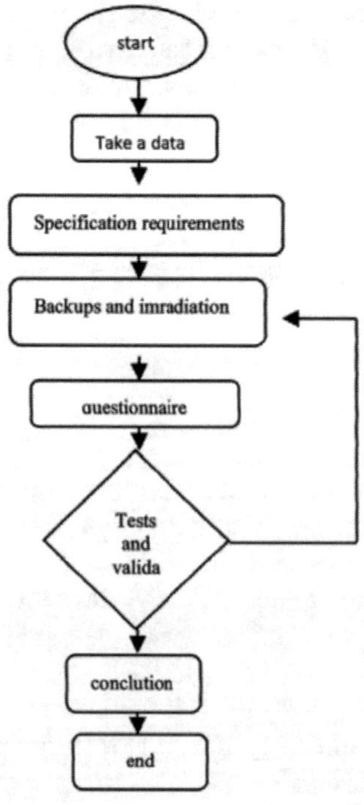

Figure 1. Flowchart Design and Testing

2.6. Visualization App Design
The development is seen in this design by putting in place detailed application specifications. At this point, the selection of media users intend to use the media accordingly, not reducing the purpose of the system.

2.6.1. Design Charts (IMAGES)
Using 3D Studio Max or SketchUp 5 The process of creating object images and 3D animation of the complex FT-UMY construction area, from creating a construction plan to a 3D building image.

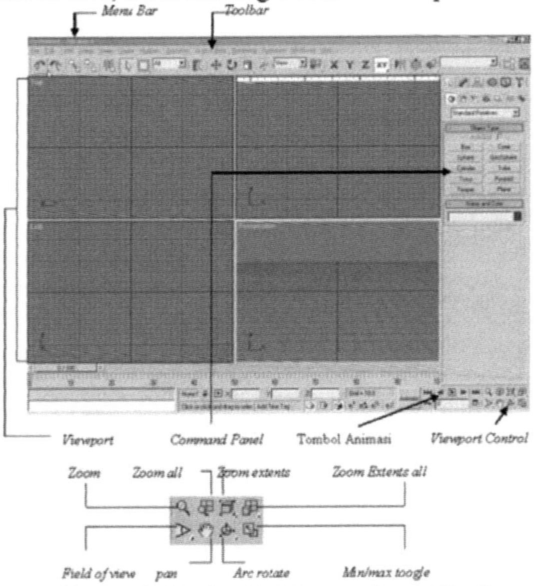

Figure 2. 3D Studio Max or SketchUp 5

2.6.2. Design of the Application Form
Display design for part of the main application (form), including:
2.6.2.1. Shape for the main menu.
2.6.2.2. Train for the search menu, which is the form search speaker, find space, find a course and find employees.
2.6.2.3. Form for the majors, that is. electro, civil and machine form.
2.6.2.4. Form for the Faculty of Engineering,
2.6.2.5. Additional forms, i.e. forms at a glance, logos, support software, on FT, thank you, and biodata admins.
2.6.2.6. Form for exit/exit, shape appears, tracks, floor plans and buildings

Figure 3. Image of design application menu

2.6.3. Database/Action Script Design

The development of the system produces parts and unifies with the system, in addition to developing application software, its functions include special activities such as editing and writing. Using 2004 Macromedia Flash MX software to create animations and database visualization of the information system with a simple action script command:

```
gotoAndStop("background",1);
}animasi : Frame 1 stop(); background
on (press) {
gotoAndStop("background",5);}
on (press) {
gotoAndStop("background",5);}
on (press) {
gotoAndStop("background",5);}
on (press) {
gotoAndStop("background",5);}
on (press) {
gotoAndStop("background1",35);
}
```

2.7. Validation and Testing of Application Functions With

Based on the test results, the quantitative analysis was to explain the ease of interpretation of the results of the data analysis. This analysis is also used to draw conclusions from the statistical state presented.The 45 respondents sampled the questionnaire from students of UMY students and non-UMY students. The 45 respondents are:

Table 3. Testing of Respondent

Major/univercity	Amount
Electro/UMY	14
Machine/UMY	4
Dentistry/UMY	2
PSIK/UMY	2
Agriculture/UMY	2
HI/UMY	3
IK/UMY	6
Management/UMY	2
Law/UMY	3
Economy/UMY	4
Accounting/UGM	1
M. Computer Science/BSI	1
AKP/UNY	1
Total	**45**

Responses from all respondents collected were classified and aggregated as required. Using the respondent's selection techniques systematically when the selected respondents actually tested the visitor guide software offered.

The type of questionnaire is a closed questionnaire, where the questionnaire consists of questions with a number of optional answers.

Table 4. App Ratings

No	Question	%			
		SS	S	TS	STS
1	easy to use.				
2	has a menu (link) that is easy to understand.				
3	has an attractive look.				
4	pictures of the building in accordance with the original building.				
5	can save time in room search.				
6	room searches are more effective than having to ask others.				
7	7 is able to provide accurate information.				
8	8 has the output/information space in the Faculty of engineering that corresponds to the location and the actual condition.				

Table 5. User Satisfaction Tests Toward Application

No	Question	Description	
		Yes	No
1	Do you know the position/layout of the rooms in the Faculty of Engineering.		
2	Do you know the position/layout of the part you are aiming for.		

All data obtained from respondents can be used as an analysis to take a conclution.

3. Result and Discussion

Flow chart that runs in the visitors guide app.

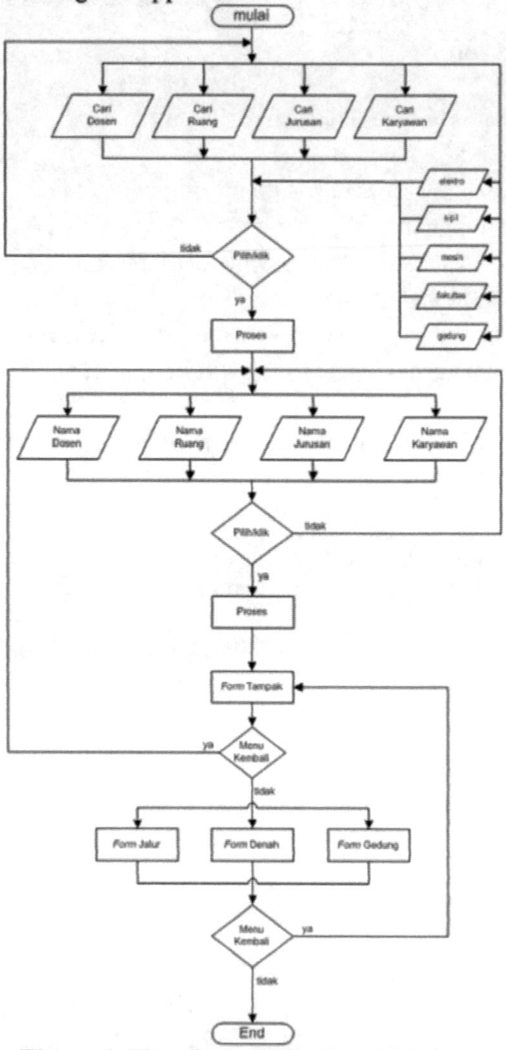

Figure 4. Flowchart of Visitors Guide App.

3.1. Complex Images of FT-UMY Building (3Dimensions)

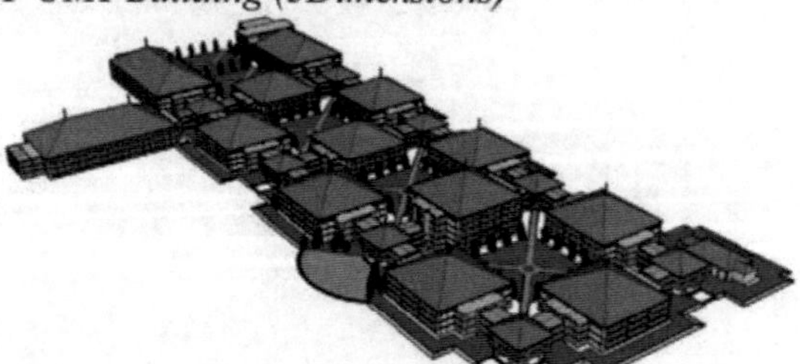

Figure 5. Complex Building of FT-UMY (3D)

The construction of the FT-UMY building has plenty of space, including: FT Conference Room, Ft Board of Directors (TU), Board of Directors (TU), Speakers Department, FT Reference Room, FT Seminar Room, Course Space Seminars, Room FT, General FT Stadium, Space SPMU-TPSDP, Space

Information Technology Reseach Group (ITRG-FT), Space Electrical English Club (EEC), PUSPER and Lab Room.

3.2. Application Menu

This Form has the main button, as follows:

a. 1 Main Menu button.
b. 4 search buttons, namely:
   Search for lecturers, search for space, search for majors, and search for Employees button.
c. 3 course buttons, namely:
   Electro, civilian, and Machine buttons.
d. 1 button Faculty of Engineering.
e. Button building, namely:
f. Laboratory link buttons, F1 Building buttons, F3 Building buttons and F4 building buttons.
g. Additional buttons, namely: Fleeting, logos button, supporting software , about FT, thank you, and biodata admin buttons.

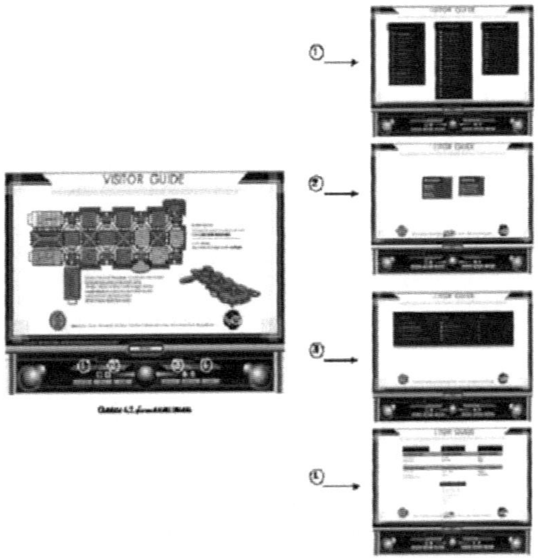

Figure 6. Application Menu

3.2.1. Lecturer Search Menu

This form is used to locate and locate the location/position of the speaker room, such as the electro speaker, the civilian speaker and the FT-UMY machine speaker.

3.2.2. Room Search Menu

This form is used to locate and locate the location/position of conference rooms and FT-UMY laboratories. This form contains two functions of the research space, namely the conference room and the laboratory. In the laboratory, the research is transformed into 3 buttons: electrical engineering, civil engineering, mechanical engineering.

3.2.3. Major Search Menu

This form is used to locate and locate the location/position of the speaker room, the administration room, the lab room, the seminar room and the main room in each major.

3.2.4. Employee Search Menu

This form is used to locate and locate the location/position of the employee's space, such as employee administration majors, laboratory employees and engineering faculty employees.

3.2.5. Faculty search Menu

Figure 7. Faculty Menu

This form contains the names of the Faculty of Engineering space, such as the Faculty of the Executive Room, the Faculty of the TU Room, the Seminar Room, the Conference Room, the EEC Space, the ITRG Space, the PUSPER Space, the Faculty of the Reference Room , The SPMU-TPSDP space, faculty area, General Stadium and teaching room.

3.2.6. Building Menu

This form is used to find the location/position and know the rooms based on the name of the building in FT-UMY.

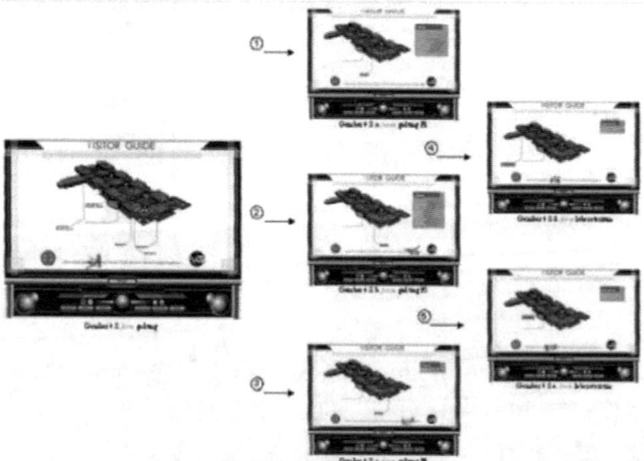

Figure 8. Building Menu

This form contains five link buttons: F1 Building, F3 Building, F4 Building and (two) laboratories.

3.2.7. Exit Form

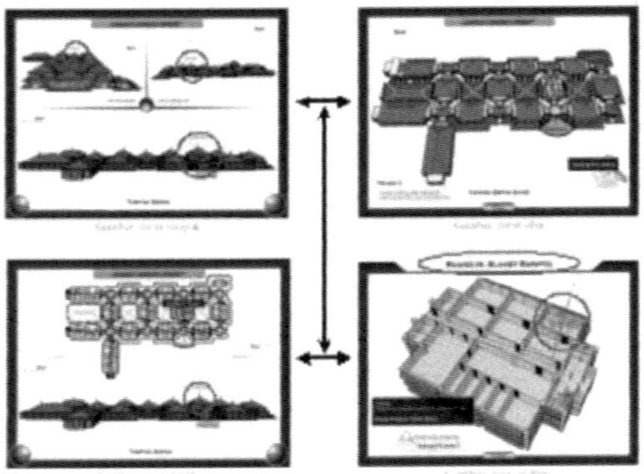

Figure 9. Exit/exit

The form that was created must be easily understood and understood by the user, and then made 4 different views, as follows:

3.2.7.1. Form Appears

Displays images of 3D buildings with 3 buildings visible, face, rearview mirror and front-facing.

3.2.7.2. Path Form

Shows a 3D building image with a path/direction to the destination point.

3.2.7.3. Floor Plan Form

Shows an image of the 2D building floor view and the front view of the building.

3.2.7.4. Building Form

Shows a more detailed 3D room image with a space location, a space name and a path/direction to the destination point.

3.3. Software Validation

This validation process is done to prove that all the functions of the software must be expected by the administrator and the user. By doing a certain component part validation process, starting from the input until the software output is expected to get a better application result. Following Table 6 Software Validation.

Table 6. Software Validation

No	Function	Output Result	Status
1	Search for lecturer	Shape appearances: Side view, rear view and front view of the building. Plane shape: Images of the 2D construction plan look up and in front of the building. Line shape: path/direction to the destination point. Construction form: View the speaker's name and the speaker room at FT-UMY and the path/direction to the destination point.	reached
2	Search for room	Shape appearances: Side view, rear view and front view of the building. Plane shape: Images of the 2D construction plan look up and in front of the building. Line shape: path/direction to the destination point.	reached

		Shape building: Conference room and display lab in FT-UMY and line/direction at point destination location.	
3	Search for major	Shape appearances: Side view, rear view and front view of the building. Plane shape: Images of the 2D construction plan look up and in front of the building. Line shape: path/direction to the destination point. Building shape: Shows the name of the space and rooms on the FT-UMY and the path/direction to the destination point.	reached
4	Search for employes	Shape appearances: Side view, rear view and front view of the building. Plane shape: Images of the 2D construction plan look up and in front of the building path form: The path/direction to the destination point location. Building form: Displays the employee's name and the employee's space at FT-UMY as well as the path/direction to the destination point location.	reached

3.4. Software Function Test by User

This test was carried out by demonstrating the visitor guide complex visualization software FT-UMY building and providing questionnaires to the user in order to know the level of satisfaction of respondents in terms of ease in the use of the software, software display, software benefits and real-world software information.

The *questionnaires* distributed to respondents up to 45 people calculated their statistical data with a different percentage of responses on each of the questions asked. The following Table 7 is a result table of questionnaire testing software functions by user.

Table 7. The Result of Questionnaire Testing Software Functions by User

No	Questions	Percentage(%) ≈ Respondent			
		SS	S	TS	STS
1	Easy to use.	22 % ≈ 10	76 % ≈ 34	2 % ≈ 1	0 % ≈ 0
2	Has a menu (link) that is easy to understand.	13 % ≈ 6	69 % ≈ 31	18 % ≈ 8	0 % ≈ 0
3	It has an attractive look.	20 % ≈ 9	69 % ≈ 31	11 % ≈ 5	0 % ≈ 0
4	Images of buildings according to the original building.	13 % ≈ 6	69 % ≈ 31	16 % ≈ 7	2 % ≈ 1
5	Can save time in room search.	22 % ≈ 10	67 % ≈ 30	11 % ≈ 5	0 % ≈ 0
6	Room search is more effective than having to ask others.	11 % ≈ 5	71 % ≈ 32	18 % ≈ 8	0 % ≈ 0
7	Able to provide accurate information.	9 % ≈ 4	71 % ≈ 32	20 % ≈ 9	0 % ≈ 0

Table 8. User Satisfaction Tests Toward Application

No	QUESTION	Description (%)≈ Respondent	
		Yes	No
1	Do you know the position/layout of the rooms in the Faculty of Engineering.	82.22 % ≈ 37	17.78 % ≈ 8
2	Do you know the position/layout of the part you are aiming for.	88.89 % ≈ 40	11.11 % ≈ 5

4. Conclusion

(1) Based on the resul of the questionnaire that distributed to 45 respondents, it can be concluded that this visitor guide software can help visitors in the process of finding a room at FT-UMY. In case this software can save time in the search room. Up to 10 respondents - 22% said they strongly agreed, 30 respondents - 67% said they agreed, in case room search is more effective than having to ask others, Up to 5 respondents, of whom 11% agree, 32 respondents, 71% responded by the acc Ord.

(2) This visitor guide software has data input, namely: 1 faculty button and has 4 search buttons, namely: button search speaker, search space, search for majors, and employee search button. It has 4/exit data consisting of: visible (side view, rear view, and front view), path, floor plan, and building.

5. References

[1] Gora S, Winastwan. Instant 3D animation using Ulead COOL 3D Studio, digital video production series. Published in collaborationAndi Yogyakarta Publisher with Deli Publishing, 2014

[2] Hendroyono, Tony. Herwibowo, Yudhi. Everything is free from the internet.Andi YogyakartaPublisher, 2014

[3] Jeprie, Mohammad. Practical application with Action Script 2.0 using Flash MX .PT Elex Media KomputindoPublisher, 2015

[4] Kurniawan, Yahya. Practical tips for mastering Action Script 2.0 Flash MX.PT Elex Media KomputindoPublisher, 2015

[5] Pramono, Andi. Creating animations with Macromedia Flash MX Professional. Andi YogyakartaPublisher, 2015

[6] Pramono, Andi. *Flash tools*, Flash support software collection. Andi Yogyakarta Publisher, 2015

[7] Zamidra Zam, Efvy. 101 tricks & techniques hidden in Windows XP, to advance.Gava MediaPublisher, 2015

[8] Campus development team, Image of UMY northern zone building;Auto CAD Files. UMY, 2007

Nearest Excellent Potential Location Using Distance Algorithm

R Purbaningtyas and A Arizal
Informatics Department, Engineering Faculty Bhayangkara University of Surabaya, Street Ahmad Yani No. 114 Surabaya, Indonesia

Email: raniubhara@gmail.com

Abstract. This study aims to find the proper distance calculation method that will be applied to the Sidoarjo on Hands (SoH) application. This study was conducted by comparing three distance algorithms namely Euclidean Distance, Manhattan Distance, and Haversine Formula. The results showed that the Euclidean Distance method was the proper method because this method had has the smallest Mean Absolute Deviation (MAD) with 1.71 in amount.

1. Introduction

Sidoarjo on Hands (SoH) is a mobile-based application which is intended to support Sidoarjo district promotion [1]. This application provides information about excellent potentials in Sidoarjo regency [2]. The SoH application was developed in two stages. The initial stage only displays potential data of Sidoarjo Regency. The second stage is equipped with features of recommended potentials of Sidoarjo Regency which is built by SoH. Furthermore, this stage matches with user's interest [3] and nearest location [4].

These two considerations are critical for users' fulfilling needs. Thus, adding an accurate location calculation relating to the potential displays is necessary.

Nowadays, various algorithms are offered to calculate distance estimation. However, in this paper, we focus on the comparison [5] of three main methods named Euclidean Distance, Manhattan Distance, and Haversine Formula. Then, we can recommend the best method to be implemented in the SoH application based on the smallest difference [6] between the calculation results and the actual distance.

2. Euclidean distance method

The Euclidean Distance method is a calculation between two points in the Euclidean space [7]. Euclidean space was introduced by Euclid, a mathematician from Greece around 300 B.C which focus on a relationship study between angles and distances. This Euclidean connects with Pythagorean Theorem and is applied in 1,2 and 3 dimensions. The formula of Euclidean Distance [8] is as follow:

$$d(x,y) = \sqrt{(x_1 - x_2)^2 + (y_1 - y_2)^2} \tag{1}$$

where x and y are latitude and longitude. The results of this calculation is still in degree units whereas we need km as a distance unit. Thus, we have to adjust the result by multiplying with 111.319 km due to one degree of earth equal to 111.319 km.

3. Manhattan Distance Method

The principle of Manhattan Distance [9] replaces squares in the previous formula by adding the absolute differences from the variables. This procedure is called an absolute block or as known as city block distance.

$$d(x,y) = L_p = i(x,y) = \sum_i^n |x_i - y_i| \quad (2)$$

4. Haversine Formula

We use the Haversine method [10] to calculate the distance between two points angles on the earth's surface using latitude and longitude as input variables. Assuming that the earth is perfectly round with radius R 6371 km, and address of 2 points in the coordinate of the ball (latitude and longitude) are lat1 lon1, and lat2 lon2, then the Haversine formula is written using this equation:

$$d = 2R\sin^{-1}\sqrt{sin^2\left(\frac{\phi_2-\phi_1}{2}\right) + \cos(\phi_1)\cos(\phi_2)\,sin^2\left(\frac{\psi_2-\psi_1}{2}\right)} \quad (3)$$

where d is the distance between two points with latitude and longitude (ϕ,ψ) and R is the radius of the Earth. The Haversine formula is an important equation in navigation [11][12][13].

5. Result and Discussion

This research was conducted in the following stages:

(1) Collecting data by defining the user's starting point to 15 different destination locations
(2) Each starting point and a destination point will be recorded what the value of the coordinates are
(3) Measure the actual distance from the origin point to the destination point
(4) Measure the distance from the starting point to each destination using the Euclidean distance method, Manhattan distance method, and also the Haversine method
(5) Calculate the difference of each algorithm by comparing the actual distance value and the calculated value of the algorithms
(6) Calculating the absolute value of each error of the three methods
(7) Define the finest algorithm based on the smallest value in the calculation error

For comparing the accuracy of calculation between three methods, a user was in Sidoarjo town square with coordinates (-7.446061,112.717707), was given 15 different locations in Sidoarjo regency:

- Location A is Indah Bordir Sidoarjo (-7.444184, 112.720208) with the actual distance from the user's location is 2.4 km
- Location B is Intako Tanggulangin (-7.505345, 112.694379) with the actual distance from the user's location as far as 9.7 km
- Location C is East Java Province Cooperative and MSME Office (-7.381197, 112.741744) with the actual distance from the user's location as far as 11 km
- Location D is Ecco Leather (-7.473713, 112.714597) with the actual distance from the user's location as far as 3.8 km
- Location E is Mitra Jaya (-7.497852, 112.696946) with the actual distance from the user's location as far as 8.2 km
- Location F is Fitrah Jaya (-7.499064, 112.69749) with the actual distance from the user's location as far as 8 km
- Location G is UD Diya Aini Jaya (-7,501568, 112.705614) with the actual distance from the user's location as far as 7 km
- Location H is Gugum Leather Jacket (-7.499108, 112.700124) with the actual distance from the user's location as far as 7.7 km
- Location I is Yokohama Sandal (-7.352664, 112.748203) with the actual distance from the user's location as far as 12.8 km

- Location J is Hasta Indah Bordir (-7.502453, 112.707734) with the actual distance from the user's location as far as 6.8 km
- Location K is Permata Tanggulangin Collection (-7.50287, 112.703678) with the actual distance from the user's location as far as 8 km
- Location L is Maju Makmur (-7.498908, 112.697412) with the actual distance from the user's location as far as 8 km
- Location M is Teratai Indah Bordir (-7.494357, 112.687915) with the actual distance from the user's location as far as 9.6 km
- Location N is Yan Kurin Collection (-7.486597, 112.710988) with the actual distance from the user's location as far as 5.1 km
- Location O is UD New Paulo (-7.404576, 112.723424) with the actual distance from the user's location as far as 6.5 km

5.1. First iteration

For the first destination location, we used Euclidean Distance method, and the result as shown:

$$d(start, A) = \sqrt{(x_1 - x_2)^2 + (y_1 - y_2)^2}$$
$$= \sqrt{(-7.446061 - -7.444184)^2 + (112.717707 - 112.720208)^2}$$
$$= 0{,}35 \text{ km}$$

Table 1 depicted the result of 15 locations using Euclidean Distance method

Table 1. Euclidean distance result

Origin	Destination	Actual Distance	Euclidean Distance Result
Start (Sidoarjo town square)	A (Indah Bordir Sidoarjo)	2.4 km	0.35 km
	B (Intako Tanggulangin)	9.7 km	7.08 km
	C (East Java Province Cooperative and MSME Office)	11 km	7.71 km
	D (Ecco Leather)	3.8 km	3.09 km
	E (Mitra Jaya)	8.2 km	6.2 km
	F (Fitrah Jaya)	8 km	6.31 km
	G (UD Diya Aini Jaya)	7 km	6.32 km
	H (Gugum Leather Jacket)	7.7 km	6.21 km
	I (Yokohama Sandal)	12.8 km	10.95 km
	J (Hasta Indah Bordir)	6.8 km	6.37 km
	K (Permata Tanggulangin Collection)	8 km	6.5 km
	L (Maju Makmur)	8 km	6.29 km
	M (Teratai Indah Bordir)	9.6 km	6.31 km
	N (Yan Kurin Collection)	5.1 km	4.56 km
	O (UD New Paulo)	6.5 km	4.67 km

5.2. Second iteration

We calculated using Manhattan Distance method and the result as below. More detail, table 2 showed the measurement of 15 locations

$$d(start, A) = \sum_i^n |x_i - y_i|$$
$$= |-7.446061 - -7.444184| + |112.717707 - 112.720208|$$
$$= 0.5 \text{ km}$$

Table 2. Manhattan distance result

Origin	Destination	Actual Distance	Manhattan Distance Result
Start (Sidoarjo town square)	A (Indah Bordir Sidoarjo)	2.4 km	0.5 km
	B (Intako Tanggulangin)	9.7 km	9.19 km
	C (East Java Province Cooperative and MSME Office)	11 km	9.91 km
	D (Ecco Leather)	3.8 km	3,42 km
	E (Mitra Jaya)	8.2 km	8,07 km
	F (Fitrah Jaya)	8 km	8.14 km
	G (UD Diya Aini Jaya)	7 km	7.52 km
	H (Gugum Leather Jacket)	7.7 km	7.85 km
	I (Yokohama Sandal)	12.8 km	13.8 km
	J (Hasta Indah Bordir)	6.8 km	7.38 km
	K (Permata Tanggulangin Collection)	8 km	7.88 km
	L (Maju Makmur)	8 km	8.13 km
	M (Teratai Indah Bordir)	9.6 km	8.68 km
	N (Yan Kurin Collection)	5.1 km	5.25 km
	O (UD New Paulo)	6.5 km	5.26 km

5.3. Third iteration

For the first destination, when we measured using the Haversine Formula (3) method, the result was 0.35 km. Thus, for all destination, if measured using the Haversine Formula method, the calculation results are follows:

Table 3. Haversine Formula Result

Origin	Destination	Actual Distance	Haversine Formula Result
Start (Sidoarjo town square)	A (Indah Bordir Sidoarjo)	2.4 km	0.35 km
	B (Intako Tanggulangin)	9.7 km	7.01 km
	C (East Java Province Cooperative and MSME Office)	11 km	7.70 km
	D (Ecco Leather)	3.8 km	3.08 km
	E (Mitra Jaya)	8.2 km	6.19 km
	F (Fitrah Jaya)	8 km	6.29 km
	G (UD Diya Aini Jaya)	7 km	6.31 km
	H (Gugum Leather Jacket)	7.7 km	6.2 km
	I (Yokohama Sandal)	12.8 km	10.92 km
	J (Hasta Indah Bordir)	6.8 km	6.36 km
	K (Permata Tanggulangin Collection)	8 km	6.49 km
	L (Maju Makmur)	8 km	6.28 km
	M (Teratai Indah Bordir)	9.6 km	6.29 km
	N (Yan Kurin Collection)	5.1 km	4.56 km
	O (UD New Paulo)	6.5 km	4.66 km

5.4. Final result determination

The finest algorithm will be determined using the Mean Absolute Deviation (MAD) method [14]. MAD is a procedure for measuring the accuracy of the results of calculations with an average error (absolute value of each error) [15]. The formula for calculating MAD values below:

$$MAD = \frac{\sum_{i=1}^{n} |x_i - d_i|}{n} \quad (4)$$

where x is the actual distance data and d is the distance data obtained from the calculation results. The results of distance computation errors with the MAD method are as follows:

Table 4. MAD calculation

Origin	Destination	Actual Distance	Error in Computation Results		
			Euclidean Distance	Manhattan Distance	Haversine Formula
Start (Sidoarjo town square)	A (Indah Bordir Sidoarjo)	2.4 km	-2.05	-1.90	-2.05
	B (Intako Tanggulangin)	9.7 km	-2.62	-18.89	-2.63
	C (East Java Province Cooperative and MSME Office)	11 km	-3.29	-1.09	-3.31
	D (Ecco Leather)	3.8 km	-0,71	-7,22	-0,72
	E (Mitra Jaya)	8.2 km	-2,00	-16,27	-2,01
	F (Fitrah Jaya)	8 km	-1,69	-16,14	-1,71
	G (UD Diya Aini Jaya)	7 km	-0,68	-14,52	-0,69
	H (Gugum Leather Jacket)	7.7 km	-1,49	-15,55	-1,50
	I (Yokohama Sandal)	12.8 km	-1,85	1,00	-1,88
	J (Hasta Indah Bordir)	6.8 km	-0,43	-14,18	-0,44
	K (Permata Tanggulangin Collection)	8 km	-1,49	-15,88	-1,51
	L (Maju Makmur)	8 km	-1,71	-16,13	-1,72
	M (Teratai Indah Bordir)	9.6 km	-3,29	-18,28	-3,31
	N (Yan Kurin Collection)	5.1 km	-0,54	-10,35	-0,54
	O (UD New Paulo)	6.5 km	-1,83	-1,24	-1,84
	Mean Absolute Deviation (MAD)		1,71	11.11	1.72

According to Table 4, Euclidean Distance method has the smallest MAD value when it is compared to Manhattan Distance and Haversine Formula. Euclidean Distance method has an average error (difference) 1.71. It can be concluded that the Euclidean Distance method is more appropriate for SoH application.

6. Conclusion

The Sidoarjo on Hands (SoH) application is a mobile-based application intended to support the promotion of the Sidoarjo Regency. This application was developed in stages. In the final development phase, will be added a feature to display information about the distance from the current location of the SoH users to the location of the potential of the intended area. This study tries to find the most appropriate distance calculation method by comparing the three distance algorithms, namely Euclidean Distance, Manhattan Distance, and Haversine Formula method. The results show that the Euclidean Distance method is the proper method to be applied in SoH application. The most noticeable reason because this method has the smallest value in the calculation error of the distance obtained compared to the actual distance value.

7. References

[1] R. Purbaningtyas *et all* 2017 *Pros. SNATIF* 291–296
[2] R. Purbaningtyas *et all* 2017 *J. Penelit. Pos dan Inform.* **7** 87-96
[3] R. Purbaningtyas *et all* 2018 *Pros. SENTIKA* 181-188
[4] R. Purbaningtyas *et all* 2019 *Probl. Inf. Technol.* **10** 70–76
[5] L. Greche *et all* 2017 *Proc. Int. Conf. on Wireless Technologies, Embedded and Intelligent Systems (WITS)* 1–4

[6] M. Cruz López-de-los-Mozos *et all* 2001 *Eur. J. Oper. Res.* **135** 184–194
[7] D. Nugraheny 2017 *Angkasa J. Ilm. Bid. Teknol.* **7** 21-30
[8] I. Dokmanic *et all* 2015 *IEEE Signal Process. Mag.* **32** 12–30
[9] W.-Y. Chiu *et all* 2016 *IEEE Trans. Evol. Comput.* **20** 972–985
[10] N. Chopde *et all* 2013 *International Journal of Engineering Research and Applications (IJERA)* **1** 162-165
[11] E. Winarno *et all* 2017 *International Conference on Innovative and Creative Information Technology (ICITech)*. 1–4
[12] M. Basyir *et all* 2018 *Emit. Int. J. Eng. Technol.* **5** 270-278
[13] Z. Arifin *et all* 2016 *Proc* 3rd *International Conference on Information Technology, Computer, and Electrical Engineering (ICITACEE)* 293–296
[14] M. Rahmani *et all* 2017 Transp. *Res. Part C Emerg. Technol.* **85** 628–643
[15] C. E. Simpkins *et all* 2017 *Ecol. Inform.* **38** 1–11

Acknowledgments
Thanks to the DRPM of the Ministry of Research and Higher Education of the Republic of Indonesia for funding this research through Individual National Strategic Research schemes in 2018.

Identifying the Quality System of Cocoa Beans to Increase Productivity Using Backpropagation Neural Network Algorithm: A Case Study at Sumber Rejeki Farmers Group, Patuk Gunung Kidul

U Lestari, R A Kumalasanti, and E S Wulandari
Department of Informatics Engineering, Institut Sains &Teknologi AKPRIND Yogyakarta, Kalisahak Street 28, Yogyakarta, Indonesia

E-mail: uning@akprind.ac.id, rosalia@akprind.ac.id, ekasepti089@gmail.com

Abstract An image is a two-dimensional function (x, y) consisting of a number of finite elements. Image processing uses Backpropagation and Principal Component Analysis (PCA) Artificial Neural Networks. The processed data is the image data of 256 x 256 pixel cocoa beans taken from the Sumber Rejeki Farmer Group (KTSR) with the results of two types of identification output, namely quality cocoa beans and non-quality cocoa beans. The identification process in the system through two stages of the process, namely the training process where the image data of cocoa beans through the training process with input parameters ANN. Image data that has been completed through the training process is used for the testing process. Based on the results of the study using the Backpropagation method with parameters alpha value = 0.35, tolerance value error = 0.05, iteration value = 1000, input neuron value = 255, hidden neuron value = 6, output neuron value = 3 produce quality image data 14 and 6 quality images that are not qualified from the test image data are 20 images of cocoa beans with an accuracy rate of 70% and an error rate of 30%. Based on the tests that have been conducted, it can be concluded that the ANN Backpropagation algorithm can produce a classification of cacao seed quality with a high degree of accuracy

1. Introduction

Cocoa is a plantation commodity which in recent decades has proven to be an adequate source of income and provides good welfare for farmers. Cocoa commodities are also able to play a role as levers of economic growth, especially in rural areas in production centers. In addition, cocoa also contributes significantly to the acquisition of foreign exchange and absorbs more than one million workers involved in the production and processing and trade sectors. Given this huge role, cocoa has been determined by the government as a priority commodity to be developed. [1]

The statistical data available at the Central Statistics Agency for the area of cocoa plantations in Indonesia in the period of 2012 was 1.77 million hectares, while in 2015 it decreased to 1.70 million hectares and in 2016 amounted to 1.72 hectares. While data on cocoa production in the Special Region of Yogyakarta in 2013 amounted to 853, in 2014 amounted to 1039, in 2015 amounted to 1121, in 2016 amounted to 1057 and in 2017 amounted to 866 with total growth of 7.93%. [2]

The fact shows that there is a continuous increase in world cocoa needs, so that in the future cocoa prices are expected to remain good so that the development of cocoa into new production centers to fill a portion of the world's needs needs to be continued. The results of the research show that in Indonesia there is still a vast area of land that is suitable for these development needs. Revitalizing existing cocoa areas and developing into new central areas is an action that must be taken by all national cocoa stakeholders to realize Indonesia as the world's largest cocoa producer. [1]

Cocoa plantations in Indonesia have experienced rapid development in the last 20 years and in 2002 Indonesia's cocoa plantation area was recorded at an area of 914,051 ha. Most of the cocoa plantations (87.4%) are managed by the people and the remaining 6.0% are large state plantations and 6.7% are large private plantations. In terms of quality, Indonesian cocoa is not inferior to world cocoa. Nevertheless, Indonesian cocoa agribusiness still faces a variety of complex problems, including low productivity, low product quality and not optimal development of downstream cocoa products. [3]

Quality is a parameter in determining a problem. In determining the quality of cocoa beans itself has several parameters so that determining the quality of cocoa is one of the important aspects in developing sustainable cocoa production and a major factor in consumer satisfaction. To be successful, cocoa farmers should understand the quality from the start. Criteria for the quality of cocoa beans with good quality include full cooking, good fermentation with dry brown seeds, sour smell, no purple or slaty if broken down, not more than 1 gram dry weight, uniform size, non-broken shells and red brownish. (Lampung Agricultural Technology Assessment Center, 2008) Sumber Rejeki Farmers Group (KTSR) is one of the farmer groups in the Patuk sub-district. The Sumber Rejeki Farmers Group was founded in 1980 as a forum for the community to strengthen friendship and sources of sharing agricultural knowledge. The growing age of the Sumber Rejeki Farmers Group also conducted training and established a business to buy wet and dry Cocoa beans. Cocoa beans collected in the Sumber Rejeki Farmer Group are then deposited with the cocoa management officer. Cocoa beans that are deposited to the management officer still use the selection process manually without using parameters but only see the condition of the cocoa beans. The Sumber Rejeki Farmer Group is expected to become a forum for the community to improve economic resources and a place for educating the community to know the parameters for selecting quality cocoa beans. Cocoa beans in the Sumber Rejeki farmer group with a variety of qualities. The cocoa beans were obtained from cocoa farmers from cocoa beans that were still wet, and then by the Sumber Rejeki Farmers Group, they processed wet cocoa beans into dried cocoa beans. The dried cocoa beans will be sold. At this time the determination of the quality of cocoa beans is still manual. Therefore, to find out the quality of cocoa beans with good quality by taking the image of cocoa beans, this study will develop the management of the quality of cocoa beans based on variable texture and color with the image of cocoa beans.

Research on the identification of digital images has been carried out by researchers with various identification objects such as facial images, fruit images, satellite images and others. Digital image processing is a process of improving image quality seen from radiometric aspects (contrast enhancement, color transformation, image restoration) and from geometric aspects (rotation, translation, scale, genetic transformation) by carrying out information withdrawal or object descriptions or object recognition contained in the image other than that there is a process of compression or reduction of data for the purpose of data storage, data transmission and data processing time [4].

The application of digital image processing is expected to be able to provide information for identifying quality with the object of cocoa beans with the results of good and bad quality cocoa beans and can minimize losses if the cocoa beans are sold. The location of the study was conducted at the Sumber Rejeki Farmer Group located in Trosari Hamlet RT / RW 019/006, Salam Village, Patuk District, Gunungkidul Regency, Special Region of Yogyakarta. This application refers to data on cocoa beans in the farmer group. Several previous studies have used the object of cocoa beans and cocoa fruit with various methods and application of the system.

The research discusses the implementation of the Backpropagation method to identify the types of defective cocoa beans based on the shape of the seeds. The method used uses the Backpropagation method. The image processing process by applying the method uses the main data, namely data on cocoa beans with parameters or limitations of the study of the type of defective cocoa beans based on the shape of cocoa beans or edge detection. The system created can help process knowledge from cocoa bean data. The process of identifying the cocoa beans begins by taking the image of the object using the camera, adjusting the image size, performing grayscale and threshold processes, edge detection or edge detection, image extraction and generating identification results in binary imagery, which is then used for the process data learning. In the research that has been done, the value of learning rate (α) affects the number of iterations of a pattern entered in the data learning process to achieve the iterative termination condition. The results of the research that have been done using the Backpropagation method with alpha = 0.6, hidden layer = 3 fault tolerance = 0.0001, and target = 0.9 produce a system that has an accuracy rate of (76%) and has an error rate (24%) in determining quality of cocoa beans. [5]

Research describes the identification of types of diseases in cocoa by digital image processing and K-Nearst Neighbor. This method is almost similar to clustering, in determining the type of disease in cocoa fruit there are 2 types, namely (1). fruit rot is a disease caused by P. palmiwora fungus which can attack young fruit until cooked. The affected fruit appears blackish brown patches from the base, middle or end of the fruit. (2). Helotopis is a cocoa fruit disease caused by the attack of ladybugs. This disease usually attacks cocoa fruit or young branches of cacao fruit trees. The results of the study were the identification system for cacao fruit using PCA feature image extraction method and K-NN calcification method with an accuracy rate of 86.67%, the size of the image used or the normalized image of 128 x 64 pixels. In the classification process with K-NN, the type of distance method that produces the highest accuracy is the Ciryblock method with k = 1 while the average computation time required for the pre-processing process is 0.31284 seconds, the average computational time for the image extraction process is 0.03598 seconds and the average computation time for the classification process is 0.07774 seconds. [6]

Research on the identification of the maturity quality of cocoa fruit using image processing and fuzzy logic methods has also been done before. This study uses color processing models RGB (red, green, blue) and image histograms. In the fruit classification process based on the characteristics of the same shape or color at the fruit maturity level with mature class groupings having yellow characteristics, the mature setangah class has characteristics of yellowish green and young cocoa fruit has a green characteristic. In applying the fuzzy logic method that is an appropriate way to map an input space into the output space based on the concept of fuzzy sets. The concept of fuzzy logic is used to solve problems that are uncertinly, imprecise, noisy, fuzzy logic as the logic between machine language and human language [7].

Research on pattern matching with the Backpropagation method has also been done before by taking human face objects. The comparison variables used were distance of facial features including nose, eyes and mult from the subject of facial images. In the process of facial recognition, the process of analyzing and identifying faces is done. Face detection includes separating window images with 2 classes namely face (target) and background (clutter). The results of this study are a system that displays humans with high-level facial recognition with a very high degree of accuracy [8].

The Research for describes a pattern recognition system to clarify data patterns and group them according to class also been done before. The purpose of this research is to learn various techniques in recognizing complex patterns, identifying and implementing the best techniques that are appropriate and analyzing results and levels of accuracy. The performance of the pattern monitoring system in this study depends on the type of approach used to identify and clarify the patterns and inputs. If the network model is based on an unsupervised learning approach, then there is no a priori knowledge about a set of classes that has been set and input data differentiated into groups based on structural characteristics or other similar characteristics which can sometimes be a tedious task. If this model is based on a supervised learning approach, the system developed is more accurate and efficient because the network has been trained with the desired results. This engineering approach is better in performance than other techniques because of the high level of accuracy for complex pattern recognition, adaptive learning and better tolerance factors for errors even though more time may be needed to train the network for very complex patterns [9]

Research with the Principal Component Analysis (PCA) and Backpropagation Neural Network methods has also been carried out to identify faces The dimensions of the face image are reduced by PCA and face recognition by the Backpropagation network. This system consists of a database of a series of facial patterns for each individual with PCA characteristic features called "eigenfaces" extracted from stored images, which are combined with Backpropagation networks for facial recognition. The results of face recognition using PCA and the Backpropagation method with a simulation approach are done with matlab and tested on the ORL face database. This database has more than one individual face image with different conditions. The database is divided into two sets, namely, training databases and testing databases. The network is trained on database training and then one of the images from the testing database is entered as input to test the network [10].

The difference in this identification application with the references mentioned earlier is that this research using the method the Backpropagation method which is added by the Principal Component Analysis (PCA) method and the quality parameters of cocoa beans, namely the texture and color of cocoa beans.

2. Research Methodology

Making an application system through several steps is as follows:

2.1. Data collection

Data collection in the form of digital data or images of cocoa beans. In data collection is done by taking pictures of objects with the help of a digital camera tool. Image taking with dominant white background with image data for each criterion of 10 cocoa beans.

2.2. Resize Image

Resize or resize the image of the cocoa beans to the desired size. In this study the image size of native cocoa beans with a size of 6000 pixels x 4000 pixels and resizing the image of cocoa beans with a size of 256 pixels x 256 pixels with the help of paint desktop applications and JPG image extensions.

2.3. Image input

Image Input Process in implementing the system. Image input by moving image data on the camera into a laptop or storage with the size of the image of cocoa beans 256 pixels x 256 pixels.

2.4. Threshold

The threshold process is the process of changing the image with a degree of gray to a binary or black and white image so that it can be seen which areas include objects and backgrounds of the image clearly. The image from the threshold process is used further for object recognition and feature extraction processes.

2.5. Image Extraction with the Principal Component Analysis Algorithm

Image extraction process or simplifying image data with the Principal Component Analysis (PCA) process by changing the image data of cocoa beans through a linear transformation process. Data that has been extracted as input data in the process of implementing the Backpropagation Algorithm.

2.6. Implementation of Backpropagation Algorithm

The sixth step in the research is implementing an algorithm from the Backpropagation method. Backrpopagtion algorithm is an algorithm with the application of artificial neural networks. Backpropagation is a supervised learning method that has a target to be sought. The process of applying the Backpropagation method is used in two stages, namely the training stage, where at this stage a number of training data and targets are given and the testing phase is carried out after Backpropagation has finished learning.

2.7. Input image

The next step is image input which is the result of the image data of cocoa beans that have gone through the process of image resizing, threshold, image extraction with Principal Component Analysis and implementation of the Backpropagation method stored in the datastore. Data stored in the datastore is used as data in the process of testing the image to obtain information on the quality of cocoa beans.

2.8. Image Testing

The last step is to test the image, this step is used to determine the results of the image data that has been obtained in the input image and the application of the Backpropagation algorithm. Image testing with image data that has been stored in the data store with the image data of cocoa beans which will later in the process of testing the image will produce conclusions of the quality of the ideals of the cocoa beans. In addition, this testing process is also used to determine error tolerance and accuracy by applying the Backpropagation method.

3. Results

The results of image testing in the form of information on the image of cocoa beans that are of good quality and not good quality. In the process of determining the quality of cocoa beans there are several parameters in determining the quality used. Parameters determining the quality of cocoa beans can be seen in Figure 1 [12].

No	Picture	Name	Description
1		Full Fermentation	Brown cocoa beans are full of hollow texture. The fermentation process is complete
2		Partial Purple	Some cocoa beans are purple and some are brown. This is generally produced from an imperfect fermentation process.
3		Full Purple	Full purple cocoa beans. This is result from a short fermentation process (less than 1 weeks)
4	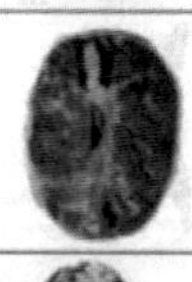	Slaty	Cocoa beans are purple and grey and have a solid texture. This results from a short fermentation process (less than 3 days)
5		Moldy	Cocoa beans are brown and there is fungus in the texture cavity of the cocoa beans. This is result from the imperfect fermentation process

Figure 1. Parameters determining the quality of cocoa beans

4. Results and Discussion

The results of the system application for identifying the quality of cocoa beans consist of two main processes, namely the Training Process and the Testing Process. Whereas button access consists of Output Target, Read Training Image, Initial Weight, Training Process, Read Test Image and Testing Process.

4.1. Training Process Stage

The initial appearance of the application for identifying the quality of cocoa beans consists of two processes, namely the Training Process and the Testing Process. The application for identifying the quality of cocoa beans begins with the Training process by determining the target output of each criterion (Slaty, Full Purple, Full Purple, Full Fermentation, Molding) then selecting the Image Read button to determine the training image data stored in the database to be processed in training process. In this process an example is taken for the target criteria of Slaty. The process of determining the target of Salty's output criteria can be seen in Figure 2. After all the criteria, namely Slaty, have finished reading the image, they are automatically stored in the database.

After the process is complete, the next step is to get values for weight V and W. By pressing the Initial Weight button then weights V and W randomly appear in the weight table V and W. This initial weight button serves to generate initial weights for weight V (weight between input and hidden layers) and W weights (weights between hidden layer and output layer). After the initial weight is raised, it will be saved automatically in the database. The next step is to determine the values of the parameters of Artificial Neural Networks which include Alpha values, Tolerance values, Iteration values, Input Neuron values, Hidden Neuron values, Output Neuron values. In this process using that Alpha value is 0.35, the tolerance value is 0.05, and the number of iterations is 1000. These values can be changed according to the requirements for each training process. The value of input neurons is 255 and cannot be changed because it is the number of eigenvalues of training images measuring 256 x 256 pixels. A 256 x 256 pixel image produces an eigenvalue of 255. The value of a hidden neuron is 6 and the output neuron value is 3 because it corresponds to the target number of each cocoa image criterion. After the process of determining parameters, the values of each parameter have been completed, the next step is to do the training process by pressing the training button.

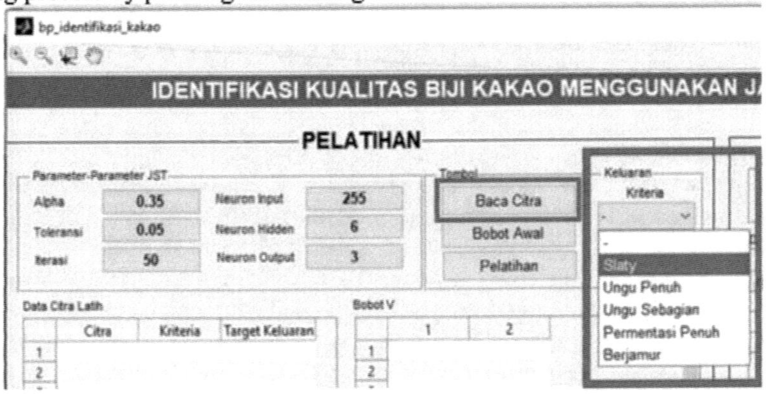

Figure 2. Process for Determining Target Criteria for Slaty

The training process will take place according to predetermined iteration and tolerance parameters. The training iteration process will continue as long as the error value of the training process is greater than the set tolerance value. If the error value of the training process is smaller than the tolerance value, the iteration of the training process will end. Conversely, if the error value of the training process is greater than the tolerance value, the training process will continue to take place in accordance with the number of iterations that have been determined.

In the dialog box, it can be seen that the number of iterations is 466 and the error value is 0.049949 which is smaller than the predetermined tolerance value of 0.05 and the computing time for the training process is 666.2426 seconds. The training process will produce a weight V and a new W weight called the trained weight term. This trained weight will be used to carry out the identification process. Weight V and weight W the training results will be displayed in the table and the results of the trained weights are automatically stored in the database.

4.2. Testing Process Stage

After the training process is completed, proceed with the Testing process, namely selecting the test image that will be used as test data in the database. This process is done by pressing the Read Image button in the testing process. After selecting the test image in the database, the results will be displayed in the test image table by displaying the image name and image size. The testing process will produce identification data which can be seen in Figure 3. Test image data in the testing process produces 14 recognizable image data according to the criteria but there are 6 data that cannot be recognized. The testing process requires computation time of 0.69412 seconds.

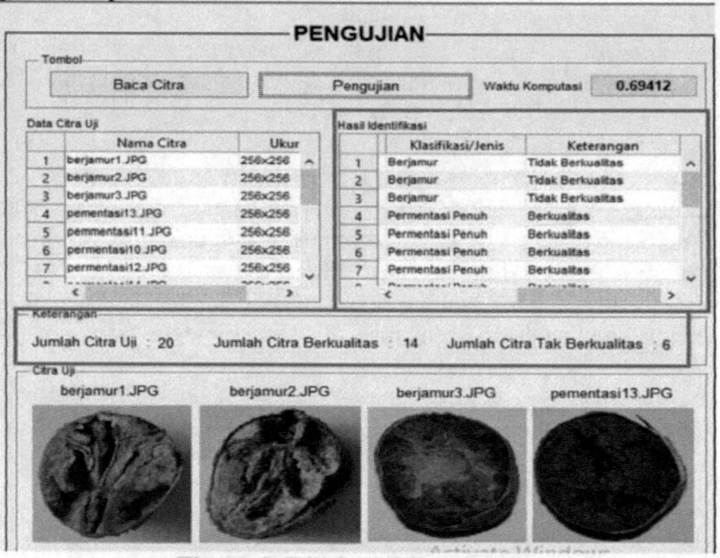

Figure 3. Display of Test Results

4.3. Analysis of System Results

The application for identifying the quality of cocoa beans using the Backpropagation method in the implementation process requires several additional methods in the identification process. In the training process or learning process in the Backpropagation method additional methods are needed, namely the image extraction process using the Pricipal Component Analysis (PCA) method. This method was added to facilitate the process of image recognition of each criterion of cocoa beans. The extraction process with PCA is used to find eigenvalues from the image of cocoa beans. In the process of extracting images with PCA the results obtained to be implemented in the Backpropagation method, namely with 256 x 256 pixel size images, an eigenvalue of 255 is obtained in the parameters of Artificial Neural Networks (ANN) as input neurons. The structured process in image extraction with PCA in the pre-training process can be seen in Figure 4.

After the image extraction process with PCA, it can be done using the Backpropagation method starting from the training process and then testing. In the training process eigenvalue data from the image of each criterion can be processed so as to produce test image data and corvegency graphs. The training process begins after the image extraction process with PCA is completed and then determines the initial weight and parameters of Artificial Neural Networks (ANN). The initial weight between weight V and Weight W is obtained randomly from weights V (hidden neurons and input neurons) and W weights (output neurons and hidden neurons). The appearance of the training process can be seen in Figure 5

The training process is carried out if the PCA process has been completed and the eigenvalues of each image are stored in the datastore. The training process begins by taking or obtaining a value of weight V and weight W obtained randomly. Furthermore, determining the parameters of Artificial Neural Networks (ANN) include alpha value = 0.35, tolerance value of 0.05, iteration value = 1000, input neuron value = 255, hidden neuron value = 6 and output neuron value = 3, After process the training was carried out in the application of the Backpropagation method and then the testing process was carried out. After the parameters have been completed, the training process can be started so that the output results in the form of a convergence graph and the training results dialog box are obtained.

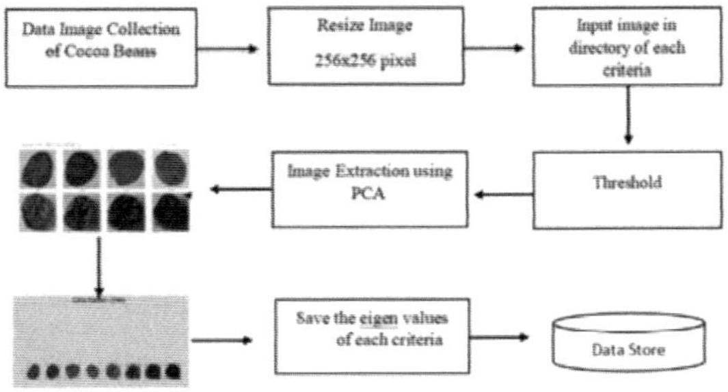

Figure 4. PCA Process

After conducting the training process, the next step is to conduct a testing process that produces identification results in accordance with the quality of each of the criteria for cocoa beans. The output of the identification results is in the form of images of good quality cocoa beans and non-quality seeds. The appearance of the testing process can be seen in Figure 8.

In the testing process begins by selecting a test image that will be carried out by the testing process with the output of a quality and non-quality image in accordance with the parameters that have been carried out in the training process. In the process of identifying the quality of cocoa beans it was carried out with 5 simulations to get the right results and accuracy in determining the quality of good and non-quality cocoa beans with the appropriate ANN parameters. The test simulation process can be seen in Table 1.

In addition to carrying out the testing simulation process, a discussion of the manual calculation process was carried out in identifying the quality of cocoa beans as follows: alpha value = 0.35, tolerance = 0, X1 = 1, X2 = 1 with hidden neurons = 3. Weight obtained randomly with a value weight V and weight W in Table 3 and Table 4.

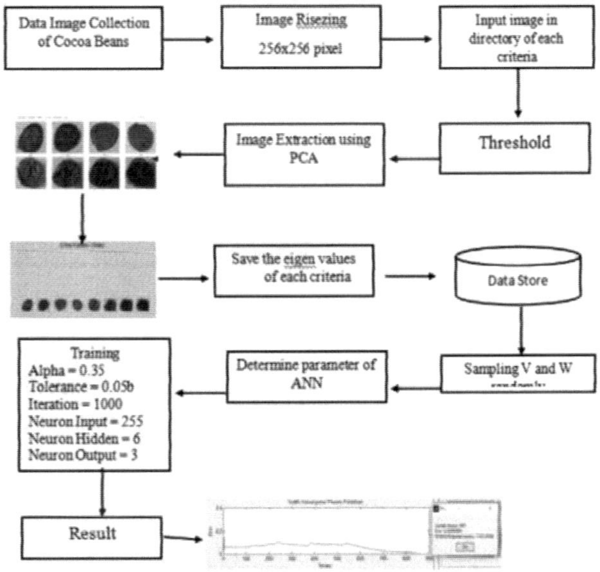

Figure 5. Training Process

Table 1. Convergence Charts and Dialog Boxes

<table>
<tr><th>No</th><th>Convergence Graph</th><th>Dialog Box</th></tr>
<tr><td>1</td><td>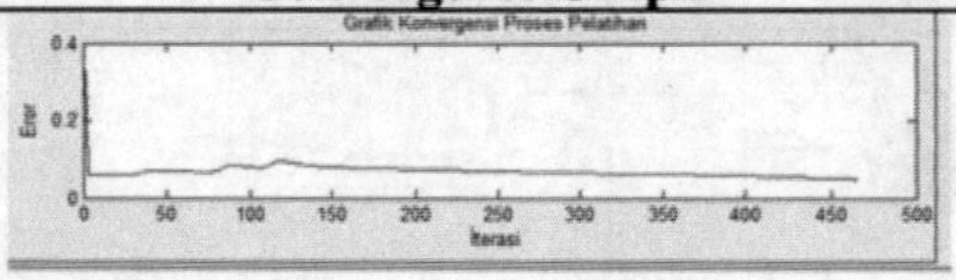
</td><td>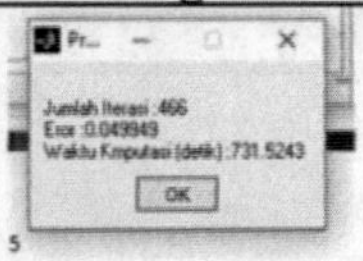
</td></tr>
<tr><td>2</td><td>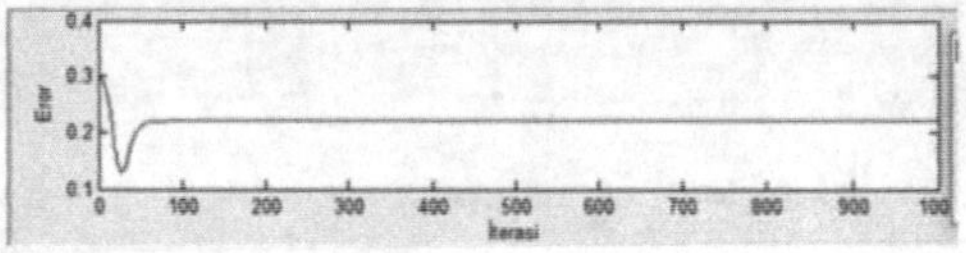
</td><td>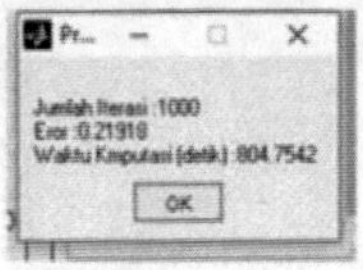
</td></tr>
<tr><td>3</td><td>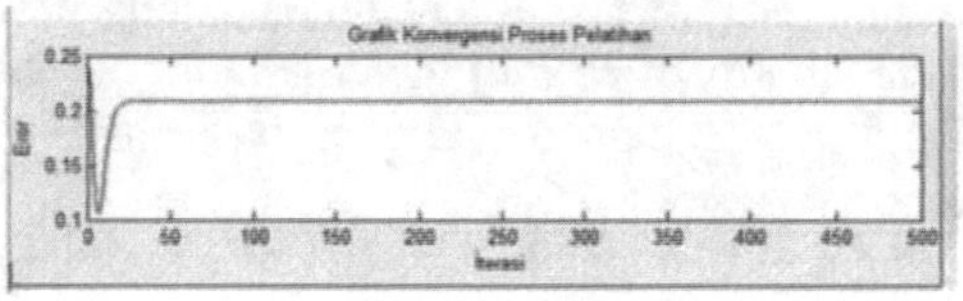
</td><td>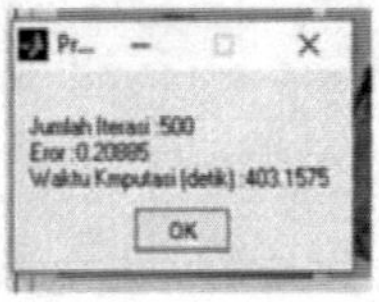
</td></tr>
<tr><td>4</td><td>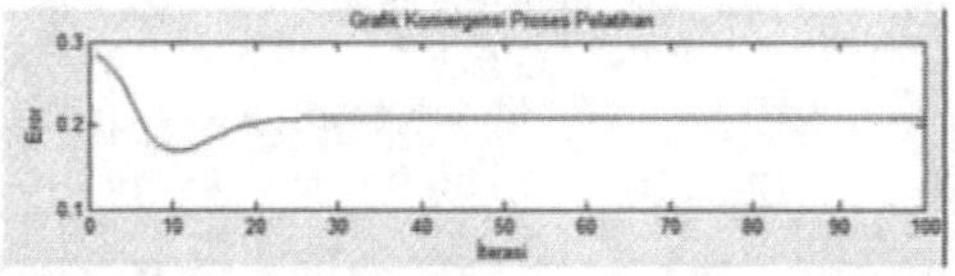
</td><td>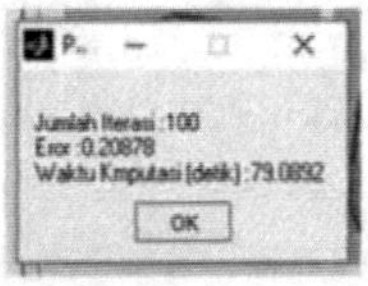
</td></tr>
<tr><td>5</td><td>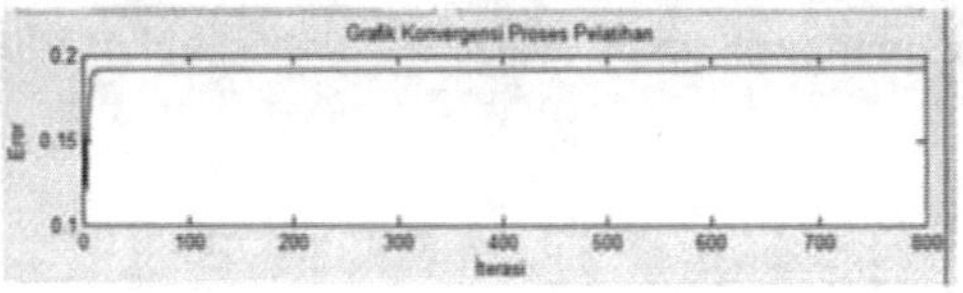
</td><td>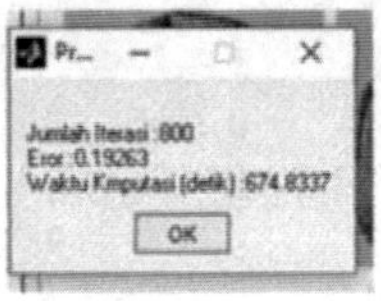
</td></tr>
</table>

Table 2. Test Simulation Results

No	Artificial Neural Network Parameters						Result (B/TB)	Accuracy
	Alpha	Tolerant	iterasi	Neuron input	Neuron hidden	Neuron output		
1	0,35	0,05	1000	255	6	3	B = 14, TB = 6	70%
2	0,02	0	1000	255	3	3	B = 0, TB = 20	0%
3	0,05	0,1	500	255	1	3	B = 0, TB = 20	0%
3	0,05	0,1	500	255	1	3	B = 0, TB = 20	0%
5	0,10	0	800	255	3	3	B = 0,TB = 20	0%

Table 3. Latest V Weight Table

	Z_1	Z_2	Z_3
X_1	-0,003	0,004	-0,002
X_2	-0,001	0,006	-0,007
1	-0,001	0,006	-0,020

Table 4. Latest W Weight Table

	Y_1	Y_2	Y_3
Z_1	0,022	0,071	0,061
Z_2	0,080	0,032	0,027
1	0,111	0,147	0,088

The implementation of the back propagation method in the application of the quality identification of cocoa beans in grouping categories or criteria from cocoa bean data is done manually based on direct observation in the field. The manual grouping process of this manual category consists of full, purple, purple, full, slaty and moldy fermentation criteria. The process of grouping categories in research with image data of cocoa beans took training samples of 8 images each from 5 criteria and 20 test images.

5. Conclusion

With the identification system using the Backpropagation method to determine the quality of cocoa beans in the Sumber Rejeki Farmer Group (KTSR) with quality and non-quality results, we can conclude:

(1) Based on the results of the study using the Backpropagation method in the process of identifying quality, it gives an output of the image of quality and non-quality cocoa beans. In the simulation process, it is obtained that with alpha parameter value = 0.35, tolerance value = 0.05, iteration value = 1000, input neurons = 255, hidden neurons = 6, and output neurons = 3, there are 14 quality cocoa beans and the image of non-quality cocoa beans is 6 out of 20 test image data with an accuracy rate of 70% and an error rate of 30%.

(2) Based on experiments or simulations that have been carried out, the value of the parameters of Artificial Neural Networks (ANN) affects the results in the process of identifying the quality of cocoa beans to produce a high degree of accuracy.

(3) By using a computerized system, the process of identifying the quality of cocoa beans is able to provide a smaller level of error and help Farmers in determining the quality of cocoa beans.

6. References

[1] Pusat Penelitian Kakao dan Kopi Indonesia 2015 *KAKAO: Sejarah, Botani, Proses Produksi, Pengolahan, dan Perdagangan* Yogyakarta Gadjah Mada University Press.

[2] Badan Pusat Statistik 2018 *www.bps.go.id.* Retrieved April 8, 2018 from Statistik Kakao Indonesia: www.bps.go.id

[3] Suryana A 2015 *Prospek dan Arah Pengembangan Agrobisnis Kakao.* Jakarta: Badan Litbang Pertanian

[4] Hermawati F A 2013 *Pengolahan Citra Digital* Yogyakarta: CV. ANDI OFFSET.

[5] Nurmuslimah S 2016 Implementasi Metode Backpropagation untuk Mengidentifikasi Jenis Biji Kakao yang Cacat Berdasarkan Bentuk Biji, Network Engineering Research Operation Journal Vol.2 No.2 2016

[6] Yogiswara G H 2016 Identifikasi Jenis Penyakit pads Kakao dengan Pengolahan Citra Digital dan K-Nearest Neighbor. *e-proceding of engginering*, 371

[7] Ardiansyah Y 2015 Aplikasi Identifikasi Mutu Kematangan Buah Kakao Menggunakan Image Processing dan Metode Fuzzy Logic. *Digital Repository.*

[8] Revathy N 2012 Face Recognition System Using Backpropagation Artifical Neural Network. *International Journal of Advanced Engineering Technology* 321-324.

[9] Kaur T 2012 *implementation of Backpropagation Algorithm :* A Neural Net-work Approach For Pattern Recognition . *International Journal of Engineering Research dan Develpoment*, 30-37.

[10] Dhoke, P. (2014). A Matlab Based Face recognition Using Pca with Backpropagation Neural Network. *International journal of Innovative Research in Computer and Comunication Engineering*, 2320.

[11] Indonesia, K. (2018). *Tampilan Tekstur Biji Kakao*, from https://www.kakao-indonesia.com

The Development of the *myKadera* System for Object Vocabulary Mastery for Students with Hearing Impairment Based on Augmented Reality

I Thalib and F Arifin
Electronics and Informatics Engineering Education Department,
Universitas Negeri Yogyakarta, Yogyakarta, Indonesian.

Email: ismail.thalib2016@student.uny.ac.id, fatchul@uny.ac.id

Abstract. This study aims to: 1) design appropriate learning media based on Augmented Reality on the introduction of object vocabulary for hearing-impaired students, 2) perform the developed system. This research belongs to research and development (R&D). This developmental research refers to the stages of learning media development by Arief Sadiman, *et al* and the stages of the development model by Borg & Gall. This research development procedure consists of three major steps that must be passed, namely planning, development and evaluation activities. The result of this Android-based myKadera system development research is that the system has some features provided including audio, sign language videos, 3D animation, and object vocabulary quizzes that were packaged more interestingly, applicative and innovative with a more real-time visual display, supported by object vocabulary learning media books with the help of AR (augmented reality) markers. The results of the test indicate that the application meets ISO / IEC 25010 standards on aspects; (1) functional suitability; all features of the application has been successful and said to be good, which means there are no functions that fail on testing, (2) application compatibility; the application is 100% compatible with sub-characteristics of co-existence, various operating systems, and various types of devices used for testing, (3) performance efficiency; the application is successfully run using the AppAchhi test device. The application is said to be good without experiencing memory leak or application damage which results in force close or launch fail. The application shows an average time behavior of 0.063 seconds/thread, an average CPU usage of 20.39%, and average memory usage of 13.08 MB, and (4) usability testing obtained a percentage of 95.50%. These results indicate that the application is very feasible and meets the usability aspect.

1. Introduction

Education is very important to be understood as the frontline in efforts to improve the quality of Indonesian human resources based on science and technology. The implementation of education in Indonesia must be equal and can be perceived by all groups, including children with disabilities. This is as contained in Law No. 2 of 1989 article 8 paragraph 1 which states that "Citizens who have physical or mental disorders are entitled to extraordinary education"[1]. It is also stated in Law No. 20 of 2003 chapter 5 paragraph 1 of the National Education System that "Every citizen has the same right to obtain good quality education despite having physical, emotional, mental, intellectual or social disabilities, are entitled to special education"[2].

Disability or disability is a term that encompasses disruption, limitations on activities and restrictions on participation. Persons with disabilities are all people who experience physical, intellectual, mental and/or sensory limitations in the long term, who are in interacting with the society can experience obstacles and difficulties to participate fully and effectively with other citizens based on equality of rights (Law No 8 of 2016 concerning persons with disabilities)[3]. Based on a global survey conducted by the United Nations Children's Fund and the University of Wisconsin, there are more than half (52%) of children in the world aged 6-9 years have at least one type of disability condition [4]. In Indonesia, there is 8.56% of the population with disabilities.

One type of disability is a condition of having difficulty/impaired speaking and or understanding/communicating with others. In Indonesia, based on data obtained from the 2015 SUPAS result for persons with disabilities who have difficulty speaking and/or understanding/communicating with others, it nationally has a percentage of 1.52% [5]. From the results of the survey, early childhood children with disabilities are a group that is very vulnerable to society treatment in the form of "socially disabled", an event that excludes children socially. This negative condition and/or treatment is dangerous for their development, their disability can become worse physically, mentally, and socially.

In this study, the focus of the research that will be carried out focuses on people with disabilities, especially related to children with disabilities as it is referred to as children with special needs. Children with special needs is "child who has something extraordinary that significantly distinguishes from children of their normal age" [6]. The type of disability that is the object of research is the disability of hearing impairment.

Hearing impairment is a condition of someone who has a hearing disability so that it prevents the process of acquiring spoken language information through hearing with or without hearing aids [7]. In this case, seeing the condition of children who have less hearing causes children's language development to stop in the process of touch so that children cannot experience the process of imitating sound when they are still little. The process of imitation is only limited to see or visual. This condition also influences in terms of receiving and capturing information from outside, both oral and written, and even greatly affects the mastery of the vocabulary possessed by children with hearing impairment.

In Indonesia, the learning process of hearing-impaired children is carried out by SDLB in each city/district. During its implementation, the learning process carried out is still experiencing obstacles, and not yet optimal. Some of the obstacles faced in the learning process of hearing-impaired children are the lack of vocabulary owned by these children because hearing-impaired children experience obstacles in hearing function so that they do not understand the sounds of language and mastery of vocabulary becomes difficult. This condition can also cause children's ability in language and communication to be lacking so that they will only get minimal information.

Vocabulary plays a very important role in language teaching since the mastery of one's vocabulary is very influential on the quality of language skills. This is consistent with the opinion of Tarigan (2011) which states that the quality of a person's language clearly depends on the quality and quantity of vocabulary they have, the greater language skills will be obtained [8]. The results of Lewton and Mackey's research in Edja Sadjaah (2005:5) show that retardation or developmental constraints of hearing-impaired children congregants have to do with language poverty, due to lack of information acquisition, causing the data abstraction and imagination experience obstacles [9].

From interviews conducted with SDLB teachers, it is known that there are 45 vocabularies that must be mastered by hearing-impaired children according to the curriculum, namely 15 fruit names, 15 vehicles names, and 15 animal names. The condition that occurs is that most children who are shown pictures of fruits, vehicles and animals are only able to name 16 to 20 names of the fruits, vehicles or animals. Referring to the demands of the 1994 SDLB curriculum, it states that children with hearing impairment must have vocabularies of at least 9000 vocabulary words for children with hearing-impaired grade IV. The reality is quite the opposite, there are many hearing-impaired children have a little vocabulary, both for mastering the vocabulary of objects and other vocabularies. Hearing-impaired children only understand concrete vocabulary and basic vocabulary. Based on the information obtained, during the learning process, children who have hearing impairment often forget the vocabularies. This causes the learning outcomes to be below the KKM (Minimum completeness criteria). Hearing-impaired children have difficulty remembering vocabularies that have been taught and it is only limited to surrounding objects which are often encountered by children.

Several studies related to the development of learning processes for children with hearing impairment have been conducted. Dhoni, Herman and Agi (2018) in his research entitled "*Android Board-Based Communication Board Application Design as Learning and Communication Media for Deaf Children, explains the Prototype Application PankoTuli*" which is designed to help reduce the difficulties faced by hearing and speech impaired people. The results of this study stated that the Prototype Application "PankoTuli" provides a method of Learning and Communication of Indonesian Sign Language for the hearing and speech impaired quickly and practically because it uses the Android-based application tools on smartphone devices [10]. Furthermore, Ardhiyani and Bachtiar (2014) researched on analysis of user interface for learning vocabulary introduction for hearing-impaired children. The results of this study are interface designs in the form of mockups and interface model simulators for learning media of vocabulary recognition for hearing-impaired children that built using Balsamiq mockup and Microsoft Tools Expression Blend 4 + sketchflow. By applying the total communication component in the delivery of vocabulary material and conducting every step of the UCD method, it can be concluded that the interface design produced is appropriate with the needs and conditions of hearing-impaired children and provides convenience to understand abstract vocabulary [11]. In Ilona research, which is about sign language learning application for hearing-speech impaired using Android-based image and Audio Visual Media. This research is in the form of Sign Language Learning Application for hearing-speech impaired patients using image-based audio and audio visual media can provide information and learning and it can help users to get sign language learning, especially in word recognition [12]. In addition, other research by Kautsar, Bordan and Sulistyawati (2015) regarding Sign Language Learning Applications for hearing-impaired-based Android users with the Bisindo method. In this research, it was built with a method of developing a multimedia system to produce multimedia-based products in the form of applications that can display learning and games that are equipped with pictures, audio and video sign language [13].

From the studies that have been described, the researcher observes that the studies conducted have not provided new innovations for the hearing-impaired learning media. Based on research reviews, the researchers offer a new system innovation in the form of an android-based learning media application, with the use of Augmented Reality technology. Augmented Reality technology is a breakthrough and innovation in the developing multimedia and image processing fields. This technology is able to lift an object that is previously flat or two-dimensional as if it is real or in real-time to be united with the surrounding environment. According to Ronald T Azuma from research published in a journal with the title "A Survey of Augmented Reality", Augmented Reality is a variation of the Virtual Environment or better known as Virtual Reality. Augmented Reality technology which is a development of Virtual Reality has a different concept [14]. Through the use of Augmented Reality technology, the presentation of learning media, especially vocabulary, is more real or in real-time. Therefore, the use of Augmented Reality technology has the potential to be developed into learning media using Android as support for hearing-impaired children in the process of object vocabulary learning.

Furthermore, in this study, the researchers packaged the learning media system application with a new media called the myKadera application (Recognizing Object Vocabulary based on Augmented Reality). myKadera is a system application development that is packaged as an android-based learning media by combining Augmented Reality technology with features provided including audio, sign language video, 3D animation, and object vocabulary quizzes that packaged more interestingly, applicable and innovative with a display visual is more real-time, supported by the object vocabulary learning media books with the help of AR (augmented reality) markers. myKadera is expected to be an alternative for teachers as a learning medium to foster children's memory and children's interactive media as relevant learning resources for hearing-impaired students to improve their vocabulary mastery of objects.

myKadera is a system application development that is packaged as an android-based learning media by combining Augmented Reality technology with features provided including audio, sign language video, 3D animation, and object vocabulary quizzes that are packaged more interesting, applicable and innovative with a display visual is more real-time, supported by the object vocabulary learning media book with the help of AR (augmented reality) markers. myKadera is expected to be an alternative for teachers as a learning medium to foster children's memory and to be children's interactive media as relevant learning resources for introducing hearing impaired students to improve their objects vocabulary mastery.

2. Research Method

Method used in this study is a research and development (R&D) method by Borg & Gall which consists of 10 development steps, namely: 1) Research and information collecting; 2) Planning; 3) Develop preliminary form of product; 4) Preliminary field testing; 5) Main product revision; 6) Main field testing; 7) Operational product revision; 8) Operational field testing; 9) Final product revision; and 10) Dissemination and implementation. In this study, the researchers also adopted several expert opinions including the stages of developing learning media by Arief Sadiman *et al* and also the stages of developing learning media by Borg & Gall. The procedures carried out in this study go through 3 important stages, namely planning, development, and evaluation. Development procedures include (1) planning phase; the activity carried out includes a needs analysis by doing a class observation to determine the most essential material to be developed and research planning by determining the research population. (2) development phase; the activity was carried out by doing a product development planning which consists of making a Storyboard and Interface Design to illustrate the workflow and detailed details in the display and to develop a measuring instrument of success/instrument, product development/program making, pouring the planned media design in the form of products and product trials through media validation by media experts, material experts, subject teachers and tested on students as users. (3) Evaluation phase; it was carried out to test the media in the experimental class and the control class.

Software testing conducted using ISO/IEC 25010 standard tests categorizes product quality into 8 (eight) characteristics, namely: Functional suitability, Performance efficiency, Compatibility, Usability, Reliability, Security, Maintainability, and Portability. Based on the eight characteristics above, in this study, the software testing phase only uses 4 of the 8 characteristics in the ISO 25010 standard from Ben David's opinion. Testing mobile software in ISO 25010 uses four aspects of testing including Functional suitability, Compatibility, Performance efficiency, and Usability [15].

3. Development and Implementation

3.1. General Description of the Application

This system application works by detecting the pattern of a marker. Markers used in designing this application are in the form of images and video barcodes. After the marker pattern is detected, a 2D or 3D object will appear directly above the marker. Besides displaying 2D or 3D objects, users can also see details of the objects and videos that are displayed. This application is equipped with instructions for its use, so that it can help the user to follow the steps in the instructions.

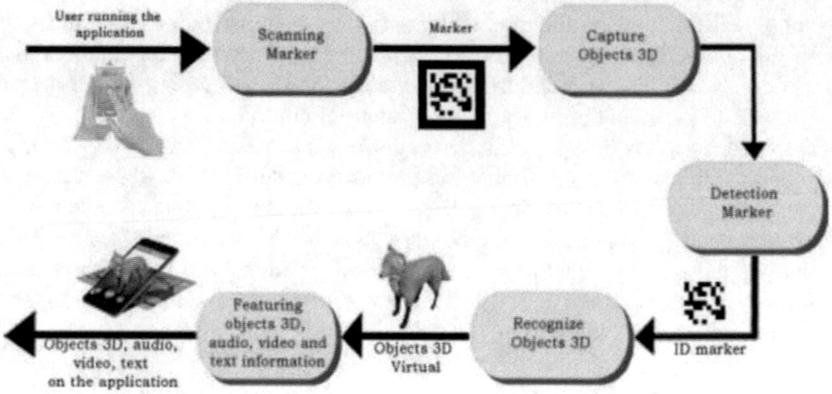

Figure 1. Block diagram of the *Mykadera* application system

The block diagram of the *Mykadera* application system, as seen in Figure 1, starts from the user running the application in the form of a scanning marker process to capture objects, followed by the detection process, ID marker, later the system will recognize data and markers in the form of 2D or 3D objects display, audio, video and information text. In the process of detecting markers, there is no internet connection needed on the smartphone used, because the database used in this application uses a local database.

3.2 Implementation
The implementation phase is the continuation stage of the system design activities to implement or create software to be feasible to operate. The activity carried out at this stage is moving or translating the logic of the program that has been made into the programming language and Java script so that it can produce interesting applications based on the planned objectives.

4. Results and Discussion
MyKadera was developed through the stage of testing the quality of the software so that it can be seen whether the software can be categorized as good software and there are no errors when used by users. At the testing stage that carried out to determine the quality of the software for this system is testing using ISO/IEC 25010 software quality standards. ISO/IEC 25010 testing becomes an international testing standard made by the International Organization for Standards (ISO) and Electrotechnical Commission (IEC) [16]. The product quality model categorizes the quality of product into 8 (eight) characteristics, namely functional suitability, reliability, performance efficiency, usability, security, compatibility, maintainability, and portability [17].

The *myKadera* system application was tested using ISO/IEC 25010 software quality testing standards by implementing 4 aspects of mobile device testing by Ben David including functional suitability, compatibility, performance efficiency, and usability. Applications must meet all aspects of these standards in order to be declared feasible.

The results of this study are sorted by testing phase. The test kits used are 5 devices available at Firebase Test Lab. The devices are *Samsung-SM-J710F, YU-YU5010A, Huawei-FRD-L02, Samsung-SM-J710FN*, and *OnePlus-ONE A2003*. The devices were chosen because they meet the operating system, memory, and device diversity criteria. In addition, the application also uses a virtual test device available at Firebase Test Lab to get general test results. Material testing is done by observation to assess whether the application content is up-to-date or not. The result of the material test is that the application is feasible to use with the addition of several explanations of several terms. The application functional suitability test was conducted by 4 programmers from various IT professions who understand the systematic development of software.

The test results are 100%. These results indicate that all application functions can run well so that it meets the functional suitability aspects. Co-existence sub-characteristic test was done by observation. The types of applications used in this test include *Play Store, Google Chrome, Joox, Google Maps, Facebook, Grab, Instagram, Lightroom, WhatsApp,* and *Telegram*. All applications were run together with the myKadera system application in one device. The results of this test are 100% so that it can be concluded that the myKadera application can run with other applications and meets the criteria very feasible on the aspect of compatibility co-existence. Testing the compatibility of the operating system was done using AppAchhi. The operating system used is Android 4.4 KitKat up to Android 8.0. Oreo The results of this test are 100%, which means the myKadera system application meets the compatibility aspect of the operating system. The compatibility testing of various types of devices was done the same as the compatibility test devices of various operating systems. The test results are 100%.

Results of performance efficiency testing used the same device as the previous test and performed automation using tools from AppAchhi. The results of the application performance efficiency test showed an average time behavior of 0.063 seconds/thread, an average CPU usage of 20.39%, and an average memory usage of 13.08 MB. From these results, the myKadera system application was declared well without experiencing a memory leak or damage resulting in a force close or launch fail. Usability testing used USE Questionnaire which amounts to 10 items and was given to 10 respondents. The test results obtained by the percentage of 95.50%. These results indicate that the application is very feasible and meets the usability aspect.

5. Discussion

Based on the results of research conducted by previous researchers, there are still some shortcomings, including: on the design of Communication Board Applications Based on Android Tablets as Learning and Communication Media for hearing-impaired children, explaining the Prototype Application "PankoTuli" which is designed to help reduce the difficulties faced by the hearing and speech impaired. Communication Board application is already based on an android tablet that displays sign language text and video by entering vocabulary in the form of sign language, however, the application has not been able to display vocabulary in the form of images to help memory and mastery of children. It has not provided complete information to children with hearing impairment related to the desired vocabulary and the design of the application is still standard, not yet sophisticated and less attractive. Other research is on user interface analysis of learning media for vocabulary recognition for hearing-impaired children. The results of this study are interface designs in the form of mockups and interface model simulators for learning media for vocabulary recognition for hearing-impaired children which are built using Balsamiq mockup tools and Microsoft expression blend 4 + sketchflow. This vocabulary recognition application for children with hearing impairment implements three total communication namely sign video, finger alphabet and speech reading on the delivery of vocabulary recognition material. However, this application still has shortcomings because the tools used are outdated to be done on children with hearing impairment. It has not been able to display images or animations on the interface design, so it has not provided simplicity in vocabulary understanding, it is less interactive and it has less attractive features.

Previous research is on Sign Language Learning Applications for Hearing Speech impairment Using Android-based Image and Audio Visual Media. This research is in the form of Sign Language Learning Application for Hearing-Speech impaired Patients Using Android-Based Image and Audio Visual Media. It can provide information and learning and it can help users to get sign language learning, especially for word introduction. This application has weaknesses in the resulting video as the video is still less clear, lack of learning the words displayed, the menu design and application features are still not interesting, the explanation of sign language is still in the form of images, quizzes provided by the application are less varied and not user friendly. Another research is about sign language learning application for hearing impaired people based on Android with the Bisindo method. In this research, it was built with a method of developing a multimedia system to produce multimedia-based products in the form of applications that can display learning and quizzes that are equipped with pictures, audio and sign language video. However, the application menu design is still standard and less attractive, bisindo letter video list view is less interactive and does not yet provide features that are actual and more informative, and the use of technology used is still not optimal for the innovation of a learning media (not actual).

Based on the problems above, the researchers provide the learning media system application with new media, namely *myKadera* (Knowing the Vocabulary Objects based on Augmented Reality). The application of augmented reality technology in the development of the Android-based *myKadera* system is seen from the features developed in the form of audio, sign language videos, 3D animations, and object vocabulary quizzes that are packaged in an interesting, applicable and innovative way with a more real-time visual display. It is the same with the provision of media vocabulary learning books which are supported by the help of AR (augmented reality) markers. *MyKadera* system is effective in the learning process and allows children to learn by looking at objects or two-dimensional images that are lifted up (augmented). This technology seems as if it eliminates the three-dimensional virtual world, integrated with the real world.

In the future, it is hoped that this system can become a breakthrough in the education world, especially for children with special needs to help to master the vocabulary of objects for hearing-impaired students. Generally, *myKadera* is expected to be an alternative for teachers as a learning media to help develop children's memory and as a relevant children's interactive media, as a learning resource for introductory learning to improve object vocabulary mastery of hearing-impaired children.

6. Conclusion

Based on the results of research and discussion, it can be concluded that the design of the myKadera system application can be used and has been running well on several android devices with different hardware versions and specifications. The myKadera system application is designed by utilizing augmented reality technology based on android in the introduction of object vocabulary to be more interesting, appropriate and innovative, because it can help hearing impaired children to recognize and see object vocabulary in real-time.

Based on the performance of the system developed, the myKadera application is feasible from the test results, through analyzing the quality of the application, the test results show that the functional suitability aspects of the application features have been successful and said to be good. Meanwhile, the compatibility aspect is very feasible because the application can run in conjunction with other applications. It can run on various operating systems and on various devices. In the aspect of performance efficiency, application is declared to be good without experiencing a memory leak or application damage resulting in force close or launch fail, with an average calculation of 0.013 second/thread time behavior, an average CPU usage of 20.39%, and average usage 13.08 MB memory.

Finally, the usability aspect test of the application has met the very feasible criteria with a percentage of 96.50%. Therefore, in general, it can be concluded that the application of the myKadera system is very interesting, easy to operate, efficient, innovative, and very suitable for use in the learning process for mastering vocabulary objects for hearing impaired students.

7. References

[1] UU RI. No. 20 tahun 1989, *tentang Sistem Pendidikan Nasional dan Penjelasannya.*

[2] Republik Indonesia. (2003). *Undang-Undang RI Nomor 20, Tahun 2003, tentang Sistem Pendidikan Nasional.*

[3] Undang-Undang Nomor 8 Tahun 2016. *tentang Penyandang Disabilitas.*
[4] UNICEF, *The State of World's Children 2013*: Children With Disabilities. 2013.
[5] Badan Pusat Statistik (BPS). *Profil Penduduk Indonesia Hasil Supas 2015.* Jakarta: BPS. 2015.
[6] Astati. Wardani, I. A. Hernawati, T., & Somad, P., *Pengantar Pendidikan Luar Biasa*. Jakarta: Universitas Terbuka. 2007.
[7] Heri, Purwanto, *Ortopedagogik Umum*. Yogyakarta: IKIP Yogyakarta. 1998.
[8] Tarigan, Henry Guntur, *Pengajaran Kosakata.* Bandung: Angkasa. 2011.
[9] Edja Sadjaah, *Pendidikan Bahasa Bagi Anak Gangguan Pendengaran dalam Keluarga*. Jakarta: Depdiknas. 2005.
[10] Setyawan, D.I., Tolle, H., & Kharisma, A. P., Perancangan Aplikasi Communication Board berbasis Android Tablet Sebagai Media Pembelajaran dan Komunikasi Bagi Anak Tunarungu. Vol.2, No. 8, Agustus 2018.
[11] Ardhiyani, J. and Bachtiar, A. M., *Analisis User Interface Media Pembelajaran Pengenalan Kosakata Untuk Anak Tunarungu* Jurnal Ilmiah Komputer dan Informatika (KOMPUTA). 2014.
[12] Ilona, D., Moch., A., *Aplikasi Pembelajaran Bahasa Isyarat Bagi Tunarungu Wicara Menggunakan Media Gambar Dan Audio Visualberbasis Android.* Teknik Informatika: Universitas Muhammadiyah Sidoarjo.
[13] I. Kautsar, R. I. Borman, and A. Sulistyawati, *Aplikasi Pembelajaran Bahasa Isyarat Bagi Penyandang Tuna Rungu Berbasis Android Dengan Metode BISINDO*, in SEMNASTEKNOMEDIA ONLINE, 2015.
[14] Azuma, Ronald, *A Survey of Augmented Reality.* California: Hughes Research Laboratory. 1997.
[15] David, Assaf Ben, *Mobile Application Testing (Best Practices to Ensure Quality)*. Amdocs. 2011.
[16] ISO/IEC, *System and software engineering - Systems and software Quality Requirement and Evaluation (SQuaRE) - System and software quality models,* 1. (I. J. 7, Editor). Diakses dari ISO: http://iso.org/obp/ui/#iso:std:iso-ec:25010 pada 22 September 2018.
[17] Wagner, S, *Software Product Quality Control.* Springer, 60-64. 2013.

Acknowledgments
I would like to express my gratitude for the reviewers for the comments and advice. I would like to thank Mr. Fatchul Arifin, my supervisor, for all the guides and supports. I would also like to express my gratitude for the school, teachers and students who participated in this research.

Augmented Reality As Android Based Learning Media for Wood Field Laboratory Work

B Wulandari[1], G N I P Pratama[2], N Hasanah[1] and N Yuniarti[3]
[1]Electronics and Informatics Engineering Education Department
[2]Civil Engineering and Planning Education Department
[3]Electrical Engineering Education Department
Universitas Negeri Yogyakarta, Indonesia

Email: bektiwulandari@uny.ac.id

Abstract. The objectives of this research were: (1) producing Android-based AR products in the form of a Wood Field Laboratory Work module in Department of Civil and Planning Engineering Education, Faculty of Engineering UNY to facilitate students' visualization abilities; (2) knowing the quality of Augmented Reality products for the Wood Field Laboratory Work module so that it is suitable for use by students and teaching staff; and (3) addressing the learning problems that have so far been present in the Wood Field Laboratory Work. The method used was the Software Development Life Cycle software development model which was described by the v-model waterfall. The v-model included: (1) the stage of specification of user requirements; (2) system architecture design; (3) designing components; (4) writing program codes; and (5) testing. The test results using ISO 25010 indicated that the application of learning media was stated to be very feasible in functional suitability, compability, and usability aspects. While from the performance efficiency aspect, the application was considered feasible because there was no memory leak on the device that run it, and CPU usage by five different devices reaches 10%, which is still below the safe limit (15%).

1. Introduction

One of the demands of special skills that must be possessed by students, especially in the Study Program of Civil Engineering Education (CEE), Department of Civil and Planning Engineering Education (CPEE), Faculty of Engineering UNY is wood engineering expertise obtained in the Wood Field Laboratory Work course, where the learning outcomes include students having the ability to design furniture, plan and implement wood construction and finishing wood furniture. Wood Work is very important in the development process, where Wood Work is a construction work involving wood work or wood-using work. Wood work can support/facilitate the development process on the project or to support the contents of the building in the form of furniture/furniture, cabinets, chairs, tables, scaffolding and so on. Currently furniture and building materials made of wood are increasingly needed by the community. And as we know wood is one of the most abundant construction materials in nature and was first used in the history of mankind. Wood is chosen as a construction material in addition to the reason it is easy to get, the price is relatively cheap and has a high aesthetic value.

Based on the observations made, interesting things that were encountered in the Wood Field Laboratory Work course in the wood workshop were (1) students were still difficult to visualize the pictures contained in the Wood Field Laboratory Work module so that students did not have a clear picture when plunging into practice (2) students tend not to read the practice module before implementing the Wood Field Laboratory Work (3) the material in the Wood Field Laboratory Work module is quite complex and dominated by reading text. Therefore, the research team felt the need for innovative and creative solutions to learning media in the Wood Field Laboratory Work, Department of CPEE, Faculty of Engineering UNY.

One of the breakthrough solutions offered is to use Augmented Reality (AR). AR is a technology that enables people to visualize cyberspace as a part of the real world that is around effectively so as to make the real world seem to be able to connect with cyberspace and an interaction can occur [1]. AR itself is usually developed on iOS and Android platforms. Both of these mobile operating systems are of course a trend among people all over the world, especially the Android operating system that is booming. This can at least be seen based on data released by IDC (International Data Corporation) where until the third quarter of 2012, Android has controlled 75% of the smartphone operating system market share [2].

In 2016, Tim Cook (CEO, Apple) also stated that AR is a revolutionary technology that has the potential to change the world. He even views the future of AR as important as smartphones in the present. However, AR technology still needs further research and development to optimize its benefits for the community [3]. In the same year, the Vice President of Business and Operations for Google's virtual reality (VR), Amit Singh, said that in 2018 there will be hundreds of millions of Android smartphones that support AR technology [4]. The presence of AR technology is actually welcomed positively by the people of Indonesia, especially in the fields of education, advertising and game applications. As many as 66% of respondents agreed that the AR application can make a real contribution in the future, and to use the AR, the community simply uses a smartphone [5]. Until now, AR has become increasingly popular with the public. According to Senja Lazuardy (Co-Founder & IT Director, Slingshot Group) the reason AR is in demand is because of technological developments, both hardware specifications and system performance on smartphones that have begun to support the functions that exist in AR technology. At present the community thinks AR is only for game and marketing purposes. In fact, AR can also be used for education, engineering, etc. [6].

AR technology will be very appropriate if it is used in a learning module in the form of three-dimensional objects such as in the Wood Field Laboratory Work course in Study Program of CEE that requires the imagination of the user before actually doing the practice. In addition, AR will also help students to be more motivated in the learning process in a non-boring way.

The objectives of this research were: (1) producing Augmented Reality products for the Wood Field Laboratory Work module in Department of CPEE to facilitate student visualization abilities; (2) knowing the quality of Augmented Reality products for the Wood Field Laboratory Work module so that it is suitable for use by students and teaching staff.

2. Research Method

Learning media will be created using the Software Development Life Cycle (SDLC) software development model which is described by the v-model waterfall. According to Pressman [7], the v-model is a variant of the waterfall model. Figure 1 below is a v-model diagram that will be used in the proposed research.

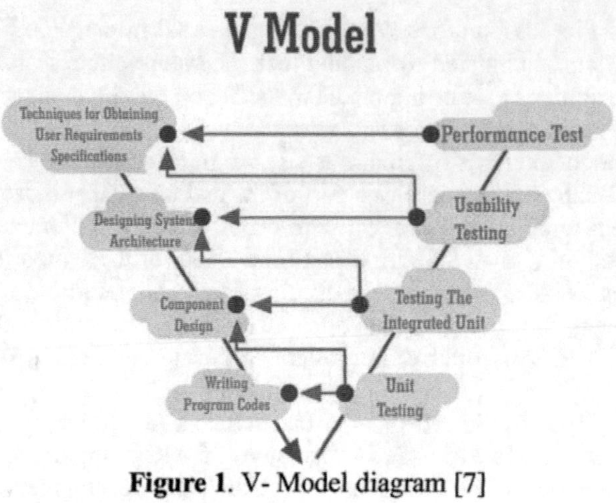

Figure 1. V- Model diagram [7]

The research and development carried out focused on making and testing the quality of learning media software for Wood Field Laboratory Work. The following are the stages of product development using the v-model.

2.1. Techniques for Obtaining User Requirements Specifications

This stage is the stage of understanding what needs are needed by users who will help interpret software features and their functions [7] in designing AR-based Wood Field Laboratory Work learning media in Wood Workshops, Department of CPEE, Faculty of Engineering UNY. In addition to observations, interviews were also needed from the head of the Wood Workshop, Wood Field Laboratory Work lecturer, and technicians to obtain information about the development of media and materials in accordance with the syllabus. After that can be analyzed the list of requests or user requirements (user requirement list) that need to be developed in the making of AR-based Wood Filed Laboratory Work learning media.

2.2. Designing System Architecture

The system architecture design stage is the stage to describe the system workflow that will be created. The system architecture design phase is made using Unified Modeling Language (UML) modeling. According to Rosa and Sallahudin [8], UML is a visual language for modeling and communication about a system using diagrams and supporting texts. Designing a system model consists of making use case diagrams, sequence diagrams, and activity diagrams.

2.3. Component Design

The component design stage is the design of the User Interface (UI) design using storyboarding and User Experience (UX) design by looking at the UML diagram that has been created, all of which are tailored to the needs of the user.

2.4. Program Codes

The stage of writing program codes is the stage to apply the design that has been made into the programming language. The software used is Unity 3D with the C # programming language so that applications can be made according to what has been planned and targeted to the user.

2.5. Testing

Testing using ISO 25010 [9] made by the International Organization for Standardization (ISO), uses four of the eight aspects in ISO 25010, they are functional aspects, performance efficiency, compatibility, and usability. The selection of the four aspects is based on the suitability of the testing aspects of the system contained in the application developed. The functional suitability aspect is the level at which software can provide the functionality needed when the software is used. Performance efficiency aspect is carried out to measure processor and memory consumption. The compatibility aspect aims to find out the extent to which the system, product, or component can be run from one device with different versions of Android and screens. While the usability aspect is the extent to which the software can be used by users to achieve certain goals effectively, efficiently, and meet satisfaction in its use.

3. Result and Discussion

3.1. Stage of Collecting User Requirements Specifications

The results obtained from the data collection stage are related to the conditions and needs of users, they are:

(1) The Wood Field Laboratory Work module consists of 2-dimensional and 3-dimensional images. Nevertheless, students still experience problems in visualizing it because the images are still static. The module also contains more text than images so that students find it difficult to translate existing images.

(2) Even though students have taken theoretical courses before carrying out the practice, students tend not to read the practice module before going straight into practice. This can cause problems in practice such as errors in the sequence of steps - practices that can inhibit or fail in a process of making wood products.

(3) Broadly speaking, Wood Field Laboratory Work aim to provide the basics of using manual carpentry tools/tools. The material provided relates to woodworking tools and assemblies, wood materials, wood joints, product design planning. The most difficult material to understand by students is about wood joints that are often used in woodwork. There are many types of connections that are learned to make beams and boards. The connection is very important role in giving strength to a construction, so that the use and placement must be really tight between one wood and the other wood.

3.2. Stage of System Architecture Designing

This stage is the design stage of AR-based work learning media architecture using the Unified Modeling Language (UML) modeling. System design is described through use case diagrams, sequence diagrams, and activity diagrams.

3.3. Stage of Component Designing

This stage is the stage of designing the learning media interface that will be developed. The interface is designed based on the results of the needs analysis. This media interface design is described using storyboards.

3.4. Stage of Programming

The process of making a program is an application development process from the design results to a programming language so that the results of the design can run according to its functionality. The tags are as follows.

3.4.1. Asset Preparation

The process of making the program begins with installing the 5.6.1f1 unit, Vuforia SDK, Java JDK, and the Android SDK. These tools are tools that must be used to create this application on the Android platform. After that, it was followed by designing markers that functioned as verification tools so that 3D objects could be raised, as well as pivots where 3D objects were located.

3.4.2. Creating an AR Camera Page

The AR Camera page will contain 3D objects. This page is the core of the application. Making 3D objects using special software called Blender 3D. The purpose of using this software is that 3D objects that are created can be as real and as similar as possible to the original. Figure 2 shows an example of making 3D objects for Door With Double Panels using Blender 3D.

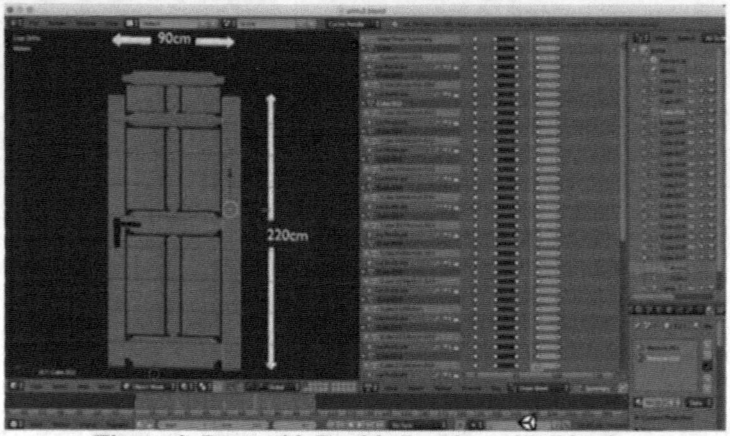

Figure 2. Door with Double Panels on 3D Blender

Then all 3D objects that have been created with Blender 3D are exported to Unity. Unity is software to create a layout of learning media that will be created. Figure 6 shows an example of creating a page layout for displaying 3D objects for Door with Double Panels.

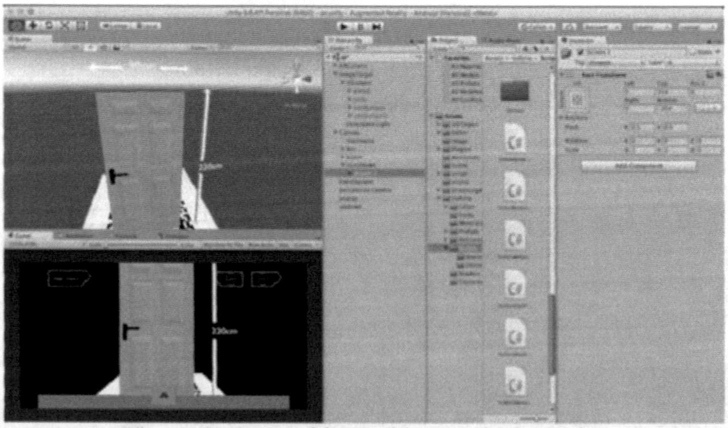

Figure 3. Door with Double Panels On Unity

Examples of several 3D objects that have been completed for learning media applications are shown in Figure 4, which is a Door with Double Panels. This 3D object consists of several wooden connections that will appear when the user zooms in at certain angles.

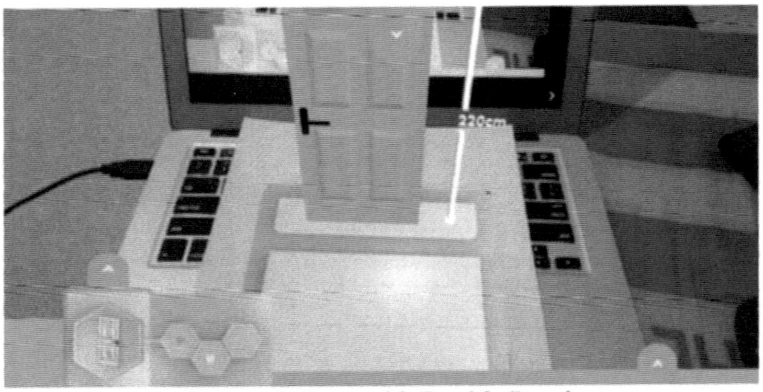

Figure 4. Door with Double Panels

When the 3D object Door with Double Panels is zoomed, the door automatically splits into several parts and wooden joints. Each connection can be zoom again so that the user can see more details about the connection.

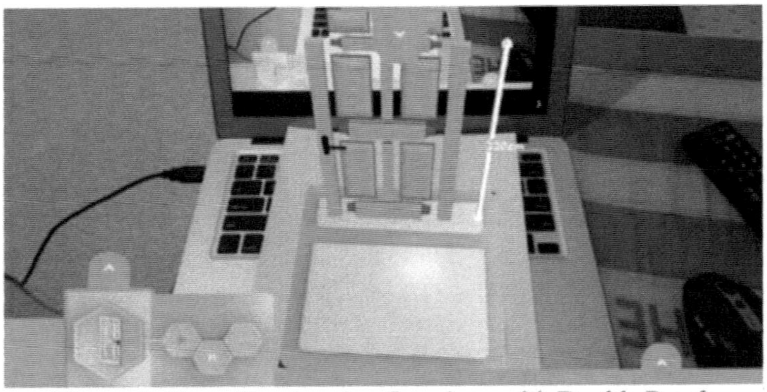

Figure 5. Door Construction Openings with Double Panels

Figure 6 shows one type of connection, it is the Dove Tail Joint. This construction is usually used on board joints beside drawers and cupboard side boards. This construction helps maintain the stability of wood and furniture from changes in shape due to shrinkage.

Figure 6. Dove Tail Joint Connection

3.5. Testing
ISO 25010 is used for testing the software.

3.5.1. Functional Suitability
Functional Suitability Tests are carried out by two experts in the field of multimedia who understand the systematics of software development. The results of Functional Suitability testing are presented in Table 1.

Table 1. Table of Functional Suitability Test Results

No	Feature	Test Score			Maximum Score
		Examiner 1	Examiner 2	Total	
1	Displaying Augmented Reality pages	1	1	2	2
2	Displaying object description of the Door with Single Panels	1	1	2	2
3	Displaying object description Door with Double Panels	1	1	2	2
4	Displaying the object description of the Door Frame	1	1	2	2
5	Displaying object description of the Dove Tail Joint	1	1	2	2
	Total	**5**	**5**	**10**	**10**

Based on the results of Functional Suitability testing, the percentage of feasibility is obtained as follows.

$$\text{Suitability Precentage (\%)} = \frac{\text{Observed Score}}{\text{Expected Score}} \times 100 = \frac{5}{5} \times 100\%$$

The calculation results of the suitability percentage are 100%, so it can be concluded that in terms of Functional Suitability, the application can be declared "Very Feasible".

3.5.2. Performance Efficiency
Performance Efficiency Test uses one of the tools for testing Android applications, it is Testdroid. Testdroid will display CPU usage and memory by the device you want to use. If memory leak occurs, the tools will provide info. Memory leak is one that can reduce the performance value of an application. If memory leak occurs, the Android system will stop the application (force close) due to lack of memory. The test results are presented in Table 2.

Table 2. Table of Performance Efficiency Test Results

No	Device	CPU	Memory	Description
1	LG Nexus 5 os 6.0.1	8% - 10%	143 MB	CPU usage under 15%, memory leak does not occur
2	Google Pixel 2	4% - 8%	180 MB	CPU usage under 15%, memory leak does not occur
3	Samsung J5	3% - 10%	90 MB	CPU usage under 15%, memory leak does not occur
4	Sony Xperia Z1	6% - 9%	90 MB	CPU usage under 15%, memory leak does not occur
5	Nokia 5	3% - 7%	150 MB	CPU usage under 15%, memory leak does not occur

Based on Table 2, the maximum CPU usage by five different devices reaches 10%. This figure is still below the safe limit set by Little Eye (mobile app analysis tools), which is 15%.

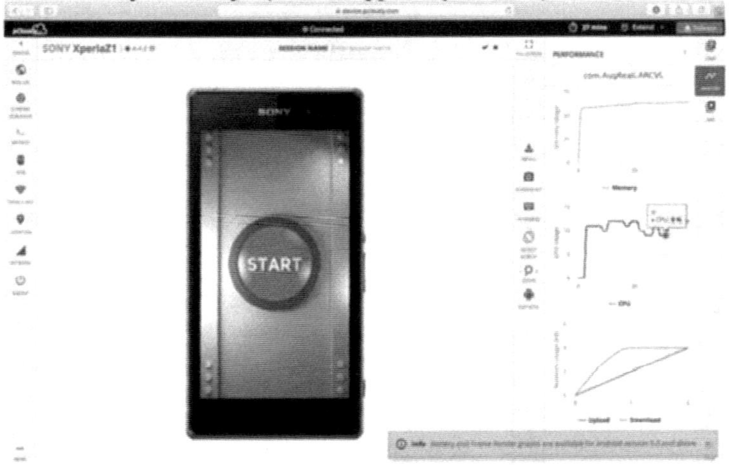

Figure 7. Testing Graph on the Sony Xperia Z1

Figure 7 shows one of the test results, it is on the Sony Xperia Z1 device. The highest CPU usage reaches 9%, which is still below the safe limit by Little Eye. The test results show that although the five devices tested use large memory for the application, memory leak does not occur, so the application can be declared "Feasible".

3.5.3. Compatibility

Compatibility test is carried out by the method of direct observation at the trial stage of installing, running, and uninstalling applications, on the Android operating system version and different screen sizes.

Table 3. Compatibility Test Result

No	Device	Android Version	Screen Resolution	Successful	Failed
1	Galaxy S8	8.0 (Oreo)	2960 x 1440	1	0
2	Galaxy S8	8.0 (Oreo)	2220 x 1080	1	0
3	Galaxy S7 edge	7.0 (Nougat)	2560 x 1440	1	0
4	Galaxy S6	6.0.1 (Marshmallow)	2560 x 1440	1	0
5	Redmi Note 2	5.0 (Lollipop)	1920 x 1080	1	0
6	Oppo A39	5.1 (Lollipop)	1280 x 720	1	0
7	Andromax R	4.1.2 (Jellybean)	1280 x 720	1	0
8	Motorola Razr	4.1.2 (Jellybean)	960 x 540	1	0
	Total			**8**	**0**

Table 3 shows that the application has been successfully running on all tested devices. The application continues successfully on eight devices with a combination of the Android version (version above Jellybean) and a different screen size. Based on the Compatibility test results, the percentage of feasibility is obtained as follows.

$$\text{Suitability Precentage (\%)} = \frac{\text{Observed Score}}{\text{Expected Score}} \times 100 = \frac{8}{8} \times 100\%$$

The calculation of the suitability percentage is 100%, so it can be concluded that in terms of Compatibility, the application is "Very Feasible".

3.5.4. Usability

Usability testing is carried out by trying applications directly to users who are students of Wood Field Laboratory Work, Study Program of CEE, Department of CPEE, Faculty of Engineering UNY. Students try the applications on each smartphone they have. The comments obtained from this test are that students like this application because of the interesting 3D object design. The most input is given so that the application supports IOS. A summary of usability testing results is shown in table 4. Based on the results of the Usability test, the percentage of feasibility is obtained as follows.

$$\text{Suitability Precentage (\%)} = \frac{\text{Observed Score}}{\text{Expected Score}} \times 100 = \frac{1415}{1500} \times 100\% = 94{,}3\%$$

The results of the feasibility calculation are 94.33%, so it can be said that this application belongs to the "Very Feasible" category.

4. Conclusion

Based on the results of the research and discussion, the research team concluded as follows:

(1) This research resulted in learning media applications using Augmented Reality technology for Android-based Wood Field Laboratory Work. Applications are built using the SDLC with the v-model waterfall method. The sequence of stages in this method includes the User Requirements Specification, System Architecture Design, Component Design, Program Code Writing, and Testing.

(2) The test results using ISO 25010 showed that in the functional suitability aspect, learning media applications are declared very feasible, because all functions in the application run 100%. From performance efficiency aspect, the application is declared very feasible, because the application can be run entirely on different screen sizes and Android versions (above the jellybean 4.1.1 version) with a percentage of 100%. From the compability aspect, the application is declared feasible because the application runs without a memory leak and CPU usage is below the safe limit (15%). From the usability aspect, the application was declared very feasible, because it obtained a value of 94.33% after being tested by Wood Field Laboratory Work of Study Program of CEE, Department of CPEE, Faculty of Engineering UNY.

5. References

[1] Jacobs, Dr. Paul E. 2012. *Augmented Reality.* http://www.qualcomm.co.id/products/augmentedreality.

[2] Tracker, IDC Worldwide Mobile Phone. (2012) *Android Marks Fourth Anniversary Since Launch with 75.0% Market Share in Third Quarter, According to IDC.* https://www.idc.com/research/container_error.jsp#.UQj817_%20tR9s

[3] Erry, FP. (2017). *Di Masa Depan Teknologi Augmented Reality Akan Sama Penting Dengan Smartphone.*https://infokomputer.grid.id/2017/02/berita/berita-reguler/di-masa-depan-teknologi-augmented-reality-akan-sama-penting-dengan-smartphone/

[4] Aziz, Lukman (2017). *Prediksi Google untuk Augmented Reality.* https://dailysocial.id/post/prediksi-google-untuk-augmented-reality-di-tahun-2018

[5] Rizki, Aditya. (2018). *Siapkah Indonesia Menyambut Augmented Reality.* http://adityarizki.net/2018/02/siapkah-indonesia-menyambut-augmented-reality/

[6] Erry, FP. (2018). *Mendunia dengan Augmented Reality.* https://infokomputer.grid.id/2018/01/profil/senja-lazuardy-mendunia-dengan-augmented-reality/

[7] Pressman, Roger S. (2010). *Software Engineering: A Practitioner's Approach.* McGraw-Hill.

[8] Rosa A. S., & Shalahuddin, M. (2011*). Modul Pembelajaran Rekayasa Perangkat Lunak (Terstruktur dan Berorientasi Objek).* Bandung: Modula.

[9] ISO/IEC. (2011). *ISO 25010.* https://www.iso.org/obp/ui/#iso:std:iso-iec:25010:ed-1:v1:en:en., accessed on July 21 2016.

This page intentionally left blank.

Journal of Physics: Conference Series **1413** (2019) 012036 doi:10.1088/1742-6596/1413/1/012036

Digital Self-Learning: Engaging Students in Studying Salat Al-Janazah (The Islamic Funeral Prayer) Based-on Digital Self-Learning

R Wardani and M J Husada
Electronics and Informatics Education Department,
Universitas Negeri Yogyakarta, Yogyakarta, Indonesia

E-mail: ratna@uny.ac.id

Abstract. As a part of Muslim society, students have a responsibility in doing Salat Al-Janazah, the ritual of prayer if there is a Muslim died. The ability to perform Salat Al-Janazah is included as part of compulsory subjects in schools. Unfortunately, there are still many students who have difficulty learning this prayer. The difficulty is due to the difference between the Salat Al-Janazah and the obligatory prayers. In addition, the prayer is not performed every day, so it is possible students learn it when in need. This problem can be solved through the technological approach. This research aims to (1) develop a digital self-learning to help the student in learning Salat Al-Janazzah by themselves; (2) Evaluate the quality of the media developed based on ISO/IEC 25010 on functional suitability, performance efficiency, compatibility, and usability aspects. In the media developed, the basic knowledge in learning Salat Al-Janazah will be transformed into digital-based material. By using digital self-learning, students can study in flexibility pattern and learn in comfortable rhythm according to their situation. Testing the quality of all aspects of ISO/IEC 25010 showing the media developed already eligible for use in learning Salat Al-Janazah.

1. Introduction

The Indonesian education curriculum contains Religion and Character Education as compulsory subjects. Islamic Education is one of the Religion and Character Education lessons in schools. In Islamic education, one of the competencies developed is the Salat Al-Janazah. Salat Al-Janazah (the Islamic funeral prayer) is one of the rituals of prayer by Muslims if there are other Muslim died. Unlike the obligatory prayer, this prayer is not performed every day. In society, a Muslim who dies becomes a collective obligation upon other Muslims in the organization of the funeral. It means if some Muslims take the responsibility of doing it, the obligation is fulfilled, but if no-one fulfills it, then all Muslims will be accountable. Salat Al-Janazah is one of the duties of Muslims associated with community responsibility. Therefore, Salat Al-Janazah becomes an important matter understood by the students as they are part of Muslims and society.

The basic competencies that must be achieved by students are basic knowledge of the Salat Al-Janazah and the implementation of the Salat Al-Janazah correctly according to the pillars and the legal procedures. Basic knowledge in learning Salat Al-Janazah contains the meaning of Salat Al-Janazah, the law of Salat Al-Janazah, the pillars of Salat Al-Janazah, and the recitation of Salat Al-Janazah.

Although Salat Al-Janazah is part of a compulsory subject in schools, many students do not understand how to do this prayer. This is because the procedure for prayer is different from the obligatory prayers. Salat Al-Janazah also has a different recitation of the dua. The main difficulty experienced by students due to the long recitation of the surah. Learning Salat Al-Janazah currently can not fully overcome this difficulty. The learning method still relies on books, text-based tutorials, the teacher gives an example of doing it, then students practice Salat Al-Janazah.

With the penetration of technology in education, conventional learning materials can be transformed into a digital-based material. In this situation, the learning strategy will change towards active learning. Student learning is more independent, where students have an environment that supports self-learning and practical training. Students can study in flexibility pattern and learn in comfortable rhythm according to their situation. The advanced technology makes it possible to make innovations in learning Salat Al-Janazah. The interactive learning media can be developed to overcome the problem of students' difficulties in learning Salat Al-Janazah.

The problem of learning difficulties can be solved through a technological approach. This approach is carried out with consideration of an increase in the number of cellular users in Indonesia. Based on the Indonesian Market Share Mobile Operating System (July 2018 - July 2019), the number of cellular users in Indonesia reached 265.3 million. Statistical data states that 93.69% of these are Android users [6]. This condition enables the development of instructional media through the use of mobile devices to overcome student learning difficulties.

The purposes of this study were to (1) develop a digital self-learning of Salat Al-Janazah to help the student in learning Salat Al-Janazzah by themselves; (2) Evaluate the quality of the media developed based on ISO / IEC 25010 on functional suitability, performance efficiency, compatibility, and usability aspects. In the media developed, the basic knowledge in learning Salat Al-Janazah which contains the meaning of Salat Al-Janazah, the law of Salat Al-Janazah, the pillars of Salat Al-Janazah, and the recitation of Salat Al-Janazah dua will be transformed into digital-based material. By using digital self-learning, students can study in flexibility pattern and learn in comfortable rhythm according to their situation.

2. Research Method

This study developed a Digital Self-Learning based on Android that is used by students to learn Salat Al-Janazah. Using the Research and Development (R&D) method, this study uses a waterfall model with variations in the V-model [5] for software development. Figure 1 describes the V waterfall model. SMA N 1 Slahung was chosen as the location for testing the quality of learning media.

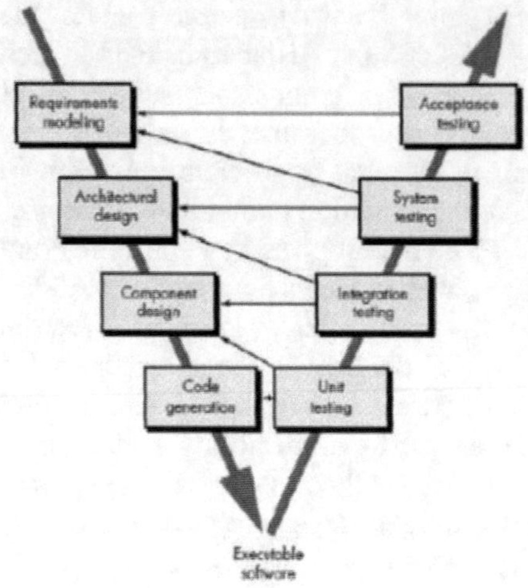

Figure 1. V Model Waterfall [5]

The data used for this study were collected by interview, observation, and questionnaire. The collection of data for quality testing of the media in this study used several research instruments. Quality testing on the usability aspect used the USE Questionnaire [3] [4] as an instrument for data collection. The functional suitability aspect used a test case to determine the quality of functions of the software developed. Quality testing for compatibility aspects carried out on various operating systems, device types, and coexistence software. Tests carried out directly on the android devices and cloud-based testing using the tools of AWS Device Farm. The testing of performance efficiency is conducted to determine the performance of an application related to the use of resources in certain conditions. In [1] stated that performance testing on mobile-based applications is a type of testing that assesses resource utilization in memory, CPU resources, and time behavior. Performance testing is performed using an automation tool from AWS Device Farm.

3. Result and Discussion

3.1. The Specification of Digital Self-Learning

The *Digital Self-Learning* of Salat Al-Jenazah contains learning material according to the syllabus. It provides an easy way for students to learn by text, motion, video, and audio. The animations help to visualize the sequence of steps in Salat Al-Janazah.

There are 3 aspects that are managed in the *Digital Self-Learning* developed, as follows:

(1) Tutorial. Tutorial with interactive methods containing reading material, demonstration of how to pray, and the discovery of reading or experience that requires verbal and written responses and tests. This tutorial program as a guidance program, which aims to provide assistance to students in order to achieve optimal learning outcomes.

(2) Drill and Practices. This feature provides training related to the subject matter that has been provided in the tutorial section. Through continuous practice, students will be able to perform Salat Al-Janazah correctly.

(3) Evaluation. This feature presents problem-based tests according to students' ability levels. When students are working on a test, the system will respond based on the answers. If the answer is correct, the program presents the next material, and if the student answers incorrectly, the program provides the facility to repeat exercises that can be given partially or at the end of the whole test.

3.2. The Design of Digital Self-Learning

This section describes the design process of Digital Self-Learning. Design process translates software requirements into models so that they can be implemented into applications at the next stage.

In order to explain the design of learning media, we use the Use Case Diagram to describe the business process and functionality specification (Figure 2). The design was elaborated from the functional specification, which is divided into 3 main functionality, namely Learning Material, Drill and Practices, and Evaluation.

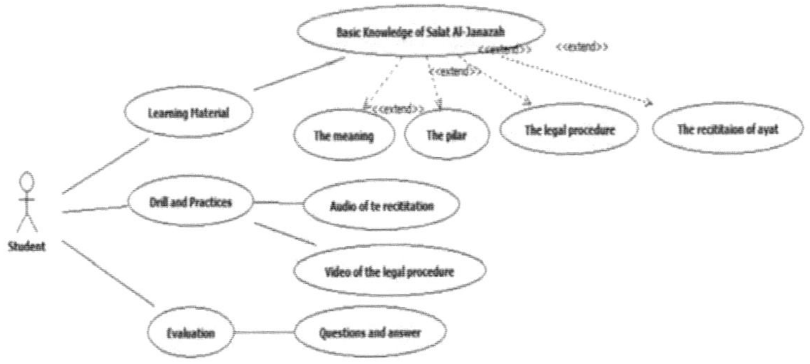

Figure 2. Use Case diagram of Digital Self-Learning

3.3. The Implementation of Digital Self-Learning

This section explains the implementation of learning media. This stage is carried out after the design process and will be implemented using a programming language. Aspects of the pre-defined functionality developed into program modules. Figure 3 shows some of the interfaces that represent the functionality of the learning media.

The stages of implementation include the following activities:

(1) Creating content Qur'an readings manually using the add-ins in Microsoft Word.
(2) Creating text-based material content with the flip view effect using the flip view library
(3) Creating the application layout using XML programming language and designer tools provided by android studio
(4) Creating a program /coding using Android Studio and Java as the programming logic.
(5) Debugging the application is one important step before the application is certified and released for widespread testing. The purpose of debugging is an internal test to find out that the application developed is running well.

Figure 3. Application interface.

3.4. The Quality of Digital Self-Learning

This section describes the results of media quality testing based on ISO/IEC 25010 on functional suitability, performance efficiency, compatibility, and usability aspects.

3.4.1. The Functional Suitability Aspect

The functional suitability of the application was tested by five experts from various professions who have software development competencies. Application testing aims to ensure that existing features can function correctly or not. Table 1 shows the results of functional suitability testing.

Table 1. Functional Suitability Testing

No	Function	Score					Total Score	Maximum Score
		1	2	3	4	5		
1	Menu	1	1	1	1	1	5	5
2	Materi	1	1	1	1	1	5	5
3	Bacaan Shalat	1	1	1	1	1	5	5
4	Video	1	1	1	1	1	5	5
5	Evaluasi	1	1	1	1	1	5	5
6	Reward	1	1	1	1	1	5	5
7	Bantuan	1	1	1	1	1	5	5
8	Tentang	1	1	1	1	1	5	5
	Total	8	8	8	8	8	40	40

Based on Table 1, the percentage of eligibility is calculated using the following formula:

$$\text{percentage of eligibility (\%)} = \frac{\text{observer score}}{\text{maximum score}} \times 100\% \quad (1)$$

$$\text{percentage of eligibility (\%)} = \frac{40}{40} \times 100\% = 100\%$$

Based on this result, the eligibility test (quality of media) states that Digital Self-Learning has a functional suitability value of 100% (very eligible), meaning that the test results show all functions can be run properly.

3.4.2. The Performance Efficiency Aspect

Performance efficiency testing using the AWS Device Farm tool in automation. There are five devices used for this test. The test results are grouped into three categories, namely CPU usage, memory, and thread execution. The results of the performance efficiency test are as follows.

Table 2. Performance Efficiency Testing

No	Perangkat	Sistem Operasi	CPU (%)	Memory (MB)	Thread (/s)
1	Google Pixel 2	8.0.0	2	104	17
2	Google Pixel	7.1.2	3	136	18
3	Huawei M8	6.0	7	46	12
4	Motorola Moto E – 2nd Gen	5.1	11	43	15
5	Samsung Galaxy S3 (Verizon)	4.4.2	9	55	13
	Average		6,4	76,8	15
	Max		11	136	18
	Min		2	43	12

Table 2 illustrates application performance using various devices that have different operating systems, CPU speeds, memory capacity, and thread execution capabilities per second. It shows that the average CPU usage, the average memory usage, and thread execution per second. CPU Utilization is the activity of the device to use CPU resources when running the application within a certain timeframe. Based on Table 2, the average CPU usage was 6.4%, where an 11% maximum usage and minimum usage is 2%. Memory utilization is the activity of the device in the use of memory resources when running applications. Table 2 shows that the average memory usage is 76.8 MB, where the maximum usage is 136 MB, and the minimum usage is 43 MB. Time behavior is the execution time of a thread in one second. Based on Table 2, the average number of threads processed is 15. Therefore the time behavior of applications learning media is 0,067 seconds / thread.

Based on performance efficiency testing, learning media applications are considered appropriate to use because there is no force closed / memory leak in the application operation.

3.4.3. The Compatibility Aspect

Compatibility testing conducted on coexistence, the suitability of the various operating systems and the suitability of the various types of devices. Coexistence testing is conducted by running the application simultaneously with other applications in one device.

The coexistence aspect (Table 3) shows the results of testing applications that run simultaneously with other applications in one device. Based on these tests, a percentage of eligibility obtained using formula (1) was 100%. It can be concluded that the application can run simultaneously with other applications in one device without a crash. Therefore, the application of Salat Al-Janazah is considered very feasible to use.

Table 3. Coexistence Testing

No	Running Application		Success	Failed
1	Learning Media of Salat Al-Janazah	*Play Store*	1	0
2	Learning Media of Salat Al-Janazah	*Mobile Legends*	1	0
3	Learning Media of Salat Al-Janazah	*Google Chrome*	1	0
4	Learning Media of Salat Al-Janazah	*Facebook*	1	0
5	Learning Media of Salat Al-Janazah	*Maps*	1	0
6	Learning Media of Salat Al-Janazah	*Youtube*	1	0
7	Learning Media of Salat Al-Janazah	*Instagram*	1	0
8	Learning Media of Salat Al-Janazah	*Maps*	1	0
9	Learning Media of Salat Al-Janazah	*Whatsapp*	1	0
10	Learning Media of Salat Al-Janazah	*Camera*	1	0
	TOTAL		10	0

Table 4. Suitability Testing in Various Devices and OS

No	Device	Setup Suite	Build-in Explorer Suite	Teardown Suite	Total
1	Google Pixel 2 (8.0.0)	1	1	1	3
2	Google Pixel (7.1.2)	1	1	1	3
3	Huawei M8 (6.0)	1	1	1	3
4	Motorola Moto E – 2nd Gen (5.1)	1	1	1	3
5	Samsung Galaxy S3 (Verizon) (4.4.2)	1	1	1	3
	Total	5	5	5	15

Compatibility testing on various devices and OS is done using the AWS Device Farm tool. The application of Al-Janazah Salat was tested using 36 android-based devices of various types and operating systems. Based on trials, 36 of these devices successfully run applications without errors.

Table 4 shows the results of testing learning media applications that are run in various operating systems and device types. The test is divided into three parts, namely the setup suite or the process when installing the application, the Build-in explorer suite or the process when running the application and the teardown suite or the process when removing the application. Based on Table 4, it can be obtained the percentage of eligibility using formula (1) of 100%.

Table 5. Compatibility Testing

No	Sub Compatibility	Score
1	Co-Existence	100%
2	Various of OS	100%
3	Various of Devices	100%

Based on the testing details (table 5) it can be concluded that the compatibility of learning media applications is 100%. These results are correlated with the score interpretation criteria and show that the compatibility of Al-Janazah Salat learning media is very feasible.

3.4.4. The Usability Aspect

Usability testing is carried out through the use of applications by 22 students SMA N 1 Slahung grade XI 2. Students try to use the application on each device for learning Salat Al-Janazah, and after that, students give their responses to a USE questionnaire. There are four aspects of usability that are tested, namely Usefulness, Satisfaction, Easy of Learning, and Easy of Use.

Table 6. Usability Testing

Respondent	Total Score	Max Score
22	2951	3300

Using the formula (1) for Tabel 6, the usability score of the learning media application of Salat Al-Janazah is 89.42%. These results interpret the learning media of Salat Al-Janazah is very feasible to use.

This study has produced Digital Self-Learning Salat Al-Janazah for high school students. This media has features for learning the material of Salat Al-Janazah, Drill and Practices for reading recitation of Salat Al-Janazah that is equipped with reading indicators and videos, and Evaluation in the form of quizzes. Media development covers all phases of development, starting from communication, planning, modeling, construction, and delivery of software to users / customers (deployment).

4. Conclusion

To conclude the study, the *Digital Self-Learning* of Salat Al-Janazah has been developed to solve the difficulty of the students in learning Salat Al-Janazah. In general, quality testing showed that the developed media is feasible to use, with detailed testing of each aspect as follows: The eligibility test (quality of media) states that Digital Self-Learning has a functional suitability value of 100% (very eligible), meaning that the test results show all functions can be run properly. The compatibility testing indicates that the application of the Digital Self-Learning is very feasible with the eligibility percentage (co-existence) of 100%, it can run on various versions of the Android operating system and can run on different types of devices. Testing on the aspect of performance efficiency is categorized as feasible, where the application can run well without the occurrence of a memory leak that can cause the application to close suddenly (force close) with an average CPU usage of 6.4% and memory usage by an average of 76.8 MB. Testing on the usability aspect was categorized as very feasible with a percentage of eligibility of 89.42%; this result was tested on 22 respondents in the school where the study was conducted.

5. References

[1] David, A.B. 2011 Mobile application testing best practices to ensure quality (Amsdoc)
[2] ISO/IEC. 2011 Systems and software engineering — systems and software quality requirements and evaluation (SQuaRE) — system and software quality models, 1. (I. J. 7, Editor) Last access June 18, 2017, (ISO: https://www.iso.org/obp/ui/#iso:std:iso-iec:25010:ed-1:v1:en)
[3] Lund A.M. 2001 Measuring Usability with The USE Questionnaire (Usability and User Experience SIG Newsletter) p.2
[4] Nielsen, J. (2012). How many test users in a usability study? Retrieved December 28, 2016, from https://www.nngroup/articles/how-many-test-users/, last accessed on Mey 5, 2019
[5] Pressman, R.S. 2015 Software Engineering: A Practioner's Approach 8th Edition (New York: MacGraw-Hill)
[6] https://gs.statcounter.com/os-market-share/mobile/indonesia , last accessed on July 21, 2019

Implementation of CBT (Computer-Based Test) System on Aptitude Test Development Using C4.5 Algorithm as Potential Detection Tool for Choosing High School Major

H Jati, R D Ristanto and Nurkhamid
Electronic and Informatics Engineering Education Department, Universitas Negeri Yogyakarta, Yogyakarta, Indonesia

Email: handaru@uny.ac.id

Abstract. This study aims to: (1) develop computer-based aptitude test software using C4.5 algorithm as an evaluation system in decision making; and (2) to reveal the level of suitability between the major desired by learners and the potential outcomes of computer-based aptitude test software with C4.5 algorithm. This research is a design and development study. The experimental subjects in this study were three high school students with a total sample of 60 students. The analysis was done by looking at the suitability of the result of the evaluation of computer-based aptitude tests that students have done. The results of this study are: (1) Computer-based aptitude test software successfully developed with several specifications namely; (a) CBT (Computer-Based Test) system model implemented using open mode; (b) Aptitude test instruments used under the Jim Barrett Aptitude Test; (c) The decision tree of the C4.5 algorithm constructed using a binary tree structure with the highest aptitude score criteria as a split value. (2) The trials conducted in three high schools obtained the average score percentage of 73.3% with the Good criteria.

1. Introduction

The phenomenon in studying or choosing the study program indicates that students who graduated from Junior High Schools or Islamic Junior High Schools entering the High School or Vocational High School are not all based on the interest of learners supported by the optimal potential, such as basic common skills (intelligence), talents, and career interests of learners. Incompatibility in the selection of study programs can lead to difficulties and a tendency to fail in learning which will have an impact on the career continuation of learners. The problem is shown in carrernews.id [1] who conducted a survey to high school and vocational high school graduates in 2015 with the theme of the suitability of majors. The survey results show that 67% of high school and vocational school graduates who continue their studies or who work feel not in accordance with the department at the time of school. In addition, the survey results also showed that 58% of high school and vocational school graduates regard a career in line with their major as important. Therefore, early guidance in major choosing, especially in preparing placement and channeling for the continuation of studies in accordance with the potential and conditions that exist in the students themselves need to be done.

The Attachment to the Regulation of the Minister of Education and Culture of the Republic of Indonesia Number 64 Year 2014 Article 1 states that major choosing is the program provided to accommodate the interests, talents and/or the ability of learners with the orientation of concentrating, extension, and deepening of subjects or vocation [2]. The term major choosing program is the same as the majoring program enacted at the senior high school level, which distinguishes only the placement and change of name system adapted to the turn of the new curriculum 2013. The major choosing itself comes from the word interest. Interest is a strong inclination or desire to develop in a focused individual and focused on the realization of a condition by considering the individual's basic capabilities, talents, interests and personal preferences [3].

The purpose of major choosing is contained in Permendikbud No. 64 Year 2014 article 2 stating that majoring in Senior High School has a purpose to give opportunity to learners develop attitude, knowledge, and skill competences of learners according to their interest, talent and/or academic ability in a group of science subjects. In general, the objectives of major choosing program are also described in the manual book of SMA-SMK Curriculum 2013 [2], to help learners to instill interest in subjects, strengthen interest in subjects, and select and set interest groups of subjects that are followed on the educational unit being pursued, career choice or further study option up to higher education level. Based on Permendikbud No.69 Year 2013 [4], the system of major choosing in high school uses three choices of major curriculum, namely: Mathematics and Natural Sciences (MIA) which contains subjects of Mathematics, Biology, Physics and Chemistry; Social Sciences (IIS) focusing on Geography, History, Sociology and Economics subjects; and Language and Cultural Studies (IBB) which contains subjects of Indonesian Language and Literature, English Language and Literature, Other Foreign Languages and Anthropology.

Implementation of the 2013 curriculum that requires the major choosing process be done from class X requires the school implement the major choosing process that is effective and efficient in the acceptance of new learners. There are several success factors of an education according to Erlin Leigh Parker [5] one of the most important is in terms of recruitment or acceptance of learners. If the recruiting tools are good, it will produce good graduates as well. Results of acceptance of learners are good, if the right tools are used in the process of acceptance. The tools used to uncover talents or special abilities possessed by a person are called aptitude tests. In KBBI (the great Dictionary the Indonesian Language), talent is the foundation (intelligence, nature, and nature) brought about from birth. Talent is a potential that must be stimulated in advance so it can be seen as a skill, knowledge, and special skills that become the provision of his life someday [6]. In his book, Suttan Surya explains that everyone is the right person, it serves as the basis for talent tests to develop the great potential of personal qualities. Aptitude Test is a test that aims to know the talents and abilities possessed by someone in the field of science [7].

Based on the book by Jim Barrett [8], Aptitude Test has nine types of problems, namely (1) Visual reasoning test (measure ability to understand reasoning expressed visually); (2) Numerical reasoning test (measure ability to understand reasoning expressed with numbers); (3) Verbal analysis test (measure ability to understand and analyze reasoning expressed using words); (4) Sequential reasoning test (measure ability to understand reasoning about process sequence); (5) Spatial recognition test (measure the ability to understand and recognize an object); (6) Three-D test (measures the ability to understand and recognize three-dimensional objects); (7) System test (measure ability in accuracy and accuracy); (8) Vocabulary test (measure ability in vocabulary use); (9) Figurework test (measure the ability to perform calculations). Each area of expertise has different characteristics based on the level or level of talent one has. In accordance with the area of expertise in high school interest, Jim Barrett [8] states that the characteristics of the skill area have a level or level of components in the aptitude test as in Table 2.

The Guidebook of Curriculum Vocational High School 2013 (2013, p.19) states that aspects that need to be considered in the major choosing of high school students include learning achievement, non-academic achievement, the value of national examination, statement of interest of learners, parental attention and potential detection of learners. Based on the observations from several high schools, major choosing process of students in high school does not match with what is specified in the guidance. There is a gap between the guidance for major choosing arranged by Kemendikbud 2013 and the major choosing process occurring in high school, that is on potential detection aspect.

One alternative to solve the problem is by applying a Computer-Based Test (CBT) system on a aptitude test instrument. Monirosadat conducted a study comparing the use of CBT with Paper and

Pencil Tests (PPT) stating that test participants were more interested in the test features presented with computers [9]. In addition, the Retnawati Study states that learners who have good technology use capabilities, CBT tend to be more accurate than PPT [10]. Jamil, Tariq, & Shami (2012) in the study of teacher perceptions (test organizers) in CBT (Computer-Based Test) vs. PBT (Paper-Based Test) stated that overall sample teachers showed a positive attitude toward CBT, but in some situations they preferred PBT especially for less experienced teachers in the CBT system , while teachers with experience in CBT systems tend to prefer CBT [11]. Some of the factors that cause difficulties in the use of CBT applications, the most influential factors are those derived from "users" and "technology use" itself [12].

There are four forms of computer-based test models and the internet namely (1) Open Mode, can be followed by anyone and without any supervision. Test participants do not need to register participants; (2) Controlled Mode, similar to an open-ended test model, but no test taker is registered by entering a username and password; (3) Supervised Mode, there is a supervisor who identifies the test participants to authenticate and validate the test-taking conditions. For tests in this mode requires the administrator to log and confirm the participants that the tests have been completed correctly at the end of the test; (4) Managed Mode, usually the test is performed centrally [13]. Organizations that manage the test process can define and assure the performance and equipment specifications at the test center. Some of the advantages of using a computer-based test are reducing the time for a test scoring job and creating a written report, eliminating logistical work such as distributing, storing and paper-based tests, test takers can immediately find out the test results. Fagbola states that the use of CBT is an effective solution for the evaluation of education in bulk [14].

The development of C4.5 Algorithm is done in CBT (Computer-Based Test) system as an evaluation system in decision making in the form of potential learners for the major choosing process in high school. C4.5 algorithm is a group decision tree algorithm that has input in the form of training samples and samples. Training samples are sample data that will be used to build a verified test tree. While samples are data fields that will be used as parameter in doing data classification [15]. C4.5 algorithm is one method to make decision tree based on training data that have been provided. The C4.5 algorithm was created by Ross Quinlan which is the development of ID3 which was also created by Quinlan [16]. Some of the development done on C4.5 is as among others can overcome the missing value, can overcome the continue data, and pruning.

There are three stages of C4.5 algorithm (1) Selecting the attribute as the root node and creating a branch for each attribute value, in the C4.5 algorithm to define attributes as root nodes, it is based on the highest gain ratio of the attributes; (2) Divide the case into the branch originating from the node. The way used in C4.5 algorithm is to calculate the number of cases divided based on attributes that can be the root node of the highest attribute value. After that calculation of gain ratio values for each attribute; (3) Repeat the process for each branch until all the cases on the branch have the same class [17]. In the C4.5 algorithm decision tree construction to determine the attribute as the root node, is based on the highest gain ratio value of the existing attributes. The attribute with the highest gain ratio value is selected as a split attribute for the node. The formula of the gain ratio is as follows:

$$Gain\ Ratio(a) = \frac{gain(a)}{split\ info(a)} \tag{1}$$

Where gain (a) is the information gain of attribute a for the set of sample X and split info (a) denotes the entropy or potential information obtained at the division of X into n subsets based on the study of attributes a. While gain (a) is defined as follows:

$$Gain(a) = info(X) - info_a(X) \tag{2}$$

The reason for using the gain ratio (a) in the C4.5 algorithm (not the gain (a)) as the criterion in the selection of attributes is that the gain has a bias attribute to the attribute having many unique values. For the split formula info (a) is as follows:

$$Split\ Info(a) = -\sum_{j=1}^{k} \frac{|X_1|}{|X|} \times log_2 \left(\frac{|X_2|}{|X|}\right) \tag{3}$$

Where X_1 denotes the 1st subset of sample X. In other words the formula for calculating the gain ratio value to be selected as an attribute of the existing node is as follows:

$$Gain\ Ratio(a) = Entropy(X) - \sum_{j=1}^{k} \frac{|X_i|}{|X|} \times Entropy(X_i) \tag{4}$$

Early idea developers about the entropy of the random, Shannon suggest that entropy is a measure of the average uncertainty of a data set when it does not know the outcome of an information source [18]. It also shows the average amount of information to be received from the information sources. To get a gain ratio value in the formation of a decision tree, it is necessary to calculate the value of information in units of bits from a collection of objects. The form of calculation for entropy is as follows:

$$Entropy\ (X) = \sum_{j=1}^{k} p_j \times log_2 \frac{1}{p_j} = -\sum_{j=1}^{k} p_j \times log_2 p_j \tag{5}$$

Where:
X = Set of cases
k = Number of partitions X
p_j = Proportion of X_j to X
The split entropy that divides X by n records into sets X_1 with n1 rows and X_2 with n2 rows is:

$$E(X_1, X_2) = \frac{n1}{n} E\ (X_1) + \frac{n2}{n} E\ (X_2) \tag{6}$$

The value of Entropy (X) indicates that X is a more random attribute. On the other hand, the smaller attribute of the Entropy (X) value implies this slightly more randomly significant attribute for data mining. The value of entropy reaches a minimum value of 0, when all other $p_j = 0$ or are in the same class. The value reaches a maximum of log_2 k, when all pj values are equal to 1/k.

2. Research Method

This study was conducted with two approaches. The first approach is to develop software with waterfall models [19]. The second approach is to use the Dick and Carey approach model to gain reinforcement of educational-related research conducted [20]. Aptitude test software development model described into five stages namely; (1) Analysis; (2) Planning; (3) Modeling; (4) Construction; and (5) Testing.

Needs analysis is done by collecting software requirements that meet user needs. Needs analysis phase is done with literature study on aptitude test development and field study to know the major choosing process in high school. At this stage of needs analysis is done by observation and direct interview with the school. The observation is done by finding the source of reading related to the major choosing process for the curriculum of 2013 in high school. Interviews are needed to find out the major choosing process undertaken in high school and find an existing gap. The purpose of the planning stage is that the research process can run effectively and efficiently, so that the resulting product can be completed on time. Planning is done by making scheduling on software development process. The development schedule includes the use of the time required to develop the software up to the timing of the software testing itself. Specification needs obtained from the needs analysis, studied and converted into software design. In designing the system used UML modeling language (Unified Modeling

Language) which serves to describe the system boundaries and system functions in general. UML is a modeling language for systems or software that paradigm 'object-oriented'. Modeling is actually used for simplifying complex issues in such a way that it is easier to learn and understand [21]. The construction phase involves the actual development of the system by developing a graphical user interface, applying models using HTML and creating database systems using SQL Server. The construction phase is based on software model design at the modeling stage. The creation of designed software is a computer based aptitude test application with C4.5 algorithm as evaluation system. Evaluation of aptitude test software development is done by expert testing and field trials. Expert testing involves five people who are experts in educational psychology and web-based applications. Testing is done by testing the function and work of the program thoroughly to minimize the error level of aptitude test software when used. Field trials were conducted in June-July 2017 at three high schools. The field trial subjects were 60 students divided into 20 students per school.

The sampling technique used to determine the respondent is a quota sampling. A quota sampling is a technique or a way to determine the respondents who have certain characteristics to the desired amount, while in the determination of the sample number of respondents refers to Jakob Nielsen which states that at least the number of respondents in the user test for quantitative research is 20 respondents [22]. Aptitude test software is used to detect potential learners based on the choice of major programs in high school at the time of admission of new learners. The analysis was conducted to determine the level of conformity between the choice of the program of interest desired by the learners with the result of potential heteksi from aptitude test software. From result of recapitulation, calculation done with formula percentage as follows:

$$presentage = \frac{number\ of\ correct}{total\ number\ of\ students} \times 100\% \quad (7)$$

From the calculation of the percentage value, it can be determined the criteria of conformity according to Suharsimi Arikunto [23].

Table 1. Criteria Conformity Level

Presentage Score	Interpretation
0% – 20%	Unacceptable
21% – 40%	Poor
41% – 60%	Acceptable
61% – 80%	Good
81% – 100%	Excellent

3. Experiment Results and Analysis

Aptitude test software is developed based on waterfall model stage, while the analysis and testing using Dick and Carey method. The results of aptitude test software development are described based on the following development stages:

3.1. Analysis

Based on observations in high school, the specification functional requirements of aptitude test software are: (1) Aptitude test software can be used any time and can be accessed anywhere by the user. (2) Users who can access the aptitude test software are the administrator and test participants. (3) Test participants may use the software and follow the test series without having to be registered as a participant in advance. (4) The login page is only used by school administrators.

3.2. Planning

Scheduling aptitude test software development process is done with the aim that research can run in accordance with the time specified. The staging phase is used as a timeline in the aptitude test software development process.

3.3. Modeling

From the results of needs analysis, computer-based test (CBT) aptitude test software is designed and developed using open mode. The choice of CBT architecture model with open mode is done according to requirement analysis that test participants can run the aptitude test software and follow the test series without having to register first as the test participant. That's because software aptitude tests will be applied at the time of admission of new learners, so that everyone can use the aptitude test software to assist in determining the choice of major direction. The design of the CBT system model is illustrated in the UML modeling language (Unified Modeling Language) use case diagram in Figure 1.

Figure 1. Use Case Diagram Aptitude Test Software

The aptitude test instrument model used in the software refers to the Jim Barret Aptitude Test which has nine types of problems: Visual, Numerical, Verbal, Sequences, Spatial, Three-D, Precision, Vocabulary and Figurework. The model of the instrument can reveal different areas of expertise. In accordance with the needs analysis in high school for major program, the field of expertise tailored to the needs of major programs in high school areas of expertise MIA, IIS and IBB in Table 2.

Table 2. Characteristics of Major Program in High School

Mathematics and Natural Sciences (MIA)		Social Science (IIS)		Language and Cultural Studies (IBB)	
Mathematics	Biology	Geography	History	Indonesian	English
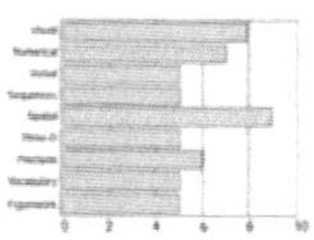	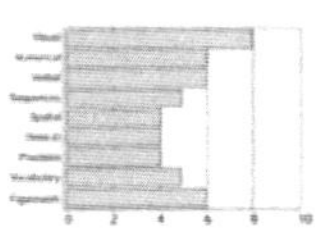	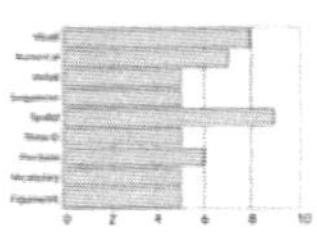	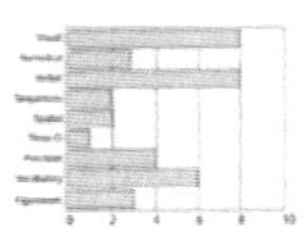	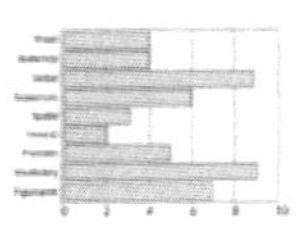	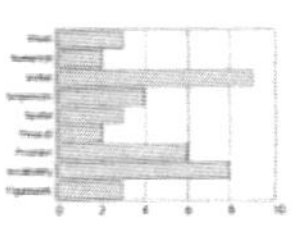
Physics	Chemistry	Sociology	Economics	Other Foreign	Anthropology
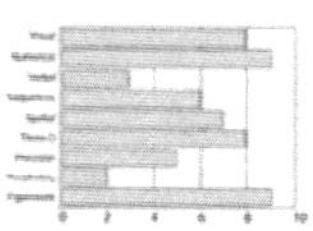	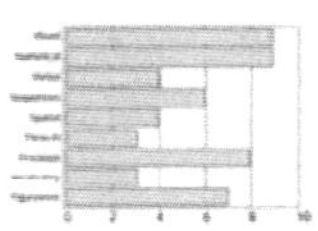	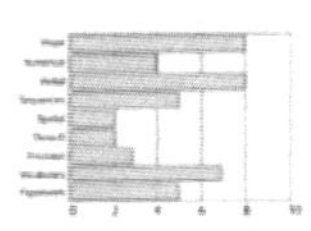	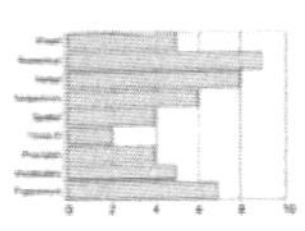	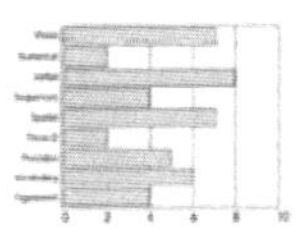	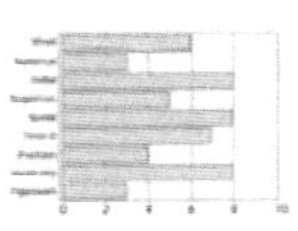

In the evaluation system of the aptitude test software, C4.5 algorithm decision tree was built using binary tree structure, i.e., every node of the decision tree has only two branches (yes or no). The reason for choosing this structure is to simplify the shape of the decision tree and to minimize the occurrence of noise that will affect the performance and speed of the process in decision making. Choosing binary tree structures to determine the potential of the aptitude test data generates three decision trees, i.e., the MIA program decision tree, the IIS program decision tree, and the IBB program decision tree which will be evaluated according to the interests of the learners. Based on the data characteristic of major program in high school program in Table 2, data as training sample for evaluation of aptitude test were obtained, as in Table 3.

Table 3. Characteristic Data Field Major Program in High School

Major Program	Indicator	Aptitude Test								
		Visual	Numerical	Verbal	Sequential	Spatial	Three-D	Precision	Vocabulary	Figurework
MIA (Mathematics and Natural Sciences)	Mathematics	6	9	5	5	3	7	5	3	9
	Biology	8	6	6	5	4	4	4	5	6
	Physics	8	9	3	6	7	8	5	2	9
	Chemistry	9	9	4	6	4	3	8	3	7
IIS (Social Sciences)	Geography	8	7	5	5	9	5	6	5	5
	History	8	3	8	2	2	1	4	6	3
	Sociology	8	4	8	5	2	2	3	7	5
	Economics	5	9	8	6	4	2	4	5	7
IBB (Language and Cultural Studies)	Indonesian	4	4	9	6	3	2	5	9	7
	English	3	2	9	4	3	2	6	8	3
	Other Foreign	7	2	8	4	7	2	5	6	4
	Anthropology	6	3	8	5	8	7	4	8	3

To create a decision tree of each major program, the major program columns in Table 3 are used as labels. When forming the MIA program decision tree, the label on MIA is changed to "correct" while the labels on the IIS and IBB programs are changed to "incorrect", otherwise it will form the IIS and IBB decision trees. After the separation of labels, from the training data sample in Table 3 calculation of the value of gain ratio to form the decision tree C4.5 algorithm. The split value determination is based on the highest aptitude test score of each attribute.

Table 4. Calculation of Gain Ratio Value on MIA Program as Root Attribute

Variable	Split	Total Case	No (S1)	Yes (S2)	Entropy	Information Gain	Split Info	Gain Ratio
Total		12	8	4	0,918			
Vis						0,143	0,414	0,347
	≥ 9	1	0	1	0			
	< 9	11	8	3	0,845			
Num						0,285	0,918	0,311
	≥ 9	4	1	3	0,811			
	< 9	8	7	1	0,544			
Verb						0,109	0,650	0,168
	≥ 9	2	2	0	0			
	< 9	10	6	4	0,971			
Seq						0,175	0,811	0,216
	≥ 5	9	5	4	0,991			
	< 5	3	3	0	0			

Spa						0,051	0,414	0,124
	≥ 9	1	1	0	0			
	< 9	11	7	4	0,946			
3D						0,116	0,811	0,142
	≥ 7	3	1	2	0,918			
	< 7	9	7	2	0,764			
Prec						0,143	0,414	0,347
	≥ 7	1	0	1	0			
	< 7	11	8	3	0,845			
Voc						0,051	0,414	0,124
	≥ 9	1	1	0	0			
	< 9	11	7	4	0,946			
Fig						0,317	0,650	0,487
	≥ 9	2	0	2	0			
	< 9	10	8	2	0,722			

The determination is based on the characteristics of a test result value, i.e., if the value is greater than the criterion value, then the value is acceptable. An example calculation of the gain ratio value to determine the attribute as a root in the formation of MIA program decision tree in Table 4. From the calculation of the gain ratio of each aptitude test attribute, the highest gain ratio is 0.487 in the figurework attribute with the split value 9. So the attribute that becomes the root node in the MIA program decision tree is the figurework with two branches of "yes" or the value of the figurework ≥ 9 and branch of "no" or the value of the figurework < 9. The same steps are performed to define attributes as branches until all instances of the branch have the same class.

Once the decision tree is completed, each algorithm from the decision tree is integrated with the interests that the learner has. From the choice of interests of the learners, then interest is adjusted to the talent of the aptitude test results so that the potential of learners can be determined appropriate for the major program in high school. The evaluation system algorithm developed in the aptitude test software to determine the potential of learners can be seen in Figure 2.

```
Algorithm for Potential Detection
Declaration:
    MIA Program Algorithm, IIS Program Algorithm, IBB Program Algorithm
    Interest 1, Interest 2, Result, Potention = string
Description:
 Interest 1
 Read (Interest 1)
 If Interest 1 = MIA then
       Read (MIA Program Algorithm)
       If Result = Correct then
             Potention ← MIA
       Else
             Return (Interest 2)
       End if
 Else if Interest 1 = IIS then
       Read (IIS Program Algorithm)
       If Result = Correct then
             Potention ← IIS
       Else
             Return (Interest 2)
       End if
 Else
       Read (IBB Program Algorithm)
       If Result = Correct then
             Potention ← IBB
       Else
             Return (Interest 2)
       End If
 End If

 Interest 2
 Read (Interest 2)
 If Interest 2 = MIA then
       Read (MIA Program Algorithm)
       If Result = Correct then
             Potention ← MIA
       Else
             Potention ← Consultation
       End if
 Else if Interest 2 = IIS then
       Read (IIS Program Algorithm)
       If Result = Correct then
             Potention ← IIS
       Else
             Potention ← Consultation
       End if
 Else
       Read (IBB Program Algorithm)
       If Result = Correct then
             Potention ← IBB
       Else
             Potention ← Consultation
       End If
 End If

 Write (Potention)
```

Figure 2. Algorithm of Evaluation System Aptitude Test Software to Determine Potential Learners

3.4. Construction

The interface of the aptitude test software was developed using a web-based application using Laravel 5.2 framework. CBT (Computer-Based Test) system model used is open mode, so everyone can take the test without prior register. The result of aptitude test software development can be seen in Figure 3.

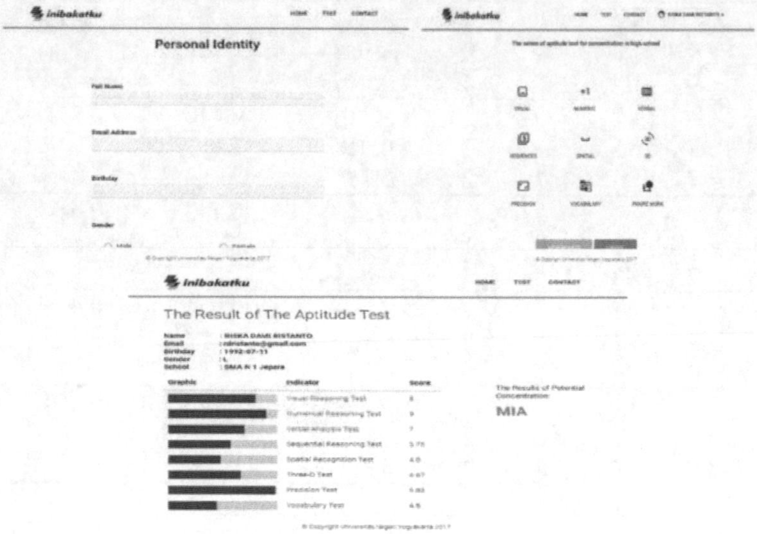

Figure 3. The Results of the Software Page View (a) Input Personal Identity (b) Test Menu (c) Test Results

3.5. Testing

Aptitude test software testing is divided into two parts: expert testing and field trials. Expert testing involves five experts in the field of educational psychology and computer-based applications. Testing is done by testing every function and performance of the program as a whole. This is done to reduce the level of error in the software when used. The test results show that the developed software has been running well and field trials can be done. Some suggestions of improvement from the experts include: (1) Need to add description list of choice of major program in high school on the form of input identity of the test participants; (2) Need to add a "see details" menu of student aptitude tests on the administrator page.

Field trials were conducted to see the suitability of the preferred choice of learners with the potential outcomes of aptitude test software. The trial was conducted involving 60 students from three high schools. The results of field trials of aptitude test software from three high schools can be concluded that the average percentage of conformity obtained a score of 73.3% with the Good criteria. More can be seen in Table 5.

Table 5. Field Test Results

School Name	Number of Respondents	Correct	Incorrect	Percentage
SMA Negeri 1 Pecangaan	20	13	7	65%
SMA Negeri 1 Welahan	20	16	4	80%
SMA Negeri 1 Jepara	20	15	5	75%
Total	**60**	**44**	**16**	
Average				**73,3%**

The development of computer-based aptitude testing software with C4.5 algorithm as an evaluation system is a potential detection tool for major choosing in high school. The development of aptitude test software is aimed at learners to determine the choice of direction of interest in accordance with the wishes and abilities possessed.

This study has some limitations as follows: (1) The potential result from aptitude test software is only as a tool to determine the direction of interest of learners , the final decision in determining the major program is done by the school; (2) Detection of student's interest is done based on the questionnaire which has been discussed with parents. The development of test instruments that detect the interests of learners is important so that the potential generated is in accordance with the interests and talents of learners; (3) Many factors influence the incompatibility of potential results. The stability of internet connection and the ability of computer use become the main obstacle experienced when conducting field trials.

4. Conclusions

The study can conclude that: (1) The computer-based talent test software was successfully developed with several specifications namely; (a) The CBT (Computer-Based Test) system model applied using open mode, so that everyone can take the aptitude test without having to register first; (b) Aptitude test instruments used by the Jim Barrett Aptitude Test which has nine types of problems: Visual, Numerical, Verbal, Sequences, Spatial, Three-D, Precision, Vocabulary and Figure work; (c) The decision tree of the C4.5 algorithm is constructed with a binary tree structure based on the characteristics of major fields in high schools as training sample data with the highest score of aptitude test scores as split values. (2) Field trials are conducted to determine the level of conformity between the choices of interests desired by learners and the potential results of aptitude test software. The results of trials conducted on three high schools obtained the average percentage of compliance score of 73.3% with Good criteria.

Suggestions for utilization of aptitude test software products that have been developed are: (1) Aptitude test software can be used as an alternative tool for schools in determining the potential of learners for the process of major in high school; (2) The potential outcome of aptitude test software can be used as a consideration of counseling teachers in determining the choice of major direction in high schools; (3) For learners, aptitude test software can be used as a tool of self-evaluation on their potential talent and ability.

5. References

[1] R. Amelia. (2015). *Lulusan SMA-SMK yang Melanjutkan Study ataupun yang Bekerja Tidak Sesuai dengan Jurusan Saat Sekolah.* Available: http://careernews.id/issues/view/3433-Lulusan-SMA-SMK-yang-Melanjutkan-Study-ataupun-yang-Bekerja-Tidak-Sesuai-dengan-Jurusan-Saat-Sekolah

[2] T. Penyusun, "Salinan lampiran permendikbud no. 69 tahun 2013 tentang kurikulum SMA-MA," *Jakarta: Permendikbud,* 2013.

[3] H. Kamaluddin, "Bimbingan dan konseling sekolah," *Jurnal Pendidikan dan Kebudayaan,* vol. 17, no. 4, pp. 447-454, 2011.

[4] D. P. Pendidikan, T. Kependididikan, D. J. P. Menengah, K. Pendidikan, and K. R. Indonesia, "Peminatan Peserta Didik SMA dan SMK," ed: Jakarta: Kemendikbud, 2013.

[5] E. L. Parker, "Factors that contribute to a successful secondary vocational education program in the state of Mississippi," 2008.

[6] B. Lucy, "Mendidik Sesuai Minat dan Bakat Anak (Painting Your Children's Future)," *Jakarta: PT. Tangga Pustaka,* 2010.

[7] S. Surya and M. Hariwijaya, "Tes bakat dan kepribadian," *Yogyakarta: PT. Citra Aji Parama,* 2012.

[8] J. Barrett, *Aptitude Personality and Motivation Tests: Analyse Your Talents and Personality and Plan Your Career.* Kogan Page Publishers, 2009.

[9] M. Hosseini, M. J. Z. Abidin, and M. Baghdarnia, "Comparability of test results of computer based tests (CBT) and paper and pencil tests (PPT) among English language learners in Iran," *Procedia-Social and Behavioral Sciences,* vol. 98, pp. 659-667, 2014.

Journal of Physics: Conference Series **1413** (2019) 012037 doi:10.1088/1742-6596/1413/1/012037

[10] H. Retnawati, "The Comparison of Accuracy Scores on the Paper and Pencil Testing vs. Computer-Based Testing," *Turkish Online Journal of Educational Technology-TOJET,* vol. 14, no. 4, pp. 135-142, 2015.
[11] M. Jamil, R. Tariq, and P. Shami, "Computer-Based vs Paper-Based Examinations: Perceptions of University Teachers," *Turkish Online Journal of Educational Technology-TOJET,* vol. 11, no. 4, pp. 371-381, 2012.
[12] H. V. Leeson, "The mode effect: A literature review of human and technological issues in computerized testing," *International Journal of Testing,* vol. 6, no. 1, pp. 1-24, 2006.
[13] D. Bartram and R. K. Hambleton, "Computer-based testing and the Internet," *Issues and advantes. England: John Wiley and Sons,* 2006.
[14] T. M. Fagbola, A. A. Adigun, and A. O. Oke, "Computer-based test (CBT) system for university academic enterprise examination," *International journal of scientific & technology research,* vol. 2, no. 8, pp. 336-342, 2013.
[15] S. Sunjana, "Aplikasi Mining Data Mahasiswa Dengan Metode Klasifikasi Decision Tree," *Jurnal Fakultas Hukum UII,* 2010.
[16] J. R. Quinlan, *C4. 5: programs for machine learning*. Elsevier, 2014.
[17] I. H. Witten, E. Frank, M. A. Hall, and C. J. Pal, *Data Mining: Practical machine learning tools and techniques*. Morgan Kaufmann, 2016.
[18] C. E. Shannon, "A mathematical theory of communication," *Bell system technical journal,* vol. 27, no. 3, pp. 379-423, 1948.
[19] R. S. Pressman, *Software engineering: a practitioner's approach*. Palgrave Macmillan, 2005.
[20] M. Gall, W. Borg, and J. Gall, "Case study research," *Educational research: An introduction,* pp. 123-163, 2003.
[21] A. Nugroho, *Rekayasa perangkat lunak berorientasi objek dengan metode USDP*. Penerbit Andi, 2010.
[22] J. Nielsen, "Usability 101: Introduction to usability," ed, 2003.
[23] S. Arikunto, "Dasar-dasar Evaluasi Pendidikan. Jakarta: Bumi Aksara. 2010," *Penelitian Tindakan Kelas (Classroom Action Research),* 2007

Journal of Physics: Conference Series **1413** (2019) 012038 doi:10.1088/1742-6596/1413/1/012038

Analysis of the Use of Thesis Guidance Information System Using the Technology Acceptance Model

Priyanto and A T Ramadhan
Electronics and Informatics Engineering Education Department,
Faculty of Engineering, Universitas Negeri Yogyakarta, Indonesia

Email: priyanto@uny.ac.id

Abstract. This study analyzes the implementation of thesis guidance information system named SIBIMTA for students of Faculty of Engineering, Universitas Negeri Yogyakarta. This study was classified as correlational research with a quantitative approach. Data were collected from from 161 respondents in Faculty of Engineering who were working on their theses. Data collection techniques used a questionnaire with 32 questions. Data were taken using the Likert scale with 4 alternative options. The results of the study were: (1) the system quality had a positive effect on perceived usefulness; (2) the system quality had a positive effect on perceived ease-of-use; (3) the information quality had a positive effect on perceived usefulness; (4) information quality had a positive effect on perceived ease of use; (5) service quality had a positive effect on perceived usefulness; (6) service quality had a positive effect on perceived ease of use; (7) the perceived ease of use had a positive effect on perceived usefulness; (8) perceived ease of use had a positive effect on attitudes toward using a new technology; (9) perceived usefulness had a positive effect on attitudes toward using a new technology; (10) the perceived usefulness did not have a positive effect on the intention to use the new technology; (11) the attitude of using technology had a positive effect on the intention to use the new technology; (12) intention to use the new technology had a positive effect on actual use.

1. Introduction

The development of technology impacts all aspects of humans lives and it has made daily activities easier and better. Technology changes and improves economy, culture, social, and educational aspects. Since the internet was discovered, activities performed manually has become more simple and practical. In the education field, technology accomodates communication between teachers and students, problems with materials, tests, homework, guidance and grading which previously had to be discussed face to face, now it can be done online through internet networking. Technology improves all aspect of daily actitivies and assists people to work faster and easier.

The observation data from the students in Faculty of Engineering, Universiats Negeri Yogyakarta, showed that before the end of 2017, students completed their thesis using a manual system. The students should meet face to face with the thesis supervisors. They should contact the lecturers first and make an appointment. All of required documents and letters were also processed manually by the academic staff. Manual academic service is not longer effective. It takes more efforts and time. Thesis guidance process will be interrupted if the guidance schedule is delayed or canceled due to certain reasons. The amount of time wasted because the manual system is considered a significant problem.

Based on these problems, the academic service system at Universitas Negeri Yogyakarta was replaced by an online-based service system particularly in the activities of thesis guidance management, which originally used a manual system and then now was swifted to an online based system. In 2016, Universitas Negeri Yogyakarta created an information system to support the thesis guidance management named SIBIMTA. The purpose of SIBIMTA is to assist the students who are working on the thesis to receive guidance from their supervisors. The students who need thesis guidance do not need to meet the lectures directly, they are only required to upload the thesis files to the information system. The existence of SIBIMTA is expected to improve the thesis guidance management to be easier, faster and effective.

A new technology in an organization usually has positive or negative impact. A positive impact is indicated by the role of the new implemented technology which improve the organization's activities. Conversely, if the new technology bring poor organization's activities, it means it has negative impact. Negative impact can occur due to several factors, for example if the users are not ready to implement the new techology. To better predict, explain, and increase user acceptance, Davis, F.D. [4] states that It is important to examine the factors influence end users in adopting a new technology, to better understand why people accept or reject computers. The theory that supports this statement is the theory of reasoned action (TRA) by Fishbein and Ajzen [8]. The theory states that users who get the benefits (positive outcomes) will continue to utilize the technology. One model that can be used to measure users' acceptance of new technology implementaion is Technology Acceptance Model (TAM).

TAM is one of the theoretical models of technology acceptance approaches. TAM was introduced by Davis. It is a derivative of TRA. According to Davis [4] the purpose of TAM is to describe the factors that determine the process of receiving information-based technology in general, explaining the behavior of end users of information technology with a wide enough variation along with the user population. TAM suggests two factors that have influence on the acceptance of a new technology, namely perceived usefulness and perceived ease-of-use. Both of the factors affect users' attitude towards using the new technology, which is directly proportional to the behavioral intention to use the technology which ultimately determines the actual usage.

Technology acceptance is also influenced by several other factors besides the impact of the technology namely the ability of the technology to run or assist the process of activities. These factors in the theory of TAM are categorized into four groups. One of the factors is system characteristics, which consist of various sub categories, including: (1) system quality, (2) information quality, and (3) service quality.

According to Afiani and Priyanto [1], Perceived usefulness (PU) is the degree to which a person believes that using a particular technology would enhance his or her job performance. In this case, it is related to how the students are working on their thesis. Perceived ease-of-use (PEOU) is defined the degree to which a person believes that using a particular technology would be free from effort. Attitudes towards the use (ATU) of technology can be interpreted as individual feelings both positive or negative about the behavior of using the technology. The behavioral intention to use (BIUS) technology here is the student's intention to use SIBIMTA.

TAM suggests two factors that have an influence on the acceptance of a new technology, namely PU and PEOU. These two factors will influence the desire to use the technology. While the external variables that will be used in TAM based on the study consisted of three variables including system quality, information quality and service quality. This study will analyze the acceptance and the use of the online thesis guidance information system based on aspects that exist in TAM.

2. Research Method

This study is classified as an exploratory descriptive study using a correlational approach which is part of the expose facto research, because it has the aim of describing the circumstances of events that occur in the field. Explorative research is expected to explore the relationship between variables, test hypotheses, develop generalizations, and develop theories that have universal validity without controlling and manipulating variables. The setting of the study was Faculty of Engineering, Universitas Negeri Yogyakarta in September-October 2018. The samples were determined by using a purposive sampling technique. This technique determines the sample based on certain considerations according to the research objectives. Based on the research objectives, the selected sample consisted of students who were writing theses. The reason was the respondents of the study should be the students who were writing theses and were already used SIBIMTA.

The method used in data collection was a survey method by distributing questions in the form of a closed questionnaire. The questionnaire consisted of the respondent's identity, instructions for filling the questionnaire and statements in accordance with the research construct. The objects of this study were influential constructs for the acceptance of the use of SIBIMTA in the academic activities of engineering students at Universitas Negeri Yogyakarta. These constructs were the constructs mentioned in the TAM theory and the DeLone and McLean information system success model (2003). The data analysis technique used was Partial Least Square (PLS).

3. Results

3.1. Inner Model Design

The design of the inner model between constructs in this study was based on the research hypothesis. The design of the inner model used smartPLS software 3. Figure 1 presents the results of the inner model design.

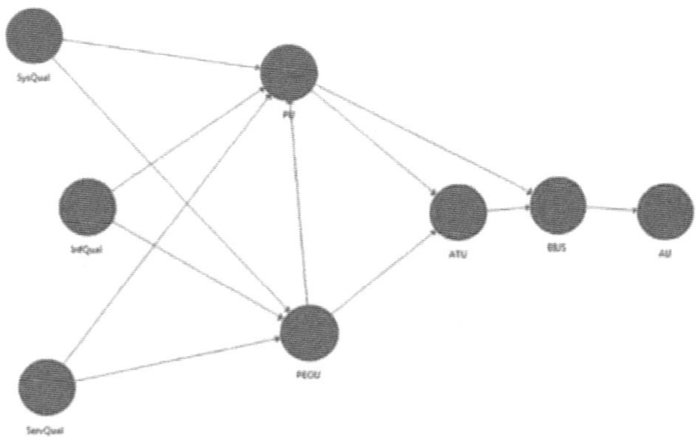

Figure 1. Inner Model Design

The design of the first inner model from Figure 1 is H1 system quality (SysQual) influencing PU followed by the second H2 system quality (SysQual) influencing the PEOU variable. H3 information quality (InfQual) affects PU followed by H4 information quality (InfQual) influences PEOU. Then H5 service quality (ServQual) influences PU continued by H6 service quality (ServQual) that influences PEOU. Then H7 PEOU influences PU, followed by H8 PEOU influencing ATU. Then H9 PU influences ATU followed by H10 PU influences BIUS. Then H11 ATU influences BIUS. Finally H12 BIUS influences AU.

3.2. *Outer Model Design*

The indicators in each construct of this study were reflective, so the arrows in each construct were headed towards the indicators. The design of the outer model used the help of smartPLS 3 software which can be seen in Figure 2.

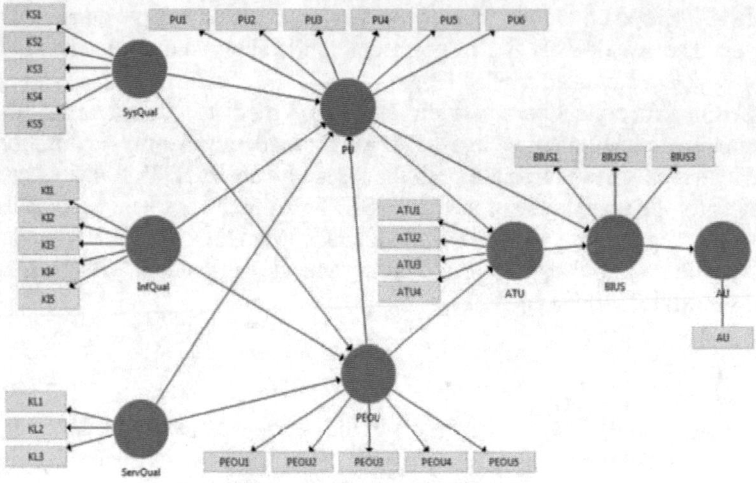

Figure 2. Outer Model Design

From Figure 2, it can be seen SysQual or system quality has indicators of KS1, KS2, KS3, KS4 and KS5. The quality of the system has an influence on PU and PEOU. InfQual or information quality has indicators of KI1, KI2, KI3, KI4 and KI5. Information quality has an influence on PU and PEOU. ServQual or service quality has indicators of KL1, KL2 and KL3. Service quality has an influence on PU and PEOU. PU or perceived usefulness has indicators of PU1, PU2, PU3, PU4, PU5 and PU6. Perceived usefulness has an influence on ATU and BIUS. PEOU or perceived ease of use has indicators of PEOU1, PEOU2, PEOU3, PEOU4 and PEOU5. Perceived ease of use has an influence on PU and ATU. ATU has indicators of ATU1, ATU2, ATU3 and ATU4. The attitude of using new technology has an influence on BIUS. BIUS or intention to use technology has the indicators of BIUS1, BIUS2, and BIUS3. The intention to use technology has an influence on the AU. AU or actual use had AU indicators.

3.3. Model Estimation

The method used in assessing parameters (estimation) in this study was an algorithm from smartPLS v3.0. To measure the inidimensionalitas of each construct in terms of the convergent validity of each construct indicator. The category for indicating individual reflexive correlation is high if it has a score higher than 0.70 with the construct being measured. For scores that have a loading factor of 0.50 to 0.60, they can still be used for models that are still under development [3]. The results of testing the model using the PLS algorithm can be seen in Figure 3.

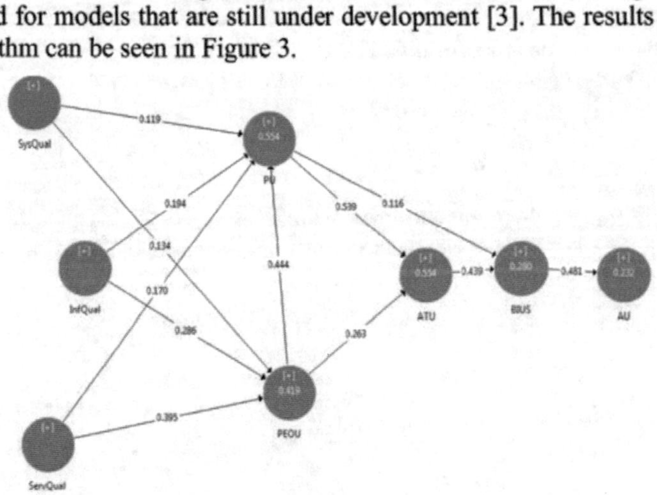

Figure 3. Results of Model Testing

The results of loading factors of the model testing, including: H1. SysQual or system quality had an influence on PU by 0.119; H2. SysQual or system quality had an influence on PEOU of 0.134; H3. InfQual or quality of information had an influence on the PU of 0.194; H4. InfQual or information quality had an influence on PEOU of 0.286; H5. ServQual or service quality had an influence on PU by 0.170; H6. ServQual or service quality had an influence on PEOU of 0.395; H7. PEOU or perceived ease of use has an influence on PU of 0.444; H8. PEOU had an influence on ATU of 0.263; H9. PU has an influence on ATU of 0.539; H10. PU has an influence on BIUS of 0.116; H11. ATU has an influence on BIUS of 0.439; H12. BIUS has an influence on AU of 0.481.

3.4. Model Evaluation

3.4.1. Testing the Outer Model

There were 3 categories in evaluating the outer model in this study. Using reflective indicators namely convergent validity (convergent validity score is the loading factor score for latent variables with the indicators. Expected score> 0.70). Discriminant validity (the score of discriminant validity is the score of cross loading factor which has the order to know whether the construct has sufficient discrimination by comparing the loading factor score on the intended construct that must be greater than the loading score with other constructs). Composite reliability (Data that has composite reliability> 0.8 has high reliability). Convergent validity from model calculations with reflexive indicators can be viewed from the relationship between item / indicator scores and the construct (loading factor) that can be assessed from the output outer loadings.

	ATU	AU	BIUS	InfQual	PEOU	PU	ServQual	SysQual
ATU1	0.871							
ATU2	0.829							
ATU3	0.862							
ATU4	0.749							
AU		1						
BIUS1			0.813					
BIUS2			0.861					
BIUS3			0.841					
KI1				0.749				
KI2				0.739				
KI3				0.749				
KI4				0.745				
KI5				0.754				

Figure 4. Outer Loading Results of ATU, AU, BIUS, and Information Quality.

	ATU	AU	BIUS	InfQual	PEOU	PU	ServQual	SysQual
KL1							0.792	
KL2							0.801	
KL3							0.853	
KS1								0.769
KS2								0.813
KS3								0.822
KS4								0.785
KS5								0.73
PEOU1					0.809			
PEOU2					0.836			
PEOU3					0.718			
PEOU4					0.775			
PEOU5					0.804			
PU1						0.865		
PU2						0.777		
PU3						0.741		
PU4						0.815		
PU5						0.761		
PU6						0.708		

Figure 5. Outer Loading Results of PEOU, PU, System Quality, and Service Quality.

An alternative way to measure discriminant validity is to equalize the square root score of the Average Variance Extracted (AVE) of each construct with the correlation score between constructs and other constructs (latent variable correlation). The score of discriminant validity is considered adequate if the model has a square root score of AVE for each construct higher than the score of the latent variable correlation. The results of AVE and latent variable correlations of PLS Algorithms can be reviewed in Table 1 and Figure 6.

Table 1. Results of AVE and AVE Square Root

	AVE	Akar AVE
KS	0.615	0.784219
KI	0.558	0.746994
KL	0.666	0.816088
PU	0.608	0.779744
PEOU	0.623	0.789303
ATU	0.688	0.829458
BIUS	0.703	0.838451
AU	1	1

	ATU	AU	BIUS	InfQual	PEOU	PU	ServQual	SysQual
ATU	1.000	0.334	0.523	0.534	0.632	0.719	0.425	0.471
AU	0.334	1.000	0.481	0.143	0.228	0.254	0.129	0.306
BIUS	0.523	0.481	1.000	0.365	0.370	0.432	0.278	0.363
InfQual	0.534	0.143	0.365	1.000	0.534	0.571	0.452	0.524
PEOU	0.632	0.228	0.370	0.534	1.000	0.685	0.552	0.365
PU	0.719	0.254	0.432	0.571	0.685	1.000	0.527	0.418
ServQual	0.425	0.129	0.278	0.452	0.552	0.527	1.000	0.206
SysQual	0.471	0.306	0.363	0.524	0.365	0.418	0.206	1.000

Figure 6. Output of Latent Variable Corelation

Based on the equation of the Table and Figure above, it can be concluded that the square root score of each construct was greater for the relationship between constructs and other constructs. For example, the AVE square root of the PU construct of 0.779 is greater than the correlation score of PU with ATU of 0.719, greater than the correlation score of PU with BIUS of 0.432, greater than the correlation score of PU with PEOU of 0.685, greater than the correlation score of PU with KS equal to 0.418, greater than the correlation between PU and KL by 0.527, and greater than the correlation between PU and KI of 0.571. Therefore it can be concluded that all constructs of the estimated model have discriminant validity criteria.

Besides testing the construct validity, a construct reliability test was assessed using two categories, namely composite reliability and Croncbach alpha of the indicator block assessing the construct. The construct is considered to be reliable when the composite reliability or Cronbach alpha scores are above 0.70 [9]. The composite reliability and Cronbach alpha output can be viewed in Table 2 and Table 3. It can be concluded that each construct had good reliability.

Table 2. Output of Composite Reliability

	Composite Reliability
KS	0.847
KI	0.748
KL	0.856
PU	0.870
PEOU	0.848
ATU	0.847
BIUS	0.798
AU	1

Table 3. Output of Cronbachs alpha

	Cronbachs Alpa
KS	0.889
KI	0.863
KL	0.856
PU	0.902
PEOU	0.892
ATU	0.898
BIUS	0.876
AU	1

3.4.2 Testing the Inner Model

After evaluating the measurement (outer) model, the structural and inner model tests were then carried out by reviewing the R-Square score in the endogenous construct. Structural models that indicate that good, moderate and weak models has an R-square result of 0.67, 0.33, and 0.19 respectively [9]. The R-square score of each endogenous construct from the estimated model is presented in Table 4.

The results of the R-square output in Table 4 indicated that there were two constructs that fall into the "moderate" criteria and one construct that falls into the "weak" criteria. The interpretation of the R-square output can be interpreted as follows: (1) The R-square score of the PU endogenous construct for this research model was 0.54. This meant that the KS, KI and KL constructs could only explain the PU construct of 54% and the rest was explained by other variables outside the model, (2) R-square score of PEOU's endogenous construct for this research model was 0.40. This meant that the KS, KI and KL constructs could only explain the ATU construct of 40% and the rest was explained by other variables outside the model. (3) The R-square score of the endogenous ATU construct for this research model is 0.55. This meant that the PU and PEOU constructs could only explain the ATU construct of 55% and the rest was explained by other variables outside the model. (4) RUS square score of BIUS endogenous construct for this research model was 0.27. This means that the PU and ATU constructs could only explain the BIUS construct by 27% and the rest is explained by other variables outside the model, (5) R-square score of the AU endogenous construct for this research model was 0.23. It meant that the BIUS construct only described the AU construct of 23% and the rest was explained by other variables from outside the model.

Table 4. Output R-Square (R^2)

	R-Square (R^2)
PU	0.554
PEOU	0.419
ATU	0.554
BIUS	0.280
AU	0.232

Hypothesis testing results showed that: (1) the first hypothesis concluded that the proposed H1 was rejected. The output path coefficients display the results of the statistical t score on the construct of the system quality to the construct of PU under 1.960 which was 1.754 so that the impact produced by the system quality on the construct of PU was not identified as significant. (2) Based on the second hypothesis testing, it was concluded that the proposed H2 was accepted. The output path coefficients displayed the results of the t statistic score on the system quality construct toward the PEOU construct above 1.960 which was 2.015 thus the impact of the system quality on the PEOU construct was significant, (3) Based on the third hypothesis testing, it was concluded that the proposed H3 was accepted. Output path coefficients display the results of the t score of statistics on the construct of information quality to the construct of PU above 1.960 which was 2.828 thus the impact produced by the quality of information on the construct of PU was identified as significant, (4) Based on the fourth hypothesis testing, it was concluded that the proposed H4 was accepted. The output path coefficients display the results of the t score of statistics on the construct of information quality on the PEOU construct above 1.960 which is 3.535 thus the impact produced by the quality of information on the PU construct is significant, (5) Based on the fifth hypothesis testing, it was concluded that the proposed H5 was accepted. The output path coefficients display the results of the t score of statistics on the construct of service quality to the construct of PU above 1.960 which was 2.204 so that the impact produced by the quality of service on the construct of PU was identified as significant. (6) Based on the sixth hypothesis testing, it was concluded that the proposed H6 was accepted. The output path coefficients displayed the results of the statistical t score on the service quality construct of the PEOU construct above 1.960, which was 6.543 so that the impact produced by the quality of the service on the PEOU construct was significant, (7) Based on the seventh hypothesis testing, it was concluded that the proposed H7 was acceptable. The output path coefficients displayed the results of the statistical t score of the PEOU construct on the PU construct above 1.960 which was 5.843 thus the impact produced by the PEOU on the PU construct was significant. Based on the eighth hypothesis testing, it was concluded that the proposed H8 was accepted. The output path coefficients displayed the results of the statistical t score of the PEOU construct on the ATU construct above 1.960, which is 3.004 thus the impact produced by the PEOU on the ATU construct. (9) Based on the ninth hypothesis testing, it was concluded that the proposed H9 was acceptable. The output path coefficients displayed the results of the statistical t score of the PU construct to the ATU construct above 1.960 which was 7.647 thys the impact produced by PU on the ATU construct is identified as significant. PU has the biggest influence on ATU, this is in accordance with other researchs [10][1][7][2]. (10) Based on the tenth hypothesis testing, it was concluded that the proposed H10 was rejected. The output path coefficients displayed the results of the statistical t score of the PU construct to the BIUS construct below 1.960 which is equal to 0.914 such that the impact produced by PU on the BIUS construct was not significant. (11) Based on the eleventh hypothesis testing, it was concluded that the proposed H11 could be accepted. The output path coefficients displayed the results of the statistical t score of the ATU construct on the BIUS construct above 1.960 which is 4.031 so that the impact produced by the ATU on the BIUS construct was identified as significant. (12) Based on the twelfth hypothesis testing, it was concluded that the proposed H12 was accepted. The output path coefficients displayed the results of the statistical t score of the BIUS construct against the AU construct above 1.960 which was 8.619 thus the impact produced by BIUS on the AU construct was identified as significant.

4. Conclusion

Based on the data obtained from this study, there were several tests that have t arithmetic scores below the t table, thus they do not have a significant impact on the dependent construct. This was because some students have not used SIBIMTA completely to do their theses because it was not recommended by their supervisors or the supervisors do not want to use it at the initial stage of implementing the system. The following details the results of testing the model, including: (1) system quality had a positive influence on perceived usefulness; (2) system quality had a positive influence on perceived ease-of-use; (3) information quality had a positive influence on perceived usefulness; (4) information quality had a positive influence on perceived ease-of-use; (5) service quality had a positive influence on perceived usefulness; (6) service quality had a positive influence on perceived ease-of-use; (7) perceived ease-of-use had a positive effect on perceived usefulness; (8) perceived ease-of-use has positive effect on the attitude to use technology; (9) perceived usefulness had a positive influence on attitudes using technology; (10) perceived usefulness did not have a positive influence on intention to use the technology; (11) attitude to use the technology had a positive influence on intention to use the technology; (12) intention to use the technology had a positive effect on actual use.

5. References

[1] Afiana, Rifatuliha and Priyanto. 2018. *An Analysis of the Acceptance of CBT VHS Application Using Technology Acceptance Model (TAM) and Theory of Planned Behavior (Tpb) Integration. J.Phys.: Conf. Ser. 1140 012026.*

[2] Ardhiani, Lisa Noor. 2015. Analisis Faktor-Faktor Penerimaan Penggunaan Quipperschool.com dengan Menggunakan Pendekatan *Technology Acceptance Model* (TAM) dan *Theory Of Planned Behavior* (TPB) di SMA Negeri 7 Yogyakarta. Jurnal Elektronik Pendidikan Teknik Informatika Volume 4, Number 2, February 2015.

[3] Chin, W. W. (1998). The Partial Least Squares Approach to Structural EquationModeling. Modern Methods for Business Research, 295-336.

[4] Davis, F.D. (1986). A Technology Acceptance Model for Empirically Testing New End-user Information Systems: Theory and Result. In Doctoral dissertation Sloan School of Management MIT.

[5] Davis, F.D. (1989). Perceived Usefulness, Perceived Ease of Use, and User Acceptance of Information Technology. MIS Quarterly (13:3), pp. 319-339.

[6] DeLone, W., and McLean E.R. (2003). *The DeLone and McLean Model of Information System Success: A Ten Year Update. Journal of Management Information Systems. Journal of Management Information Systems / Spring* 2003, Vol. 19. No .4, pp. 9—30.

[7] Destiana, Bonita. 2012. Analisis Penerimaan Pengguna Akhir terhadap Penerapan Sistem E-Learning dengan Menggunakan Pendekatan Technology Acceptance Model (TAM) di Sma N 1 Wonosari. Jurnal Elektronik Pendidikan Teknik Informatika 1,Volume 1,Nomor 1, September 2012.

[8] Fishbein, M., & Ajzen, I. (1975). Belief, attitude, intention, and behavior: an introduction to theory and research. Reading, Massachusetts: AddisonWesley

[9] Ghozali, I. (2011). *Structural Equation Modeling*, Metode Alternatif dengan *Partial Least Square*. Semarang: Badan Penerbit Universitas Diponegoro.

[10] Priyanto. 2017. *The Determinants of E-Learning Usage by Teachers Of Vocational High Schools in the Yogyakarta Special Region*. Jurnal Pendidikan Vokasi Volume 7, No 1, Februari 2017 (1-13)

[11] Ratri, Saras Maretra. 2016. Analisis Faktor-Faktor yang Memengaruhi Penggunaan E-Learning Moodle oleh Guru SMK Negeri 2 Yogyakarta dengan Pendekatan Technology Acceptance Model (TAM). Jurnal Elektronik Prodi Pendidikan Teknik Informatika Vol 5, No 4 (2016)

Simtenismeja.com Features as an Information System in the Implementation of Table Tennis Matches

D B Hertanto
Electrical Engineering Education Department, Universitas Negeri Yogyakarta, Yogyakarta, Indonesia

Email: denybudi@unt.ac.id

Abstract. The application of Information Technology has helped many business processes in various fields. Table tennis as a sport still has many problems. The process of implementing the match up to now still relies heavily on manual methods. The use of computers in the competition is still limited to documentation. Match schedules, match charts and validation of participants are examples of problems that make it difficult for the competition committee. Simtenismeja.com is an application that has been developed to overcome the above problems. This system is based online, so all processes can be done anytime and anywhere using all possible devices. This application has facilities for recording data on participant athletes, preparation of schedules and charts of matches automatically, and recording and publication of results of matches. This system has been implemented in a table tennis match. User responses have been recorded to show whether the system is proven to make it easier to do table tennis matches or not. The response test results state that all buttons, interface pages, and menus can function properly. The next process is comparing the implementation of matches manually with the implementation of matches assisted by simtenismeja.com application. In general the use of applications in matches is much more efficient in terms of time and accuracy.

1. Introduction
Table tennis is a popular sport in the world. This can be seen by the inclusion of table tennis as a sport that is contested in the Olympics. The very rapid development of table tennis through these championships made the number of participants even more so that the organizers or organizers of the game worked extra especially in terms of making the match system schedule. Ricky Aditya in the Journal of Mathematics and Science (2013), said that a good match scheduling method will affect the quality of organizing a sports tournament. According to Sonia et al. in Ejournal UNY (2015), competition issues cause table tennis athletes to tend to have high levels of anxiety before competing. According to Wasul Mualip et al. in UNY's E-Journal (2013), making a match system model is very important to do to help smooth the implementation of the match.

2. Simtenismeja.com Features
Some things in table tennis matches that must be prepared include the match schedule, match table, match chart, referees, referees, score papers, match reports and match athlete data. Making schedules, charts and match reports are the most difficult things to do. If the number of participants in a match exceeds 100 athletes, preparing a schedule and chart manually requires a lot of time and accuracy to avoid mistakes. After finishing the competition, the committee still has to report the results of the match per table. The hassles can be easily overcome using information technology.

Simtenismeja.com is an application that has been built to overcome the above problems. The application has several special features. The first feature is managing matches. To use this feature, application users must specify the date, start time, end time, rest time, number of tables, and duration of the match from the preliminary to the final. For more details, consider Figure 1 below.

Figure 1. Match Feature

After changing the configuration of the match feature, the user must set which athletes will be involved and or must be removed from the match. Furthermore, the user can access the match schedule feature. In this feature, users can choose the athlete to be seeded. The application does not limit the number of seeded, although in a good match system the maximum number of seeded is 4 athletes / team. The next process is generating a schedule. As a result, the game schedule will appear in 2 formats, which are the sequential schedule format from the earliest and the match chart format. Both formats can be printed and enlarged to be pasted on the bulletin board. In addition, athletes who compete can also access the schedule as a general user at www.simtenismeja.com.

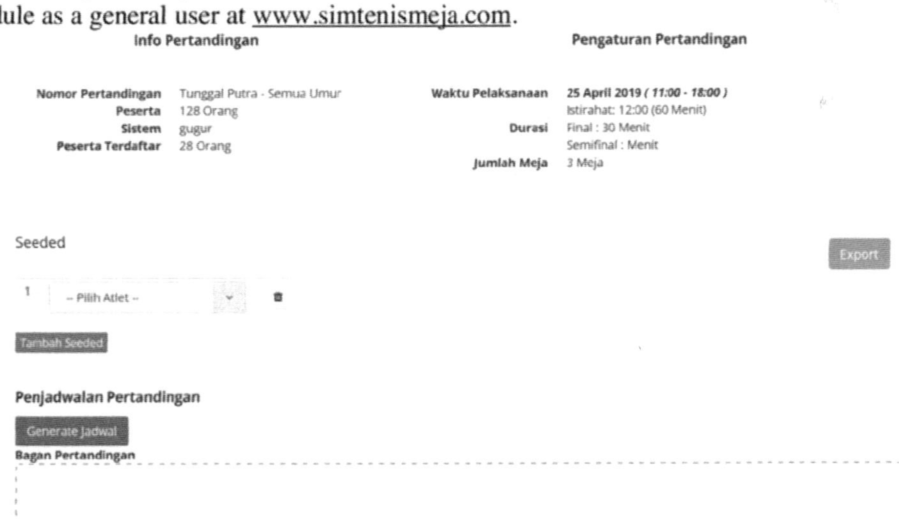

Figure 2. Arrange Match Schedule

After the schedule has been generated, the results can be seen in Figure 3. The figure shows the overall schedule of the match by displaying the date, time, table, round or round that is being passed along with the athlete involved.

Journal of Physics: Conference Series **1413** (2019) 012039 doi:10.1088/1742-6596/1413/1/012039

Tunggal Putra - Semua Umur | Informasi Pertandingan | Peserta Pertandingan | Jadwal Pertandingan | Bagan Pertandingan

Jadwal Pertandingan

No.	Waktu Pertandingan	Kelompok Pertandingan	Ronde	Meja	Peserta 1	Skor	Skor	Peserta 2	Aksi
1	25 Apr 2019, 11:00	Tunggal Putra (Semua Umur)	16 Besar	Meja 1	A. Prasetya Wigig Pamungkas	3	0	Sutirman	
2	25 Apr 2019, 11:00	Tunggal Putra (Semua Umur)	16 Besar	Meja 2	Tarto	3	0	Endra Murti Sagoro	
3	25 Apr 2019, 11:00	Tunggal Putra (Semua Umur)	16 Besar	Meja 3	Sunardianta	3	0	Edy Purnomo	
4	25 Apr 2019, 11:30	Tunggal Putra (Semua Umur)	16 Besar	Meja 1	Mutaqin	2	3	Joko Purwoko	
5	25 Apr 2019, 11:30	Tunggal Putra (Semua Umur)	16 Besar	Meja 2	Winarno	2	3	Tri Utoro	
6	25 Apr 2019, 11:30	Tunggal Putra (Semua Umur)	16 Besar	Meja 3	Parjana	2	3	Maryanto	
7	25 Apr 2019, 13:00	Tunggal Putra (Semua Umur)	16 Besar	Meja 1	Ernawan Dwi Cahyo	1	3	Martanto	
8	25 Apr 2019, 13:00	Tunggal Putra (Semua Umur)	16 Besar	Meja 2	Suprihanto	0	3	Hadwi Prihantanta	

Figure 3. Match Schedule

In this interface, the match table committee can input the match results using the action button located on the right in each row of the match schedule. The score that has been inputted can be seen by anyone and at any time at www.simtenismeja.com. The form for inputting match results is shown in Figure 4.

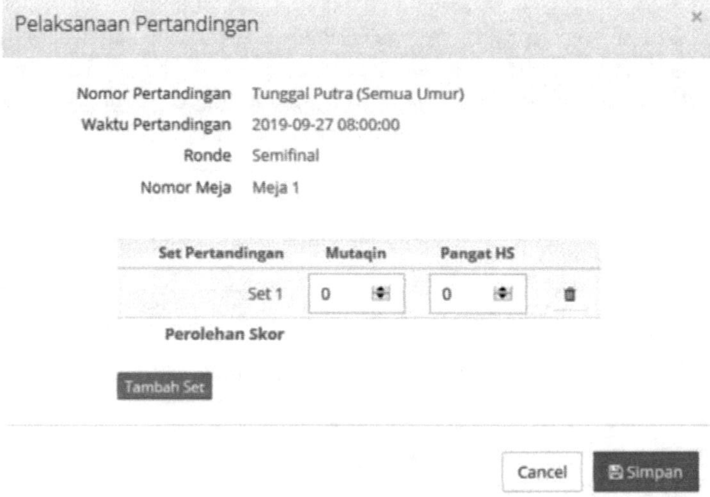

Figure 4. Input Match Results

3. Research Method

The information system built in this article refers to the Research and Development method in Pressman (2012). This paper discusses the superior features of the application by proving whether the application of the system shows far better results compared to the manual way, in terms of time, data accuracy and accessibility.

Features discussed include: athlete / team data input, manage matches, set match schedules, input match results, and report match results. Comparison of the application of the match process between the manual and the application is limited only to the superior features above. Data was taken directly at a match using direct observation and interviews.

4. Results

The results of observations and interviews in the process of implementing the match with 50 participants of the match are presented in the following description. At the game registration stage, there is no significant difference between manual registration and application use. The only difference is the amount of paper used in manual registration, while the application does not require paper at all, but requires a gadget and internet access.

At the stage of organizing the match, the system requires the same time and accuracy in the manual way. Manually, arranging matches is only stated and written down on a piece of paper to be agreed upon by all parties. This manual process is also carried out by the system that is inputted into the match management form.

Table 1. Comparison of the Process of Matching the Match between the Manual and the System

No.	Stages for 30 participant	Time Needed		Accessibility		Data Accuracy	
		Manual	System	Manual	System	Manual	System
1	Registration	5 minutes	2 minutes	offline	internet	60%-90%	99%
2	Making a Competition	10 minutes	3 minutes	offline	internet	100%	100%
3	Arrange a match	30 minutes	10-15 minutes	offline	internet	100%	100%
4	Making a match schedule	5-12 jam	2 minutes	offline	internet	60%-90%	100%
5	Making a match chart	1 jam	2 minutes	offline	internet	60%-100%	100%
6	Input result of a match	Depends on a match duration	Depends on a match duration	offline	internet	100%	100%
7	Report result of a match	30 minutes	1 minutes	offline	internet	100%	100%

At the stage of setting the schedule, the system can do it quickly. After entering the match participants and determining the player's seeded, based on the match management form, the system can quickly (only 2-5 seconds) create a match schedule as well as create a match chart. Meanwhile, manually, the committee made a chart of the match based on the number of participants. Then participants will be drawn to be included in the chart. After all participants have entered, the next step is to schedule a match. Manually, the process of drawing and entering participants requires 2-3 hours. To create a match schedule, the committee takes 5-12 hours with several checks to ensure no errors occur.

After the chart and schedule are finished, the next process is the implementation of the match. All process of recording the match is done manually by the referee using a score sheet. When reporting the results of a match, a manual process is carried out by rewriting results to the chart and schedule of the match. Whereas the application process is done by inputting a score sheet into the system. The results of the match can be accessed by all relevant parties via the internet and the updated chart of the results of the match can be reprinted by the system. A summary of all processes through observation and interviews can be seen in Table 1.

5. Conclusion

Based on the results and discussion above, it can be concluded that the system is much faster, more accurate, and wider in scope compared to the manual system. By implementing the system, the implementation of table tennis matches is expected to be carried out well.

6. References

[1] Aditya, Ricky, Suatu Metode Penjadwalan Pertandingan yang "Baik" untuk Kompetisi Liga Sepakbola Menggunakan Persegi Latin, Jurnal Matematika dan Sains, Vol. 18 No. 1, April 2013.

[2] Mualif, Wasul, dkk., Penyusunan Model Pertandingan Berbasis Komputer pada Cabor Pencak Silat, E-Journal UNY, PKO-S1, Vol 1 No. 1, Mei 2013

[3] Pressman, Roger S. (2012). Rekayasa Perangkat Lunak. Yogyakarta: ANDI OFFSET.

[4] Sonia, dkk., Tingkat Kecemasan Atlet Cabor Tenis Meja Sebelum Menghadapi Pertandingan di Kejuaraan Nasional di Semarang 2014, E-Journal UNY, PKO-S1, Vol 1 No. 1, Maret 2015

Journal of Physics: Conference Series **1413** (2019) 012040 doi:10.1088/1742-6596/1413/1/012040

Visual Speech Recognition for Daily Indonesian Words Based on Combination of Double Difference and Image Projection Method

A Nasuha, F Arifin, D Irmawati and N Hasanah
Electronics and Informatics Engineering Education Department,
Universitas Negeri Yogyakarta, Yogyakarta, Indonesia

Email: arisnasuha@uny.ac.id

Abstract. In visual speech recognition, there are two approaches to lip feature extraction, which are based on appearance and shape. The appearance-based approach is usually better, because it provides visual features that not only include lip structure but also the visibility of the teeth and tongue. However, the disadvantage of this approach is that it produces too many features. Integration of double difference and horizontal-vertical image projection, is part of the appearance approach, in addition to using image projection as dimensional reduction. In previous study, the method has succeeded in recognizing 5 daily indonesian words, with data in form of videos recorded inside the room. In this study, we used the same words with 5 additional new words recorded outside the room. MLP (Multi Layer Perceptron) and SVM (Support Vector Machine) are used as classifiers. The word recognition process is evaluated using 10-fold cross validation. The method tested reached 88.92% on classification accuracy and 0.9948 on AUC (Area Under ROC Curve).

1. Introduction

The automatic visual speech recognition, consists of 3 main stages, namely data acquisition, feature extraction, and introduction. Because of the characteristics of color and lip movement: lip color does not always contrast enough with the surrounding environment, and the shape of the lips keeps changing during the speech process [1], it is necessary to use an extraction method that is suitable for both characteristics.

The method for presenting the visual features of lips on video, based on space and time, can be linked into two categories [2], namely stationary-based methods and motion-based methods. The first method will display a video divided into an image frame, technically the feature extraction that is run on each frame. This method is called the static feature method. The second method feature will be taken from the relative movement between the object and the observer, namely the velocity distribution that appears in the conversation pattern in the picture. In other words, information is related to the composition of the objects collected and the speed of changes in composition.

Another type of feature extraction that can be found in the literature is extract according to the type of information source they process [3], which is based on the shape and appearance. ASM (Active Shape Model) and ACM (Active Contour Model) are examples of extraction based on shapes, while DFT (Discrete Fourier Transform) and DCT (Discrete Cosine Transform) are appearance-based examples. Appearance-based lip extraction provides visual features that not only contain lip structures, but also about the visibility of the tongue and teeth. The disadvantage of this extraction is the huge number of features.

There is a simple method for effective segmentation and lip tracking, namely frame difference method, which is one technique in the background substraction method. This method is widely used in video surveillance and motion detection. The advantage of this method is the low computational process.

Application of frame difference and double difference algorithm for automatic visual speech recognition has studied by Nasuha et. al. [4]. In the research, frame difference is combined with horizontal-vertical image projection. Moreover, important issue of the frame difference, namely selection of the threshold value, had been discussed. Data on the research is five daily Indonesian words recorded from ten subjects in the controlled environment. This paper will discuss the application of the above method on recognition ten daily Indonesian words and data recorded outdoor.

2. Materials and Methods

2.1. Materials

In this research, we use video recorded from 18 volunteers, 12 men and 6 women, with age distribution from 19 to 21 years. All video are in color and focused around subject's mouth in frontal face. In these video, all volunteers are recorded outdoor. Each of volunteers says ten simple words in Indonesian, which is widely used in daily life, i.e. "saya" (I), "mau" (shall, will, want), "makan" (eat), "minum" (drink), "mandi" (take a bath), "tolong" (help, please), "sapu" (broom), "meja" (table), "kursi" (chair, seat), and "sudah" (already, past, done). Each word is repeated three times, therefore there are 540 video data. The original frame size is 320x240 pixels. Each volunteer is recorded in 0.75 seconds for each word, in 30 frames per second. Example of frame from each subject can be seen in figure 1. Figure 2 is example of sequence of lip images pronouncing "mau".

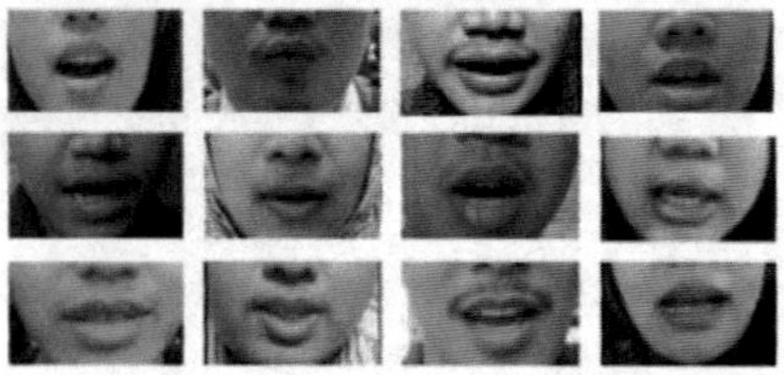

Figure 1. Examples of video frames of the subject

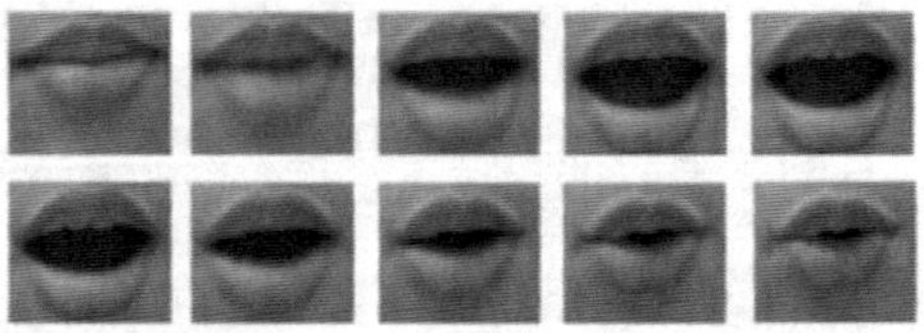

Figure 2. Sequence of lip images pronouncing "mau"

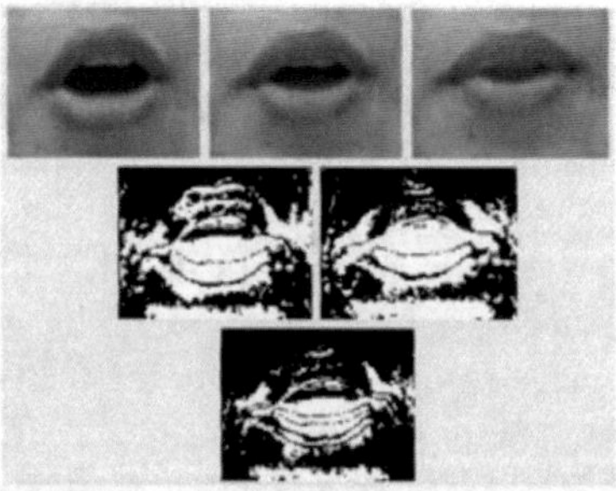

Figure 3. Illustration of double difference: three consecutive frames (top), frame difference of two consecutive frames (middle), logical AND of two frame difference (bottom)

2.2. Methods

2.2.1. Double Difference. A double difference [5] is an algorithm that places three consecutive frames as a group, calculates the difference between two consecutive frames (see equation (1) and equation (2)) and performs logical AND operations on two frame differences (see equation (3)). Wherein $I(x,y,k)$ and $I(x,y,k\text{-}1)$ are intensity value in same pixel (x,y) at the current and previous time, $dI_i(x,y)$ and $dI_i(x,y)$ are difference intensity value, T is a specific threshold value. Illustration of the algorithm can be seen in figure 3. The algorithm can better get the moving object region if the speed of the object moves and the frame speed matches.

$$dI_1(x,y) = \begin{cases} 1, |I(x,y,k) - I(x,y,k-1)| \geq T \\ 0, |I(x,y,k) - I(x,y,k-1)| < T \end{cases} \quad (1)$$

$$dI_2(x,y) = \begin{cases} 1, |I(x,y,k-1) - I(x,y,k-2)| \geq T \\ 0, |I(x,y,k-1) - I(x,y,k-2)| < T \end{cases} \quad (2)$$

$$dI(x,y) = \begin{cases} 1, dI_1(x,y) \cap dI_2(x,y) = 1 \\ 0, dI_1(x,y) \cap dI_2(x,y) = 0 \end{cases} \quad (3)$$

Double difference has been known to overcome the disadvantages of frame difference, namely foreground aperture and ghosting.

2.2.2. Horizontal-Vertical Image Projection. Image projection is a one-dimensional representation of image content. Horizontal-vertical image projections are histograms from horizontal and vertical directions of binary imagery. The vertical-horizontal image projection is defined in equation (4) and equation (5).

$$P_{hor}(y_0) = \sum_{x=0}^{M-1} I(x, y_0) \text{ for } 0 < y_0 < N-1 \quad (4)$$

$$P_{ver}(x_0) = \sum_{y=0}^{N-1} I(x_0, y) \text{ for } 0 < x_0 < M-1 \quad (5)$$

The P_{hor} and P_{ver} are image projection in horizontal and vertical way respectively, I(x,y) is intensity value in pixel (x,y), M and N are width and height of image in pixel respectively. Each row and each column of image becomes a bin in the histogram. The stored count in a bin is the number of 1-pixels that appear in that row or column. This method can extract image features quickly and easily. This method has proven successful for Amazigh handwritten character recognition [6]. Illustration of this method can be seen in figure 4.

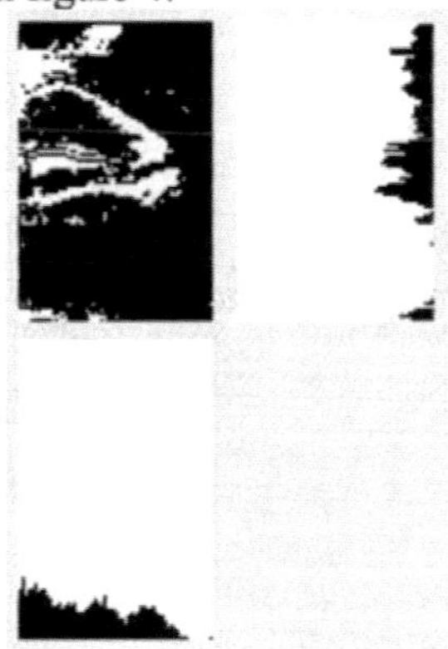

Figure 4. Illustration of Horizontal-Vertical Image Projection: Result of Double Difference (top left), Horizontal image projection (top right), Vertical image projection (bottom left)

The basic steps for the method are as follow:
1) Convert video from RGB color space to grayscale
2) Manually cropping the image in each frame to get area around the lip, from 320x240 pixels into 160x120 pixels, and down sampling to get 32x24 pixels
3) Apply double difference method, using threshold value 5
4) Folding the lip image to reduce the impact of the illumination mismatch of the left and right lip areas, therefore the size of each frame becomes 16x24 pixels, and total of data becomes doubled
5) Apply vertical and horizontal image projection, to extract the features of the lip image, which will be used as input of classifier. The results of feature extraction for each frame is (16+24) features, or 840 features for each video

6) Apply word recognition using MLP (Multi Layer Perceptron) and SVM (Support Vector Machine)

For MLP classifier, we use three layers, and the number of hidden units is varied to get the best result. We use sigmoid for activation function, and also regularization to prevent overfitting. For SVM classifier, the kernel is varied, namely linear, polynomial, RBF (Radial Basis Function) and sigmoid. We use automatic parameter search [7] to get optimum parameter for learning for each kernel.

To evaluate the model of the method, we need a way to estimate the test risk. Because of limitation of the data, we use cross-validation [8] instead of classical approach, separate the data into training and testing part. In this research we use 10-fold cross-validation.

There are many ways to measure the performance of the classifier, such as accuracy and AUC. Some researchers prefer to use the AUC [9], because AUC is a better measure of accuracy based on formal definitions of discriminancy and consistency. On the other hand, some others choose accuracy, because it is more intuitive and easy to understand, especially when the number of data for each class of the same amount. For this study, we used the accuracy and AUC to get the benefits of both.

3. Result and Discussion

The result of overall experiments is shown in figure 5 and figure 6. Figure 5 shows comparison of CA and AUC for the proposed method using MLP as classifier. Figure 6 shows comparison of CA and AUC using SVM as classifier.

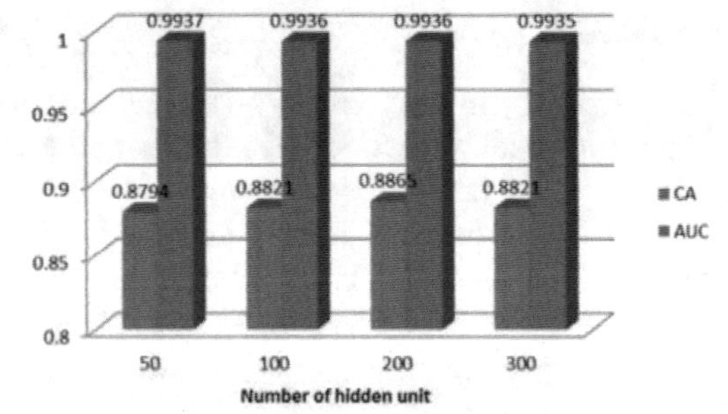

Figure 5. Classification accuracy and AUC using MLP for several hidden units

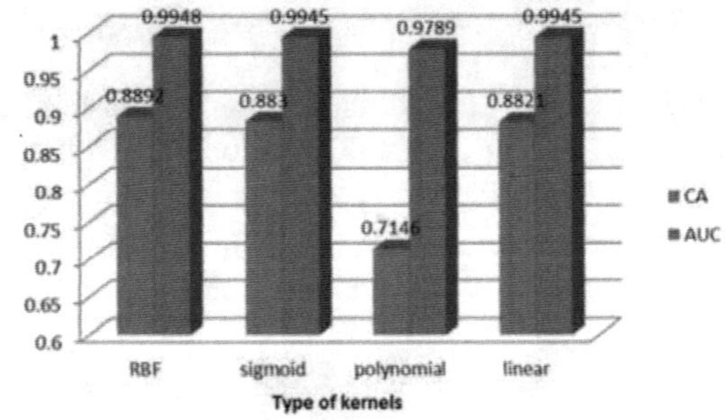

Figure 6. Classification accuracy and AUC using SVM for several kernels

In this study, when we used MLP as classifier, the best result is 88.65% on CA measurement, achieved when using 200 hidden units and 0.9937 on AUC, achieved when using 50 hidden units. When we used SVM as classifier, the best kernel is RBF. These results were 88.92% on CA, and 0.9948 on AUC.

When compared with previous studies, the results of word recognition in this study declined. This is caused by at least two things, first, namely the video data taken outside the room, where the condition is not controlled. Interference is low brightness and the presence of shadows. Secondly, the number of target classes is twice that of previous studies, while the amount of learning data available is limited.

Another difference with previous research is that the results of the SVM classifier outperform MLP in word recognition. Several studies comparing two types of classifiers have been carried out by other researchers, with results that outperform each other. Because this research did not focus on the problem, more in-depth research was not conducted.

4. Conclusion

This paper discusses the application of combination of double difference method and horizontal-vertical image projection for automatic visual speech recognition, in recognizing ten daily Indonesian words. The proposed method achieves 88.92% on CA and 0.9948 on AUC, when using SVM with RBF kernel.

Our future work in this area involves evaluating the proposed method on classifying Indonesian visemes, not Indonesian words. Other works include studying the use of adaptive or optimum method in determining the threshold value.

5. References

[1] A. Nasuha, T. A. Sardjono, and M. H. Purnomo, "Lip segmentation and tracking based on Chan-Vese model," presented at the 2013 International Conference on Information Technology and Electrical Engineering (ICITEE), Yogyakarta, Indonesia, 2013, pp. 155–158.

[2] M. Li and Y. Cheung, "A novel motion based lip feature extraction for lip-reading," in *Computational Intelligence and Security, 2008. CIS'08. International Conference on*, 2008, vol. 1, pp. 361–365.

[3] G. Potamianos, C. Neti, J. Luettin, and I. Matthews, "Audio-Visual Automatic Speech Recognition: An Overview," p. 39, 2004.

[4] A. Nasuha, F. Arifin, T. A. Sardjono, H. Takahashi, and M. H. Purnomo, "Automatic Lip Reading for Daily Indonesian Words Based on Frame Difference and Horizontal-Vertical Image Projection," *JATIT*, vol. 95, no. 2, pp. 393–402, Jan. 2017.

[5] Y. Kameda and M. Minoh, "A human motion estimation method using 3-successive video frames," presented at the International Conference on Virtual Systems and Multimedia, Gifu, Japan, 1996, pp. 135–140.

[6] Y. Es-saady, A. Rachidi, M. El yassa, and D. Mammass, "Amazigh Handwritten Character Recognition based on Horizontal and Vertical Centerline of Character," *Int. J. Adv. Sci. Technol.*, vol. 33, Sep. 2011.

[7] N.-E. Ayat, M. Cheriet, and C. Y. Suen, "Optimization of the SVM kernels using an empirical error minimization scheme," in *Pattern Recognition with Support Vector Machines*, Springer, 2002, pp. 354–369.

[8] G. Seni and J. F. Elder, *Ensemble Methods in Data Mining: Improving Accuracy Through Combining Predictions*, vol. 2. Morgan & Claypool, 2010.

[9] J. Huang and C. X. Ling, "Using AUC and accuracy in evaluating learning algorithms," *IEEE Trans. Knowl. Data Eng.*, vol. 17, no. 3, pp. 299–310, 2005.

Acknowledgement

We would like to express our gratitude to Zulfa Kurniawan and Arif Puji, due to the meaningful discussion about video processing.

AUTHOR INDEX